Contributions from Biology Education Research

Contributions from Biology Education Research (CBER) is the international book series of the European Researchers in Didactics of Biology (ERIDOB). The series includes edited collections of state-of-the-art papers presented at the ERIDOB international conferences, and monographs or edited collections of chapters by leading international scholars of the domain. The aim of the series is to shed light on global issues and trends in the teaching and learning of biology by gathering cutting edge research findings, theoretical views, and implications or concrete suggestions for everyday school practice regarding biology. The books may serve as resources for (a) getting informed about the most recent findings of biology education research to possibly integrate them in new personal research, and (b) studying about the teaching and learning of biology as a pre-service or in-service biology teacher. So, the main audiences for the series range from senior to early career biology education researchers and pre- or in-service biology teachers working at all educational levels. Book proposals for this series may be submitted to the Publishing Editor: Claudia Acuna E-mail: Claudia.Acuna@springer.com

More information about this series at https://link.springer.com/bookseries/16350

Blanca Puig • María Pilar Jiménez-Aleixandre
Editors

Critical Thinking in Biology and Environmental Education

Facing Challenges in a Post-Truth World

Springer

eridob
european researchers
in didactics of biology

Editors
Blanca Puig
Facultade Ciencias da Educación
Universidade de Santiago de Compostela
Santiago de Compostela, Spain

María Pilar Jiménez-Aleixandre
Facultade Ciencias da Educación
Universidade de Santiago de Compostela
Santiago de Compostela, Spain

ISSN 2662-2319 ISSN 2662-2327 (electronic)
Contributions from Biology Education Research
ISBN 978-3-030-92005-0 ISBN 978-3-030-92006-7 (eBook)
https://doi.org/10.1007/978-3-030-92006-7

This Springer imprint is published by the registered company Springer Nature Switzerland AG
The registered company address is: Gewerbestrasse 11, 6330 Cham, Switzerland

To our partners
Manuel González Fernández de Tejada (B. P.)
Ramón López-Facal (M. P. J. A.)
For their love and support.

Foreword

Critical thinking is a concept that has been used in education and educational research for over 20 years now. With the publication of the UN-agenda Transforming our world: the 2030 Agenda for Sustainable development, in 2015, critical thinking has received renewed and increasing attention by researchers, curriculum designers, and policy makers around the world. This edited book volume offers one such example through a timely exploration of questions related to the Whats, the Whys, and the Hows of critical thinking.

Situated within the urgency of addressing current global challenges, such as climate change, sustainability, and public health, and within the context of the so called "post-truth" era, Blanca Puig and María Pilar Jiménez-Aleixandre embark on a timely and much-needed exploration of the concept of critical thinking in biology and environmental education. In its scope, however, the book volume goes far beyond critical thinking to examine emerging crucial questions related to scientific citizenship and identity, environmental action, the role of evidence in science, as well as pedagogies and practices of resistance.

Critical Thinking in Biology and Environmental Education includes 15 chapters written by 30 authors at different stages of their careers, from 10 different countries: Spain, the UK, Mexico, the USA, Cyprus, Chile, France, Greece, Norway, and Sweden. As such it presents a comprehensive set of diverse perspectives on critical thinking and provides empirical evidence rooted in contemporary research in biology and environmental education in various parts of the world. The chapters offer concrete examples of programs and curricula situated in diverse geographical contexts and learning environments, which include different populations, ranging from young students, preservice teachers, teachers, and citizens.

In the first chapter, the editors argue about the urgency and need for critical thinking and frame their argument in the context of COVID19-pandemic and the pervasiveness of disinformation and fake news. The chapters that follow travel across epistemological, theoretical, and methodological spaces that range from dialogical argumentation, aesthetics, emotions, sociolinguistics, and spatial literacy to health education and ecofeminisn. In Chap. 2, Laura Colucci-Gray and Donald Gray share contemporary perspectives on critical thinking that highlight the role of language and context as they advance the central role of visual aesthetics in the development of a critical consciousness vis-à-vis environmental issues. In the chapter that follows, Elizabeth Hufnagel explores the intersection between emotions and critical thinking in relation to anti-science stances and climate change in education. In Chap. 4, Alejandra García Franco, Lisber Farrera Reyes, and Alma Adrianna Gómez Galindo provide an example of culturally relevant educational practices that integrate indigenous knowledge into the school science curriculum, and which aim to facilitate students' dialogue between their own ways of knowledge and scientific ways of knowing.

Part II consists of five chapters reporting research about critical thinking in biology and health education. In Chap. 5, Ravit Golan Duncan, Veronica L. Cavera, and Clark A. Chinn report on the findings of a study that analyzed students' critical evaluation of evidence to decide between explanations of phenomena. Kalypso Iordanou, in the next chapter, presents the findings of a study that aimed to examine a group of young students' critical thinking following their engagement in an intervention program based on dialogic argumentation and reflective activities genetically modified food. In Chap. 7, Blanca Puig and Noa Ageitos report on the findings of a study that aimed to examine the intersections between preservice teachers' knowledge and critical thinking skills regarding specific claims in the context of vaccination. In the next chapter, Marida Ergazaki builds on empirical evidence to discuss the thinking strategies that biology students may employ when exploring the procedure of making and using a DNA library. In Chap. 9, Araitz Uskola reports on the results of a research program that analyzed the opinion and justifications posed by students about the effectiveness of homeopathic products.

Part III consists of five chapters reporting research about critical thinking in environmental and sustainability education. In Chap. 10, Corina González-Weil, Valeria León, Delia Cisternas, Gabriel Caro, and Roberto Morales build on the findings of a multiple case study with teachers framed within theoretical perspectives around pedagogical of resistance and critical action to argue about the importance of formative and collaborative spaces for teaching in contexts of environmental degradation. In Chap. 11, Asli Sezen-Barrie, Joseph A. Henderson, and Andrea L. Drewers report on the findings of a study that aimed to examine how students develop critical thinking skills about climate change evidence, impacts, and solutions to make local and individual meaning of what is ultimately a global collective action problem. In Chap. 12, Pablo Brocos and María Pilar Jiménez-Aleixandre report on the findings of a study that examined students' capacity to develop independent positions and challenge social or peer's ideas, related to citizenship and social practice. In the chapter that follows, Kévin De Checchi, Gabriel Pallarès, Valérie Tartas, and Manuel Bächtold present empirical evidence from a case study with students in the context of a socioscientific environmental debate to argue that epistemic beliefs might serve as a means of understanding critical thinking. In the last chapter of this part, Eli Munkebye and Niklas Gericke argue about the importance of addressing critical thinking in primary school through an empirical study that examined teachers' understanding of critical thinking in the context of education for sustainable development.

In the chapter that closes the book, Blanca Puig and Marilar Jiménez-Aleixandre connect the main ideas and contributions emerging across chapters and suggest further directions for the integration of critical thinking in biology and environmental education.

Collectively, these chapters synthesize important and useful insights that promise to deepen and expand our understanding of critical thinking and provide concrete examples of related programs and approaches implemented in diverse contexts. Despite the differences in their goals, theoretical framing, methods, and contexts, a consensus exists across the chapters about the value of placing critical thinking at the forefront in efforts to promote meaningful dialogue for social change.

The world is changing and societies are transforming in a fast pace, possibly faster than ever before. The question then becomes one of how might education lead sustainable change in this fast-paced world? This book volume offers a first step towards that direction. As evident in the work done by the authors, biology and environmental education hold a privileged position to addressing current socioscientific challenges through (re)conceptualizing critical thinking for sustainability. When do we start? As Blanca and Marilar rightfully state in the first chapter, *the time is now.*

University of Groningen
Groningen, The Netherlands

Lucy Avraamidou

Contents

Part I
Perspectives on Critical Thinking

Chapter 1
Educating Critical Citizens to Face Post-truth: The Time Is Now

María Pilar Jiménez-Aleixandre and Blanca Puig

1.1 Introduction: The Need for Critical Thinking

The development of critical thinking (CT) is a consensual objective for educators in all times. But serious crises, as may be climate change and the COVID-19 pandemic, give it an added urgency. Taking a distance from particular topics, a serious concern for education, as well as for the general public, is the rise of post-truth, which McIntyre (2018) describes as "a corruption of the process by which facts are credibly gathered and reliably used to shape one's beliefs about reality." (p. 11). More worrying, according to this author, is that truth is being challenged as a mechanism for asserting political dominance. Therefore, critical thinking, rather than an ability or a practice related to areas such as science education or philosophy, should be viewed as necessary for all the citizenry. As Abrami et al. (2008) pointed out, thinking about problems within disciplinary areas has relevance for academic purposes, but thinking about social, political and ethical challenges of everyday life is relevant for students' future as functional and contributing adults:

> At a broader societal level, a democracy composed of citizens who can think for themselves on the basis of evidence and concomitant analysis, rather than emotion, prejudice, or dogma, is a plus—in fact, it sustains, builds, and perpetuates the democracy (Abrami et al., 2008 p. 1103).

A socio-scientific issue that we are currently experiencing, related to biology and health education, is the COVID-19 pandemic, which illustrates a few points relevant for critical thinking:

M. P. Jiménez-Aleixandre (✉) · B. Puig
Facultade Ciencias da Educación, Universidade de Santiago de Compostela, Santiago de Compostela, Spain
e-mail: marilarj.aleixandre@usc.es

B. Puig, M. P. Jiménez-Aleixandre (eds.), *Critical Thinking in Biology and Environmental Education*, Contributions from Biology Education Research,
https://doi.org/10.1007/978-3-030-92006-7_1

- First, COVID-19 underscores the *pervasiveness of fake news*, either about the origin of the disease or about bogus remedies such as injecting disinfectant. Certain pieces of disinformation were even promoted by some governments. The abundance of disinformation, coupled with cognitive bias, may explain the difficulties that citizens are experiencing for unravelling rigorous information from unfounded opinions.
- Second, this pandemic highlights the *complexity of socio-scientific issues* (SSI) in which multiple dimensions from different fields interact. In the case of COVID-19 this is so, not only because of health, economic, social or political impacts, but also about the causes involved in cross-species transmission, including virus biology, environmental disruption and livestock production, as the Intergovernmental Platform on Biodiversity and Ecosystem's (IPBES, 2020) report made clear. Zheng-Li Shi, world leading expert on coronavirus diseases, and her colleagues in Wuhan, made the case for the relationships between disruption of wild habitats, intensive breeding practices and epidemics. In several papers, written before COVID-19, Shi and colleagues anticipated new outbreaks, predictions that unfortunately came true, and proposed to keep in mind the *One Health* concept, meaning human, animal and environment health, endorsed by the World Health Organization (WHO). In *Nature* Zhou et al. (2018) highlighted the importance of identifying coronavirus (CoVs) diversity in bats "to mitigate future outbreaks that could threaten livestock, public health and economic growth." (p. 255). Fan et al. (2019), alerted that bat-borne CoVs would likely re-emerge to cause the next disease outbreak. Cui et al. (2019) stated: "The constant spillover of viruses from natural hosts to humans and other animals is largely due to human activities, including modern agricultural practices and urbanization." (p. 190). They suggested that the most effective way to prevent outbreaks is keeping the barriers between natural reservoirs of wildlife and human society. Combining concerns about misinformation and her scientific expertise, Shi warned, on February 2nd 2020 on her WeChat, towards conspiratorial theories about coronavirus: "The 2019 novel coronavirus is a punishment by nature to humans' unsanitary life styles."
- Third, it reveals people's *struggles with uncertainty*, as for the public it is not easy to accept that science doesn't know enough about COVID-19, that scientists do not have all the responses and may sometimes disagree with one another, and that it could take many months for vaccines to be available. People and authorities showed resistance to admit that perhaps we all should change our lifestyles, or that plans for the next years were dependent of how the pandemic, as well as governments' responses to protect their citizens would evolve.
- Fourth, it lay open *social inequalities when facing disease*: COVID-19 is striking countries with feebler or non-existent public health systems harder. Within countries, infection is far more frequent in districts where people live packed together, as for instance immigrant labourers working in slaughterhouses or picking fruit in European countries. This is a lesson that historians of science have told about tuberculosis, but that seems forgotten.

Critical thinking is not going to speed up the scientific quest for medical solutions, for COVID or any other future diseases. However, we argue that it may equip citizens to discard fake news, to use appropriate criteria to evaluate information, and perhaps to consider lifestyle changes. It could support the identification of multiple dimensions in SSI, including the acknowledgement of the uncertainty inherent to science-in-the-making (Kampourakis, 2018). Critical thinking, in the approach that we propose, is oriented to action, so it could promote responsible behaviour, such as keeping distances, avoiding crowded events or wearing a mask, even when it means some sacrifice and could be seen as benefiting others rather than oneself. Engaging in action could also mean recognizing that not everybody has the same opportunities to overcome disease, that social priorities should shift towards guaranteeing food and healthcare for all.

In this chapter we first review literature about CT characterizations and knowledge building dynamics in relation to post-truth; second, we discuss our characterization of the components of CT, which is a revision of a previous one (Jiménez-Aleixandre & Puig, 2012); third, we make the case for biology education and environmental education as privileged learning environments to develop critical thinking, which is particularly timed for current crises and post-truth challenges

1.2 Critical Thinking and Criticality: From a Focus on Skills to a Focus on Practice

While CT is considered a valuable goal for education, it is also a complex notion, not easy to define and to study, and that poses difficulties for operationalizing interventions (Abrami et al., 2008, 2015). There are different views about CT meaning, components, and about how to promote it. The rationale discussed in this section draws from two strands: first, revised characterizations of CT, including criticality and the relevance of identities and emotions; second, literature about post-truth, pseudoscience, uncertainty, and science denial. Because critical thinking is understood in different ways, both by scholars and by the general public, we begin this section by briefly addressing some stereotypes, which may obscure what is CT and what is not.

First, engaging in CT does not consists in criticizing everything, in considering only negative aspects of a given issue although, as Davies and Barnett (2015) note, for some people CT means a propensity for finding fault. As Diane Halpern (1998) pointed out, the word "critical" is not meant to imply "finding fault" but evaluation or judgment.

Second, CT involves taking into account empirical evidence, but it has also other components, some cognitive or metacognitive, as self-regulation, some dispositional, as willingness to reconsider and revise views, and affective. In other words, understanding CT as only the examination of evidence is partial.

Third, CT could be developed in the context of a range of disciplines and subject matters courses in education. However, it has been often claimed that philosophy is the only discipline teaching or promoting CT. This "proprietary" assertion was examined by Álvarez Ortiz (2007), who found that there is insufficient evidence to back the claim that studying philosophy improves CT any more than other disciplines. This chapter is framed in a perspective considering that science education can be an appropriate context to develop CT, and that biology and environmental education are privileged learning environments to do so.

Fourth, we argue that the development of CT requires engagement in its practice. In other words, that it cannot be developed by just receiving instruction, in a transmissive approach, something already pointed out by Facione (1990).

1.2.1 Critical Thinking: Criticality, Identities and Practice

During three decades studies about CT have taken as a starting point the definition developed by a Delphi panel of the American Philosophical Association (APA): "purposeful, self-regulatory judgment, which results in interpretation, analysis, evaluation, and inference, as well as explanation of the evidential, conceptual, methodological, criteriological, or contextual considerations upon which that judgment is based" (Facione, 1990, p. 2). The panel was formed by 46 experts although only two of them, Anita Silvers and Carol Tucker, were women, a bias that nowadays would probably not happen.

The panel found CT to include both cognitive and dispositional dimensions. Regarding the first, they identified six core cognitive skills: interpretation, analysis, evaluation, inference, explanation, and self-regulation, about which a strong consensus, ranging from 95% to 87%, was found. There were, however, fewer consensuses about the dispositional dimension: although there was agreement on cognitive dispositions that can be correlated to each cognitive skill, there was a division about the affective ones. 61% of the experts held that CT includes a reference to certain affective dispositions, while one third of them restricted it to only cognitive ones. Examples of affective dispositions are flexibility in considering alternatives and opinions, honesty in facing one's own biases, prejudices, stereotypes, egocentric or sociocentric tendencies, or willingness to reconsider and revise views where honest reflection suggest that change is warranted.

Although this APA definition has been highly influential, there is a recent shift from a focus on *CT as skills* to a focus on *CT as practice*. In their introduction to the *Handbook of Critical Thinking in Higher Education* Davies and Barnett (2015) pointed out some of the shortcomings of the APA definition: first, that it does not lend itself easily to educational implementation; second that it is rooted in one kind of CT, namely CT as argumentation and judgment formation, leaving out concerns about the nature of criticality – discussed below–; third, that the dimensions in the definition include skills, reflective judgment formation, and dispositions, but action is not mentioned. They concluded: "It is in principle possible to meet the stipulated requirements of the definition and not *do* anything." (Davies & Barnett, p. 11; authors' emphasis).

In a recent paper Deanna Kuhn (2019) argued for a sharpening of the construct of CT, proposing to consider it as discourse, as a *dialogic practice*:

> Extending this [dialogic] view of thinking more specifically to the construct of critical thinking, critical thinking is a dialogic practice people commit to and thereby become disposed to exercise, more than an individual ability or skill. Critical thinking as dialogue is engaged initially interactively and then with practice in interiorized form" (Kuhn, 2019, pp. 148–149).

The author noted that this dialogic view is proposed as a framing of CT, rather than a restrictive definition. She suggested that it may help to bring together separate strands of work regarding CT as a theoretical construct, a measurable skill, and an educational objective. This perspective has consequences for how to promote its development in classrooms for, as Kuhn pointed out, CT is not a fixed attribute, an individual ability that qualifies someone for admission to the shared practice, but rather a dynamic activity that is developed through engaging in the practice.

This perspective is in accordance, first, with conceptualizations of epistemic cognition as a practice (Kelly, 2016), as epistemic practices are constructed in social interaction, and they include interactionally accomplished understandings of knowing; second with the approach, framing the Next Generation Science Standards (NGSS Lead States, 2013) viewing science as consisting of a set of scientific practices.

Davies and Barnett (2015) outlined three perspectives about CT, not entirely separable: *Philosophical*, *educational* and *socially active*, the last itself a complex of positions concerned with the transformation of society; it encompasses critical pedagogy, which the authors conceive as educating to promote political activism, and critical citizenship, which they view as cultivating a critical citizenry. For the purposes of our proposal about the components of CT, the differences in interpretation of the meaning of "critical", discussed by them, are of relevance. For the philosophical perspective it means "criticism", for instance identifying weaknesses in claims or arguments, while for the critical pedagogy it rather means "critique", in other words identifying dimensions that might be missing or concealed behind claims or arguments. Davies and Barnett discussed the shift from a view of CT as only composed of skills and dispositions, to *criticality*, a view including *action*. Criticality comprises critical thinking, critical reflection and critical action.

Another shortcoming of the APA definition is that it did not adequately acknowledge affective disposition and much less emotions, a criticism that it has also been raised to criticality. Danvers (2016) claimed that "Re-imagining criticality through specifically *feminist* engagements with relations, affects, bodies and materialities... allows us to ask a different set of questions about how critical thinking is conceptualised" (p. 283, author's emphasis). Danvers' focus would be on the entanglement of unequal power, viewing CT as emerging "both *through* the web of social, material and discursive knowledge practices that constitute criticality and *with* the different bodies that enact it" (p. 283, author's emphasis).

Many biology, environmental and health education topics, besides being controversial, are emotionally laden, which may influence judgments and decisions about them, and increase difficulties for facing post-truth. Some instances are anthropogenic climate change, genetically modified organisms (GMO), genetic diagnostic, evolution, vaccination, water access, use of fertilizers or food options. Elizabeth Hufnagel (2015, 2019; Chap. 3, this book) is carrying out a program of research on emotional sense-making in climate change, pointing out that emotions provide a lens to understand how learners engaged personally with this issue. She showed how climate change – as may be the case with others mentioned above – requires, besides science understanding, critical evaluations of the ways in which the issue is construed in science education (Hufnagel, 2015). She highlighted the role of *identities* to CT, as reference points for emotions. The focus on science (and engineering) identities is of relevance for the perspective seeing criticality as involving commitment to social justice. Lucy Avraamidou (2020) advocates the adoption of postcolonial, critical race theories and feminist approaches to science identity research for the purposes of addressing classroom inequalities and promoting social change. Kelly et al. (2017) faced the challenge of promoting underrepresented students' recruitment, because girls and African-American students did not identify themselves as engineers. Their work examined how engagement in engineering provided opportunities for building new identities.

Our revised proposal of characterization of the components of CT is framed in these perspectives: considering CT as a dialogic practice and as criticality, comprising attention to identities and to critical action.

1.2.2 Knowledge-Building Dynamics: Post-truth and Science Denial in Biology and Environmental Education

Because CT has been characterized as purposeful judgment, we argue that it is connected to judgments about post-truth (we drop the quotation marks to facilitate reading), science denial and pseudoscience or, in other words, that CT is necessary to face all of them. Pseudoscientific claims and therapies, such as homeopathy or the anti-vaccination movement, have existed for a long time. However, their recent diffusion at a larger scale and the fact of being sometimes endorsed by social or political leaders are causes of concern, leading to the term "post-truth era" (McIntyre, 2018). When choosing post-truth as their word of the year in 2016, the Oxford Dictionaries defined it as "relating to or denoting circumstances in which objective facts are less influential in shaping public opinion that appeals to emotion and personal belief." This concern is reflected in the 2020 special issue of the journal *Educational Psychologist* focusing on how post-truth problems related to scientific and socio-scientific issues may be addressed in education. In their introduction to the issue, Barzilai and Chinn (2020) discussed five trends associated with the post-truth condition, which may help to make sense of this educational and social threat:

(1) Increasing prevalence and influence of *misinformation* (incorrect information without intention to mislead) and *disinformation* (inaccurate information shared with an intent to manipulate or harm); (2) Increasing rejection of well-established claims, which does not mean critique or skepticism, but rather *disagreement about verifiable facts* or claims strongly supported by evidence, such as vaccine safety, evolution or anthropogenic climate change; (3) *Placing personal belief and experience above facts and evidence*, supporting judgments on emotions or on anecdotal reports, rather than on systematic data; (4) *Declining trust in institutional sources of information* such as science and journalism, and increased consumption of news from social media and partisan sources; (5) *Increased fragmentation and polarization* of information consumption.

Barzilai and Chinn also reviewed four approaches or educational lenses to address the post-truth condition: (1) *Not knowing how to know*, how to critically deal with information; (2) *Fallible ways of knowing*, cognitive biases and limitations; (3) *Not caring (enough) about truth*, not committing oneself to the pursuit of epistemic ideals; (4) *Disagreeing about how to know*, a loss of shared epistemology.

Discussing cognitive bias, McIntyre (2018) pointed out how we often reason within an emotional context, and how feeling psychological discomfort may lead to accommodate our beliefs to our feelings. This attitude is termed motivated reasoning, an expectation of data confirming one's beliefs or hoping that what one believe is true. This goal is accomplished through mechanisms such as confirmation bias, by which people interpret information so it confirms their preexisting beliefs. Both have been studied in a range of educational contexts. Rebekka Darner (2019) examined *science denial*, understood as the unwillingness to consider evidence that contradicts one's desired conclusion, and deemed it a form of pseudoscience. She drew from work about the backfire effect (Nyhan & Reifler, 2010), showing how the presentation of refutatory evidence may reinforce mistaken beliefs, as may occur with negationists of climate change. This effect is even stronger when counter-evidence threatens people's identity or worldviews, as may be the case with evolution, highlighting the role of identities, as discussed above. Darner suggested engaging students in critical evaluation of novel explanations through plausibility appraisals. The problem of rejection of information that contradicts individuals' worldview is also discussed by Sharon and Baram-Tsabari (2020), in their work about the identification of misinformation. They suggested teaching open-mindedness in science classrooms.

Asli Sezen-Barrie and colleagues have addressed denialism of climate change and how to deal with it (Sezen-Barrie et al., Chap. 11, this book; Sezen-Barrie et al., 2019, 2020). In doing so, they draw from the idea of critical sensemaking as a lens to look at power relationships in classroom settings. Sezen-Barrie et al. (2020) explored what teachers identify as sources of ambiguities and uncertainties when teaching climate change and scaffolding students' use of epistemic tools. The scholars designed professional learning environments to provide opportunities for teachers to learn how to support students in grappling with uncertainties and in acquiring epistemic agency.

The relationships between post truth and science identity, and the role of social identities, are examined by Lapsley and Chaloner (2020), and we will return to this issue in the next section.

Feinstein and Waddington (2020), drawing from climate change examples, focused on how science education can help people work together to make appropriate use of science in social contexts, to help people survive and sustain democracy in the post truth era. These authors warned against the "neoliberal trope that individuals can and should take on responsibilities that would otherwise fall on institutions" (p. 156), for instance sustainable consumption instead of environmental regulation. Therefore education alone cannot offer solutions to post truth, without elimination of structural factors that exacerbate polarization. They argued that accurate scientific knowledge does not necessarily lead to particular courses of action, so we should be more concerned with public support of policies that produce social and environmental catastrophe than with deep understanding of climate science. Feinstein and Waddington concluded that "for science to be an important part of civic discourse, civic discourse – including its more pluralistic, creative, and chaotic forms – must become an important part of science education." (p. 161).

We draw from these ideas about the post-truth challenges and the relevance of civic discourse for our characterization of CT discussed next.

1.3 A Revised Characterization of Critical Thinking

In a previous work (Jiménez-Aleixandre & Puig, 2012) we proposed a holistic characterization of CT, including four types of components or dimensions, one set of two related to argumentation, and another set of two "related to social emancipation and to citizenship" (p. 1006). Our argument, then and now, is that while almost all definitions of CT take into account the first dimensions related to purposeful judgment, there is another set of dimensions about a second meaning for *critical*, drawing from critical theorists, such as Habermas, Bourdieu or Fairclough, as well as from critical educators, such as Célestin Freinet or Paulo Freire, which should also be included. This second sense, acknowledged in the *criticality* approach, was anticipated in our characterization.

In this chapter we present a revised version of the characterization, including the two sets, with refinements in the definition of the components in light of current perspectives about CT discussed in the previous section.

Set A, *Purposeful judgment*, commitments to epistemic criteria and to evidence: (1) Cognitive and epistemic skills; (2) Critical character, dispositions.

Set B, *Civic participation and social justice*, commitments to independent thinking and civic action: (3) The capacity to develop independent opinions and to challenge socially and culturally established ideas; and (4) The capacity to criticize inequalities and discourses that justify them, aligned with critical discourse analysis (Fairclough, 1995).

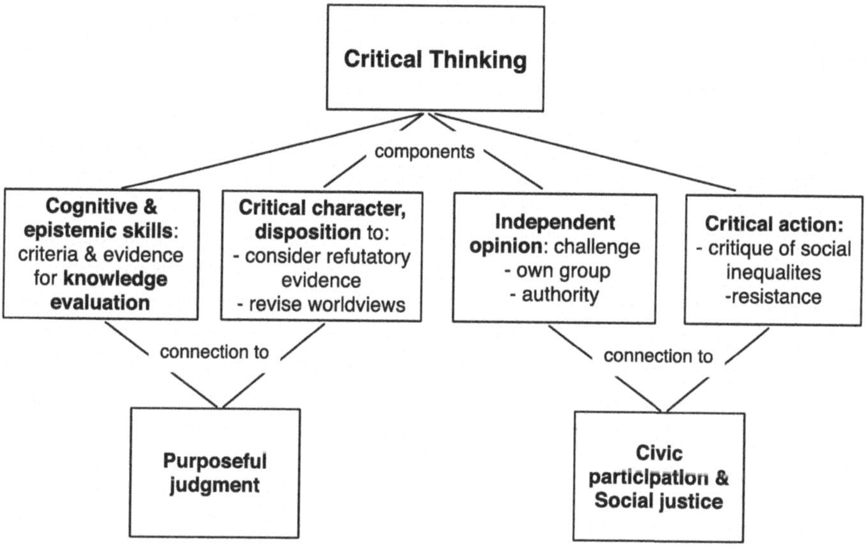

Fig. 1.1 A revised characterization of components of critical thinking

Figure 1.1 summarizes the revised characterization and its components or dimensions, with examples from each dimension that should not be considered an exhaustive list. Next we discuss each component, with more attention to the second set, which is a distinctive feature of our proposal.

A first component, *cognitive and epistemic skills*, involves developing and using epistemic standards or criteria in knowledge building and knowledge evaluation. A core post truth challenge, according to Chinn et al. (2020), is the educational response to disagreements about appropriate ways of knowing. Unpacking these deep epistemic disagreements, they made a case that an apt response to them, and to post-truth, requires making epistemic assumptions visible, justifying and negotiating them, and developing shared commitments to appropriate standards and processes of reasoning, in sum discussing ways of knowing together with others. In order to develop these meta-epistemic abilities, they proposed educational practices termed *explorations into knowing*, illustrated with the vaccination example. These practices involve a shift from content-focused discourse, for instance whether vaccines are harmful, to meta-epistemic discourse focused on epistemic aims, standards and processes, for instance whether systematic studies are more reliable and why, or whether fit with evidence is a more appropriate standard than fit with intuition. Duncan et al. (Chap. 5, this book) make the case for how laypeople, including students, engage in the practice of reasoning with evidence.

A second component, *critical character*, involves dispositions, for instance, to consider evidence, be it supporting one's own claims or contradicting previous beliefs; to revise views; to evaluate the reliability of sources. It requires open-mindedness, regarding a diversity of worldviews. Related to this component is engagement in the process of forming a *science identity* which, as Avraamidou

(2020) argues, is personal but also political, for there are inequalities about who is or is not recognized as a science person in specific contexts. Identities are the reference points for emotions (Hufnagel, Chap. 3, this book), and their role to CT is decisive because how people evaluate information is shaped by their identities.

A third component is the *capacity to develop independent opinions* and *to challenge* socially and culturally established ideas. The focus is on the influence of social interactions and peer's approval on CT. Although a great deal of work about CT focuses on individual learners, studies about the sociology of knowledge show that knowledge and beliefs about the world are socially constructed: "we may discount even the evidence of our own senses if we think that our beliefs are not in harmony with those around us. In short, peer pressure works." (McIntyre, 2018, p. 39). Social conformity, the need for peer group approval, especially in adolescence, and the difficulties experienced for challenging or overcoming socially established ideas are documented in the literature. Joan Solomon (1987) reviewed students' ideas about science through this lens, suggesting that the continual reaffirmation of social notions with other people makes them very durable and resistant to change, as well as rather opaque to rational analysis. She pointed out that too different a viewpoint would exclude from social intercourse, a price that few are ready to pay. The role of identity-protective mechanisms in the preservation of belief systems was examined by Lapsley and Chaloner (2020) in the context of post-truth challenges. They argued that belief systems are relatively immune to correction by contradicting information when they support social identities. In order to face these challenges, they suggested that science education should include the internalization of science identity, as well as a focus on intellectual virtue – cognitive excellence or habits of mind – related to CT. It could be noted that social identities may have been manipulated through appeals to emotions. Brocos and Jiménez-Aleixandre (Chap. 12, this book) examine the obstacles hindering the adoption of decisions challenging culturally established practices.

The capacity to challenge authority or peers' ideas should not be understood as a lack of consideration to different views, on the contrary. It involves a careful evaluation of the information provided by different sources, of the assumptions behind them, of the extent of their support by evidence. We suggest that a crucial disposition in this component is to be prepared to challenge the mainstream ideas of one's own group or community. As an example (Jiménez-Aleixandre & Puig, 2012), it denotes a higher degree of independent thinking to be against capital punishment in some contexts and countries where it is legal, that to be against it in other countries where death punishment was abolished years ago. Darwin delayed for twenty years the publication of *The origin of species* because he was afraid of its controversial nature. His journals and notebooks document his reluctance to make public these ideas, in particular about the origin of man, his fears of conflicts both with the socially dominant creationism and with the religious beliefs of his wife (Desmond & Moore, 1992).

A fourth component is *critical action*, which has, among others, two facets. On the one hand *critical consciousness*, the capacity to analyse and criticize inequalities and discourses that justify them, aligned with critical discourse analysis (Fairclough,

1995). The issue of critical consciousness about environmental issues is explored by Colucci-Gray and Gray (Chap.2, this book), in a work framed by the enactivist approach, paying attention to embodied understanding and the coordination of social action. On the other hand *critical participation*, which in a criticality approach is related to engaging in action, resisting asymmetrical relations of power. We draw from Jürgen Habermas' (1981) notion of critical theory as a form of self-reflective knowledge that expands the scope of autonomy, reducing domination. Habermas' theory of communicative action assigns people the potentiality to develop actions directed to social change. In the context of teaching about climate change, Sezen-Barrie et al. (2020) use critical sensemaking as a lens to look at power relationships in classrooms, where teachers, canonical views of science and assessment systems have more power than students do. Drawing from a decolonial perspective, García-Franco, Farrera-Reyes and Gómez-Galindo (Chap. 4, this book) discuss the contributions of traditional and local knowledge, in the Mayan Highlands in Mexico, to the development of CT relevant to students' lives, and to science learning. Science teachers' forms of resistance in environmentally degraded areas in Chile are analysed from an eco-feminist perspective in González-Weil et al. (Chap. 10, this book). Teachers' sensemaking of educating in sciences is mediated through the relationship between science teaching and territory. In particular actions oriented towards citizen participation about the use and quality of water are discussed. These studies provide examples of CT connection to social emancipation and to citizenship.

1.4 Biology Education and Environmental Education as Privileged Contexts for the Development of Critical Thinking

The perspective of critical thinking oriented to action may apply to a range of educational settings in science, social sciences, language or arts. However, we suggest that biology and environmental education are learning environments particularly relevant for developing CT. In this section we briefly address some reasons that support this view and show the potential of biology and environmental education as privileged contexts for the practice of critical thinking.

First, controversies in biology and environmental education are issues of major concern in a range of life settings. Climate change and its consequences on human and natural systems is an outstanding example. Resources that we depend upon – such as water and energy –, agriculture or human health, are experiencing the effects of global warming. Reducing our vulnerability to these impacts depends not only upon our capacity to understand the scientific issue of climate change and its implications, but also upon us taking critical action to reduce greenhouse emissions or to change intensive breeding.

Second, issues in biology and environmental education have a social impact and are close to students' interests and needs. For instance, issues related to health are personally relevant as they have a direct impact on people. Individuals can be affected by health problems, and what we do to prevent them, can also affect the health of others, as the pandemic of COVID-19 shows. Health and environmental issues offer an opportunity to open up science to individually relevant questions (Zeyer & Kyburz-Graber, 2012), so they can contribute to the development of critical thinking. For instance, understanding resistance to antibiotics will introduce students to the science behind the problem. According to Schulz and Nakamoto (2012), it would not be sufficient to provide students with proper knowledge regarding the effects and possible risks of antibiotics, but also the judgment skills regarding the proper use of antibiotics.

Third, biology and environmental education provide real world situations in which individuals are expected to make reasoned and independent decisions. Health issues such as vaccination, gene therapy, and the COVID-19 pandemic provide students' opportunities to make critical choices based on evidence. Making thoughtful decisions and taking action about environmental problems require independent thinking. For instance, recycling and reducing waste materials is a personal choice made by many people to protect the environment, but not always encouraged by government policies and their communities. Both personal commitment and civic action putting pressure on governments in order to modify structural injustice are needed.

Fourth, environmental and health issues are complex problems, with no clear-cut solutions, that present a degree of uncertainty, so they do not only involve understanding the scientific notions, but also the processes of knowledge construction and knowledge evaluation in science. For instance, environmental problems as the decline of bees allow students' engagement in the critical examination of the different perspectives about diverse causes and consequences of this problem (Puig & Evagorou, 2020); but also on the role of science on problem-solving attempts. These questions may help to raise awareness that science does not provide definitive truths, and offer an opportunity to get students familiar with scientific ways of thinking and researching showing the potential and limits of scientific endeavours (Zeyer & Kyburz-Graber, 2012).

Fifth, SSIs in biology and environmental education include elements of critique and social action. Issues such as vaccination, food consumption, use of land and energies, involve potential consequences of action or inaction that present risks both to the wellbeing of the human society and of the environment. The World Health Organization (WHO) released a campaign call "Vaccines work for all" in April 2020 to promote the use of vaccines to protect people of all ages against disease. CT is necessary for making responsible decisions as vaccination and for understanding the social consequences of our actions. Furthermore, serious crises as the COVID-19 pandemic reveal and magnify existing social inequalities. Recent studies indicate that Black and Latino populations are experiencing higher rates of infection and COVID-related death than their white counterparts across USA (Davila et al., 2020). Critical thinking is crucial for addressing these social inequalities when facing diseases and other biology and environmental problems in the classroom.

Sixth, environmental and health problems are multifaceted problems that require multi-sectorial responses, thus collaborative work to achieve better solutions. The "One health" approach promoted by WHO, means that human health cannot be understood without the health of other living beings and the environment. Many of the same microbes infect animals and humans, as they share the eco-systems they live in, so efforts by just one sector cannot prevent or eliminate the problem (Hitziger et al., 2018).

Seventh, socio-scientific issues in biology and environmental educations are value-laden problems that involve informal reasoning and in many cases emotions. They provide a background of opportunities for value judgment, as for instance when deciding what to eat and the adequacy of different diets according to environmental and nutritional criteria (Brocos & Jiménez-Aleixandre, Chap. 12, this book); when dealing with this issue in the biology classroom, conflicts emerge in the debate, and the acknowledgement that it is not always possible to reach a solution that meets all interests. This acknowledgement is part of the development of critical thinking about complex, real life issues. In order to address SSIs, according to Lombard et al. (2020), cognitive empathy is a significant skill; what means understanding others' emotional reactions in the process of developing CT. Future biology teachers expressed that they feel unequipped to manage students' reactions and to guide these debates (Evagorou & Puig, 2017). While encouraging students to consider evidence-based alternative explanations is of primary importance, it is equally important that teachers are equipped for this purpose.

Eighth, teaching of SSIs in biology and environmental education presents challenges related with pseudoscience and the post-truth era as the anti-vaccination and climate change denial movements and the increase of fake news during the pandemic reveal. Although fake news are not new, the easy way they spread through social media and the Internet makes difficult to control them (Willingham, 2008). The amount of information in social media paradoxically does not help to value and assess other opinions; rather it obstructs CT, particularly perspective-taking (Lombard et al., 2020). Furthermore, the pervasiveness of fake news is more frequent in relation to biology and environmental topics – homeopathy, vaccination, and climate change – than in other science areas. Learners need to develop CT to assess health messages, distinguish scientific facts from opinions, and make personal decisions on immunization and medical treatments, amongst other issues (Puig & Ageitos, Chap. 7, this book; Uskola, Chap. 9, this book).

For students to truly learn how to make argued and responsible decisions on SSI they will encounter in life as those discussed above, they need practice, which aligns with the call for developing CT in biology classrooms. While research on content knowledge and different modes of reasoning on SSI is vast in biology education (Ratcliffe & Grace, 2003), CT development has been understudied. This is the gap that the chapters in this book seek to fill. The timing is appropriate, as the twenty-first century has brought significant changes in all aspects of life, including education (Partnership for 21st Century Learning, 2002), and CT is now being considered a goal.

1.5 Concluding Thoughts

Devastating events shatter certainties and expectations. Although there were numerous wars and catastrophes in many countries after World War II, none of them reached planetary dimension until COVID-19 struck. There are lessons that we, as humanity, should learn from the pandemic, and it would be wrong to strive for a return to life "just as it was before", for it was an unsustainable life and one devoid of equity and social justice. Before the pandemic humans were – they still are – causing severe damage to planet Earth, depleting resources, breaking natural balances, ignoring scientific forewarnings about irreversible effects of global warming, carrying out intensive animal breeding with attention only to money profit and not to environmental care or much less to the rights of sentient beings.

Lessons from the pandemic should be framed in the *One Health* concept, meaning human, animal and environment health. In other words, humans cannot expect to be healthy if animals are living in unhealthy conditions and the environment is deteriorating. Biology taught us those relationships between unwholesome environments and diseases, and humans need to pay heed to them, before the next deadly outbreak. An important lesson is that we should place caring for the environment before benefits for big companies: far from being wishful thinking, this reverse of current values is a question of survival.

The implications for biology and environmental education are, among others, first, that the social impacts of scientific issues, rather than a footnote, should be an integral part of biology teaching; second that the complexity of SSIs need to be fully addressed; third, that a part of developing critical thinking is to be prepared to take critical action. As Colucci-Gray and Gray (Chap. 2, this book) claim, critical education is a process of cultivating consciousness for reason, action and social justice. The Lancet Migration Panel (Orcutt et al., 2020) issued a call for the inclusion of migrants and refugees in the COVID-19 response. It is a question of social justice, equity, human rights but, as these authors pointed out, it should not be overlooked that from a self-interest perspective the control of the outbreak will only be successful if all populations are included. The case of infections beginning with immigrants forced to live in unsanitary conditions is revealing. Fourth, biology education and environmental education should involve learning to use adequate criteria to assess information. The more citizens in a given society are capable of discarding fake news, of distinguishing between reliable and unreliable information, the more democratic that society would be. It would be a step on the road to a world where humans identify themselves as part of the environment and not out of or above it.

Acknowledgements Work supported by the Spanish Ministry of Science, Education and Universities, partly funded by the European Regional Development Fund (ERDF). Contract grant PGC2018-096581-B-C22; and European Union (Erasmus+) project CRITHINKEDU , code, 2016-1-T01-KA203-022808.

References

Abrami, P. C., Bernard, R. M., Borokhovski, E., Wade, A., Surkes, M. A., Tamim, R., & Zhang, D. (2008). Instructional interventions affecting critical thinking skills and dispositions: A stage 1 meta-analysis. *Review of Educational Research, 78*, 1102–1134.

Abrami, P. C., Bernard, R. M., Borokhovski, E., Waddington, D. I., Wade, C. A., & Persson, T. (2015). Strategies for teaching students to think critically: A meta-analysis. *Review of Educational Research, 85*(2), 275–314. https://doi.org/10.3102/0034654314551063

Álvarez Ortiz, C. M. (2007). *Does philosophy improve critical thinking skills?* Master thesis. University of Melbourne. Accessed at https://www.reasoninglab.com/wp-content/uploads/2017/05/Alvarez-Final_Version.pdf

Avraamidou, L. (2020). Science identity as a landscape of becoming: Rethinking recognition and emotions through an intersectionality lens. *Cultural Studies of Science Education, 15*, 323–345. https://doi.org/10.1007/s11422-019-09954-7

Barzilai, S., & Chinn, C. A. (2020). A review of educational responses to the "post-truth" condition: Four lenses on "post-truth" problems. *Educational Psychologist, 55*(3), 107–119. https://doi.org/10.1080/00461520.2020.1786388

Chinn, C. A., Barzilai, S., & Duncan, R. G. (2020). Disagreeing about how to know. The instructional value of explorations into knowing. *Educational Psychologist, 55*(3), 167–180. https://doi.org/10.1080/00461520.2020.1786387

Cui, J., Li, F., & Shi, Z.-L. (2019). Origin and evolution of pathogenic coronaviruses. *Nature Reviews Microbiology, 17*, 181–192.

Danvers, E. C. (2016). Criticality's affective entanglements: Rethinking emotion and critical thinking in higher education. *Gender and Education, 28*, 282–297.

Darner, R. (2019). How can educators confront science denial? *Educational Researcher, 48*(4), 229–238. https://doi.org/10.3102/0013189X19849415

Davies, M., & Barnett, R. (2015). Introduction. In M. Davies & R. Barnett (Eds.), *The Palgrave handbook of critical thinking in higher education* (pp. 1–25). Palgrave Macmillan.

Davila, K., Abraham, M., & Seaberry, C. (2020). *Towards health equity in Connecticut*. The role of social inequality and the impact of COVID-19. DataHaven, June 2020. Accessed at: https://ctdatahaven.org/reports/towards-health-equity-connecticut

Desmond, A., & Moore, J. (1992). *Darwin*. Penguin.

Evagorou, M. & Puig Mauriz, B. (2017). Engaging elementary school pre-service teachers in modeling a socioscientific issue as a way to help them appreciate the social aspects of science. *International Journal of Education in Mathematics, Science and Technology, 5*(2), 113–123. https://doi.org/10.18404/ijemst.99074

Facione, P. A. (1990). *Critical thinking: A statement of expert consensus for purposes of educational assessment and instruction*. California State University.

Fairclough, N. (1995). *Critical discourse analysis. The critical study of language*. Longman.

Fan, Y., Zhao, K., Shi, Z.-L., & Zhou, P. (2019). Bat coronaviruses in China. *Viruses, 11*, 210. https://doi.org/10.3390/v11030210

Feinstein, N. W., & Waddington, D. I. (2020). Individual truth judgments or purposeful, collective sensemaking? Rethinking science education's response to the post-truth era. *Educational Psychologist, 55*(3), 155–166. https://doi.org/10.1080/00461520.2020.1780130

Habermas, J. (1981–1984). *The theory of communicative action*. Beacon Press.

Halpern, D. F. (1998). Teaching critical thinking for transfer across domains. Dispositions, skills, structure training and metacognitive monitoring. *American Psychologist, 53*(4), 449–455.

Hitziger, M. O., Esposito, R., Canali, M., Aragrande, M., Häsler, B., & Rüegg, S. (2018). Knowledge integration in one health policy formulation, implementation and evaluation. *Bulletin of the World Health Organization, 96*, 211–218. https://doi.org/10.2471/BLT.17.202705. Accessed at: https://www.who.int/bulletin/volumes/96/3/17-202705/en/

Hufnagel, E. (2015). Preservice elementary teachers' emotional connections and disconnections to climate change in a science course. *Journal of Research in Science Teaching, 52*, 1296–1324. https://doi.org/10.1002/tea.21245

Hufnagel, E. (2019). Emotional discourse as constructed in an environmental science course. In G. J. Kelly & J. L. Green (Eds.), *Theory and methods for sociocultural research in science and engineering education* (pp. 155–180). Routledge.

IPBES. (2020). *IPBES workshop on biodiversity and pandemics. Executive report.* Accessed at https://ipbes.net/pandemics

Jiménez-Aleixandre, M. P., & Puig, B. (2012). Argumentation, evidence evaluation and critical thinking. In B. Fraser, K. Tobin, & C. McRobbie (Eds.), *Second international handbook for science education* (Vol. 2, pp. 1001–1017). Springer.

Kampourakis, K. (2018). Science and uncertainty. *Science & Education, 27*(9), 829–830.

Kelly, G. J. (2016). Methodological considerations for interactional perspectives on epistemic cognition. In J. A. Greene, W. A. Sandoval, & I. Bråten (Eds.), *Handbook of epistemic cognition* (pp. 393–408). Routledge.

Kelly, G. J., Cunningham, C., & Rickets, A. (2017). Engaging in identity work through engineering practices in elementary classrooms. *Linguistics and Education, 39*, 48–59. https://doi.org/10.1016/j.linged.2017.05.003

Kuhn, D. (2019). Critical thinking as discourse. *Human Development, 62*, 146–164.

Lapsley, D., & Chaloner, D. (2020). Post-truth and science identity: A virtue-based approach to science education. *Educational Psychologist, 55*(3), 132–143. https://doi.org/10.1080/00461520.2020.1778480

Lombard, F., Schneider, D. K., Merminod, M., & Weiss, L. (2020). Balancing emotion and reason to develop critical thinking about popularized neurosciences. *Science & Education, 29*, 1139–1176.

McIntyre, L. (2018). *Post-truth.* MIT Press.

NGSS Lead States. (2013). *Next generation science standards: For states, by states.* The National Academies Press.

Nyhan, B., & Reifler, J. (2010). When corrections fail: The persistence of political misperceptions. *Political Behavior, 32*, 303–330.

Orcutt, M., Patel, P., Burns, R., Hiam, L., Aldridge, R., Devakumar, D., Kumar, B., Spiegel, P., & Abubakar, I. (2020, May 9). Global call to action for inclusion of migrants and refugees in the COVID-19 response. *The Lancet, 395*, 1482–1483.

Partnership for 21st Century Learning. (2002). *Learning for the 21st century.* Partnership for 21st Century Learning. Available at: https://eric.ed.gov/?id=ED480035

Puig, B., & Evagorou, M. (2020). Design of a socioscientific issue unit with the use of modeling: The case of bees. *International Journal of Designs for Learning, 11*(1), 98–107. https://doi.org/10.14434/ijdl.v11i1.24142

Ratcliffe, M., & Grace, M. (2003). *Science education for citizenship.* Open University Press.

Schulz, P. J., & Nakamoto, K. (2012). The concept of health literacy. In A. Zeyer & R. Kyburz-Graber (Eds.), *Science. Environment. Health. Towards a renewed pedagogy for science education* (pp. 69–87). Springer.

Sezen-Barrie, A., Shea, N., & Borman, J. H. (2019). Probing into the sources of ignorance: Science teachers' practices of constructing arguments or rebuttals to denialism of climate change. *Environmental Education Research, 25*(6), 846–866.

Sezen-Barrie, A., Stapleton, M. K., & Marbach-Ad, G. (2020). Science teachers' sensemaking of the use of epistemic tools to scaffold students' knowledge (re)construction in classrooms. *Journal of Research in Science Teaching.* https://doi.org/10.1002/tea.21621

Sharon, A. J., & Baram-Tsabari, A. (2020). Can science literacy help individuals identify misinformation in everyday life? *Science Education.* https://doi.org/10.1002/sce.21581

Solomon, J. (1987). Social influences on the construction of pupils' understanding of science. *Studies in Science Education, 14*, 63–82.
Willingham, D. T. (2008). Critical thinking: Why is it so hard to teach? *Arts Education Policy Review, 109*(4), 21–32. https://doi.org/10.3200/AEPR.109.4.21-32
Zeyer, A., & Kyburz-Graber, R. (2012). *Science. Environment. Health. Towards a renewed pedagogy for science education*. Springer.
Zhou, P., Fan, H., Lan, T., . . . Ma, J.-Y. (2018, April 12). Fatal swine acute diarrhoea syndrome caused by an HKU2-related coronavirus of bat origin. *Nature, 556*, 255–258.

Chapter 2
Critical Thinking in the Flesh: Movement and Metaphors in a World in Flux

Laura Colucci-Gray and Donald Gray

2.1 Introduction: Critical Thinking as the Invitation to Think Anew in Biology Education

Confronted with the global social, economic and environmental crisis, critical questions have been raised as to the purpose and practices of education vis a' vis the future. Writing about the decline of reason in everyday life, Stanley Aronowitz (1977) highlighted 40 years ago that "critical thinking is the fundamental precondition for an autonomous and self-motivated public or citizenry" (p. 49). At heart, critical education is a process of cultivating consciousness for reason, action and social justice. From this perspective, being conscious 'is a radical form of being' (Freire, 1978), in which education helps learners to understand systems of power that regulate social interaction, critically analyse their situation, and to link theory and action for positive change. In biology education, this concern for the 'critical' has manifested itself through a variegated set of research practices, ranging from the analysis of students' preconceptions of biological concepts, to supporting students' argumentation and decision-making on complex, socio-scientific and socio-environmental issues. Arguably however, scholarly work on the possible role of pedagogy in radically transforming education for social and ecological justice, and develop more sustainable imaginations, remains fragmentary at best (Osberg & Biesta, 2021).

L. Colucci-Gray (✉)
Moray House School of Education and Sport, The University of Edinburgh, Edinburgh, Scotland
e-mail: laura.colucci-gray@ed.ac.uk

D. Gray
School of Education, University of Aberdeen, Aberdeen, Scotland

B. Puig, M. P. Jiménez-Aleixandre (eds.), *Critical Thinking in Biology and Environmental Education*, Contributions from Biology Education Research,
https://doi.org/10.1007/978-3-030-92006-7_2

Starting from the more established definition of critical thinking as a skill that can be taught and applied by students to the redefinition of existing conceptual schemas, this chapter will first discuss the value and the limitations of 'teaching for critical thinking' whereby 'thinking' is separate from 'doing'. In the second part of the analysis, we will discuss recent perspectives on critical thinking highlighting the role of language tools and context in shaping the ability to formulate new thoughts. Specifically, we will focus upon enactivist theories of cognition (Gallagher & Lindgren, 2015), offering a renovated attention to the role of the body and language: not as a set of idioms for the delivery of information, but as a tool located within an inherently embodied understanding of oneself *in the world.* (Shapiro & Stolz, 2019). This positions shifts the focus of attention from critical thinking as process taking place 'in the head' to critical thinking as the emergence of conscious thought in dialogue between oneself and the environment. Methodologically, the writing of this chapter will follow the guidelines offered by Saldaña (1999) on the creation of texts unfolding in the manner of theatre, that is, through the construction of three-dimensional portrayals. Perceptions from the research literature will join in dialogue with field-notes extracted directly from instances of teaching: *"revealing how different characters react to one another"* (Saldaña, 1999 cited by Leavy, 2015, p. 185). Following the post-qualitative insights of St. Pierre (2014, p. 7), *the aim is not to offer our "participants to our readers on a silver platter"* for knowledge consumption, but to create a 'collective story' of which the researchers – and writers of this chapter – are also a part. Drawing on current thinking in the philosophy of biology, this chapter will take a critical stance over 'intellectualist' views of critical thinking in biology education with a view to re-align school science with the critical questions raised in biology at the turn of the century.

2.2 Biology at the Interface Between Environment and Society

Starting from an historical account of the evolution of the concept of 'gene', in 2000, Evelyn Fox Keller asked the question: "What will the biology of the twenty-first century look like"? Without claiming to predict the future, Keller made reference to a newly emerging lexicon, one that emphasises 'checkpoints'; 'epigenetics' and 'metabolic networks', in order to signal the profound transformation of the field and most importantly, to engage the audience in considering what *needs* are those terms expected to satisfy in the coming decades. For example, Keller (2010) details the astonishment that derived from new understandings of DNA in the cell. One of the most powerful and challenging critiques was addressed to the very notion of genes as autonomous elements, or that "to think of the development of traits as a product of causal elements interacting with one another" (Keller, 2010 p. 6), that she deemed as fiction, because development depends on the complex orchestration of multiple elements and interactions. Far from being a 'repository' of genetic information, self-sufficient and self-replicating, a more accurate portrayal of genetic transmission reflects the complex and dynamic communication amongst the different component parts of the cell, coordinating with each other to edit, proofread and repair the new molecules. Hence, changes in biological understandings reflected in language point to significant changes of understandings of the living world, caught between early mechanistic ideas to emerging views of the Planet as a living system, a complex network of material and energy transformations.

These ideas resonate with current philosophies of *process* biology (Nicholson & Dupré, 2018), rooted in the philosophical ideas of Whitehead (1925) whom, in *Science and the Modern World,* argued against the reductionist, and determinist view of nature, which dominated for most of the nineteenth and twentieth centuries, by advocating a philosophy of nature that stressed the interdependence of all things. "If science is not to degenerate into a medley of ad hoc hypotheses – argued Whitehead – it must become philosophical and must enter upon a thorough criticism of its own foundations" (Whitehead 1925, p. 25). Hence, the task for a biology of the twenty-first century is that of re-imagining a new philosophy of biology that would emancipate itself from reductionist influences, but also from obscure and obscurantist vitalistic notions, which fail to account for the dialectical tension between change and structure in biology. Another prominent exponent of processual biology, Edward Stuart Russell for example rejected the static construction of the organism as a machine to suggest instead a view of the organism as a 'something happening', as "a phase of a life-cycle" (Russell, 1930, p. 171). As he put it in a subsequent discussion "[i]t is as a lifecycle progression and not as a static organisation that the living thing is ultimately to be conceived" (Russell, 1945, p. 186).

Such recognition changes the focus of biology from the study of things or objects to the study of phenomena taking place at different time-scales, thus calling for a more sophisticated grasp of time and temporal change as different branches of biology may capture it (e.g. from the quicker pace of physiology to the slower *tempo* of inheritance and evolutionary change).

Overall, this distinctly processual approach to biology that emerged from the influence of the metaphysics of Alfred Whitehead, is one that warns against the danger of abstraction and one which rejects a preoccupation with biology as concerned with the primacy of objects and things regarded as the basic 'building blocks' of reality (Nicholson & Dupré, 2018). Rather, the dynamicity of change is primal. In this view, things do not precede processes; and processes do not exist simply as manifestations of the materiality of things undergoing some external agency. As Nicholson and Dupré (2018) remarked: "What we identify as things are no more than transient patterns of stability in the surrounding flux, temporary eddies in the continuous flow of process (p. 13). The peculiar properties of an organism originate from the interrelations amongst the parts, and the same applies to the specific components (e.g. organs; tissues, etc.), whose properties are determined by the interconnections taking place amongst all entities at play. In their process of 'becoming', organisms, parts and relations change themselves and one another (Gagliasso, 2001). Such state of affairs is exemplified by metabolic processes, which concretely demonstrate that our physical beings are a 'knot' amongst the multitude of living processes that connect the bios. Whether we feed on meat or vegetables, and the fact that we are ourselves the living environment for other species of microorganisms, are a physical manifestation and a reminder that our existence is dependent upon a 'sharing of worlds and knowing functions' with the rest of the living world.

Notably, this form thinking has influenced current developments in ecology, with recognition of the limitations of previous ideas of species interactions, which emphasised the negative, competitive aspects (i.e. competition and predation). Understanding nature as flux brings recognition of ecosystems as open, relational units (see Pierce, 2015, p. 192) for also the commonly known 'invasive' species can contribute to maintain stability in nature. For an extensive discussion of this notion of 'shared' worlds, see for example Jablonka and Lamb (2005) on the ongoing interplay of genetic, epigenetic, behavioural and symbolic dimensions in the evolution of living organisms. With reference to a series of studies with rabbits, the researchers illustrate how the intelligence of young rabbits with respect to their feeding habits is shaped by their genetic characteristics, as well as the biochemical conditions of the mother's womb, which is – in turn – continuously shaped by the mother's behaviour responding to wider environmental conditions. Far from being a diktat of their genetic programme, the 'vegetarian diet' of the young rabbits can thus be understood 'contingently' as a result and manifestation of their existence in a 'shared' world'. Interestingly, these ideas appear to resonate with thinking in education – developed in a different time and context – as Vygotsky (1994) maintained that the person is not separate from her environment; instead, there is a unity/identity of the two. To understand the person's characteristic behaviour, one needs to know the characteristics of the environment, including other people and their language.

2.3 Language and Material Demarcations

While acknowledging the growth of support for a processual view of biology influencing other realms of general culture and education, the pervasive bias towards a reductionist view of biological 'things' however is perpetuated in everyday communication through powerful and pervasive forms of demarcation, continuously separating processes from things.[1] As Keller stated (2000), the words that scientists

[1] Keller (2000) provides a critique of the central dogma of the gene according to which 'One gene-one protein' by acknowledging first that such specificity is an idea that is no longer accepted, replaced as it were by established recognition that 'one gene-many proteins'. However, the problem that persists is that there is no means to ascertain how a gene could decide over which protein to make. Keller argues that this problem lies with language underscored by a deeply mechanistic epistemology, which separates process from product: the ultimate responsibility for the product does not lie with the gene but in the complex regulatory system of the cell as a whole (for a fuller articulation of this debate see Keller, 2000, p. 63).

use play a crucial role in the way they are being motivated to action, their attention directed to particular features and questions. Hence, a first line of inquiry into critical thinking in biology starts from the use of language, and most importantly, the materials, economic and social context in which that language originates and functions.

Such understanding is critical with recognition of contemporary developments in biology. On the one hand, we assist at the ongoing reduction of biodiversity globally while on the other hand, developments in computing technologies integrating the bio with the 'micro', 'nano' and 'digital' dimensions (Seo et al., 2019) are significantly modifying original conceptions of the living, and the boundary between our humanity and our machines, including the sense of our own selves. An interesting example is the substantial investment into DNA computing, that is, machines, which incorporate DNA materials instead of the traditional silicon-based transistors, yielding the promise of increased computational power at smaller scales. However the hopes of what DNA computing might be able to achieve raise critical questions about how such technologies will or could interact with the evolutionary processes longer term; and how such developments – regardless of their desirability – call for a critical review of what we understand and want to protect about 'being human' (Facer, 2011).

Donna Haraway (2016) captures this idea through the shift from noun to verb, 'worlding', blending the material and the semiotic to remove the boundaries between subject and environment, to highlight and make possible different temporalities and modes of being. "Reality is an active verb – says Haraway – and the nouns all seem to be gerunds with more appendages than an octopus" (Haraway, 2008, p. 6). 'Worlding' stands in stark contrast with intellectualised views of man and environment, and representational views of knowledge, to assert instead an active, ontological process; a way of being in and attending to the world.

Taking a view of language as a central and dynamic response to the way in which scientists imagine their object of inquiry and asks questions of it provides a route into the process of critical and reflexive inquiry in biology. Arguably, the task of inquiring into future imaginaries of biology is not the task of linguists and philosophers alone, but one that exposes the limitations of contemporary understandings of critical thinking in biology education, which may fall short of accounting for socio-technical change in education. We argue that a contemporary analysis of critical thinking in biology cannot be disentangled from understanding the processual nature of biological entities, brought into sharp focus by the prospected trans-human future that our children might inhabit. In the following sections, we will draw upon a processual view of biology and an enactivist view of cognition to derive some insights for a pedagogy of critical thinking in contemporary biology education.

2.4 Structure and Process Supporting Critical Thinking in Biology Education

Between the years 1899 and 1910, a series of 87 futuristic pictures by Jean-Marc Côté and other artists were issued in France. Originally, in the form of paper cards enclosed in cigar boxes and, later, as postcards, the images depicted the world as it was imagined to be like in the then distant year of 2000 (The Public Domain Review). The images depict all aspects of social life at the time, from agriculture to space exploration, domestic life to education. One of such pictures related to learning and the classroom (Fig. 2.1) is presented to the reader here as a prompt for illustrating features of critical thinking. The question is quite simple: what do you see?

Two views of critical thinking in education have derived from psychological analysis. One is Gestalt psychology – a tradition that focuses on perception (see Wertheimer, 1938). In Gestalt psychology, critical thinking and problem solving are understood as the reorganisation of perception that is, through the ability to approach a problem from a different point of view. Critical thinking is thus largely associated with 'visual' experience and habits, which need somewhat to be 'reorganised' into a new configuration. For example, repeated experience of "perceiving an object in one way, such as a pair of tongs as a tool for handcraft, may hinder its perception in another way, such as seeing the pair of tongs as a possible ballast" (see Maier, 1931 in Schnotz et al., 2010, p, 11). The other is psychology of information processing, focusing on thinking as a linear process of finding the right way through a problem space. This approach functions on the basis that the students will have 'a

At School

Fig. 2.1 'At School'. Futuristic image of education in the year 2000 (as featured in Public Doman Review; public domain)

representation' of the initial state of the problem at their disposal and a rather concrete idea about the goal. The process of learning will then unfold through the setting up of the conceptual space, which will support 'intermediate states' of thinking until the goal state has been reached.

When referring back to the image in Fig. 2.1, the first immediate reaction is that of focusing on the individual components in the picture: first the heads of the students, which are located at the centre of the visual space. Then the cables connecting the heads with the systems of electrical wires above, and moving on to the image on the right hand-side: the whirring machine and the two human figures feeding it with books and human labour. Once the detail of the main subjects is uncovered, further questions then will arise: who is feeding the machine? Who is the boy turning the handle? And why is he dressed differently from the other boys? Such questions lead to a sudden realisation that perhaps the depiction of children learning in the classroom – albeit strange – is not as innocent as it might have first appeared. Some assumptions about equality, the nature of knowledge and the teacher's role are re-viewed.

As summarised by Schnotz et al. (2010), while the central tenet of Gestalt theory is sudden restructuring that leads to insight, the information processing approach relies on the search for possible paths; it is a gradual and linear process leading from the initial state to the goal state. While they may appear as two contrasting approaches, the authors maintain that these are in fact complementary to one another: one focusing on the 'structure' and 'framing' of the problem; the other focusing on the actual procedures. The 'visual' reading of the picture above illustrated this potential integration. The subjects represented in the picture could be 'framed' in Gestalt terms as a 'class' of individuals (in which the boy crouching down near the machine and the man standing up were an 'exception' with the potential to reveal the 'limits' of the frame). Alternatively, they could be perceived as a set of individual entities, showing up in rows of stationary bodies. Drawing on Ohlsson (1992), Schnotz et al. (2010) uphold the view that these different modes of thinking can and should be integrated within the process of learning, the central question being how the problem to be solved (the initial state) is represented and structured – as an image – in the mind of the learner.

Such emphasis on 'mental representations' or 'schema' has given rise to a significant amount of literature in science education, and especially research focussing on the analysis of students' conceptions and representations. The assumption underpinning this work is that learning progresses on the basis of existing representations, responsible for shaping the activation of thinking patterns that students would apply to the matter at hand. Examples of this tradition are taken from the extensive research conducted over the past 30 years on children's views: of the water cycle (Bar, 1989), of the weather (Dove, 1998) and of the shape of the Earth across multiple countries (see Diakidoy et al., 1997; Frède et al., 2011). Most of the studies conclude with making recommendations about the teaching of concepts that need to pre-exist (or to be taught) in order for the correct representation to take form. For example, understanding the weather – and specifically rainfall as a phenomenon – "requires prior understanding of evaporation, accounting for condensation and heaviness" (Bar, 1989, p. 481).

Hence, ascertaining or 'capturing' the existence of students' conceptions (or alternative frameworks, Viennot, 1979) with a view to build a different account – or a different conception – that may accommodate new evidence, has formed the basis of the established pedagogical tradition of teaching for conceptual change (Hewson & Hewson, 1984). In this view, structure and thinking are closely interrelated as learners draw upon their conceptual schema to account for a new observation or an experience and gradually modify their views. However, such interrelation is also somewhat tautological, as the construction of the 'new' representation requires *operations* to be performed on that very structure.

This recognition poses two problems; first, that choosing a new (and potentially better) representation is not as simple as 'picking' one that may be available. This is because representations do not exist per se, but they will show their representational features to the extent to which thinking operations are activated upon them. Secondly, there is the question of 'experience', which in traditional Gestalt theory is understood to be potentially limiting the opportunities for learners to change their ways of seeing and understanding reality. For example, returning to the image in Fig. 2.1, viewers may or may not question what is being depicted on the postcard; if their own experience of classrooms and teaching has been one that resonates with 'sitting at a desk and not talking', the representation on the card – albeit anachronistic – will validate that particular 'configuration' of learning and teaching.

Hence, the idea of critical thinking as a cognitive function that can be taught either in the form of information processing or visual/integrated understanding, reaches an impasse when dealing with students who may not have valid representations at hand. As remarked by Wetzels et al. (2010), learners with different levels of prior knowledge may also differ in intelligence, motivation, or interest, which in turn will affect the activation of thinking processes and learning. Such state of affairs is at odds with the notion advanced by processual biology that human existence is always in flux, for our learning is depending and contingent upon a world in ongoing transformations. The literature on conceptual change falls short of explanations for how to bridge conceptual representations with a pedagogy of critical contingency.

2.5 Embodied Cognition and Enactivism

In a revised account of the work of Vygotsky, Wolff-Michael Roth put forward a critique of classical approaches to learning in science, which are based on representational models of knowing whereby pupils' thinking can be externalised and matched against the correct/incorrect conceptions or schemas. "Studies of conceptions tend to focus on semantics [. . .]. The images that students construct are taken to 'signify' a presumed reality, thereby failing to address other aspects of knowing" (Roth, 2017, p. 256). The author identifies an essential problem with an overly intellectualised view of knowing, namely one that emphasises the mind-body gap, by reducing body movement and experience to intellect. Children's views of the world appear as if they were dead, language being separate from affect and from its materiality (e.g., Vygotsky, 1987).

Indeed, recent contributions from research on learning and cognition across a diverse array of discipline areas, such as philosophy, psychology, linguistics, neuroscience, and computer science, have challenged traditional cognitivist accounts of the mind (Shapiro, 2011). The view that is emerging instead is known as embodied cognition, which can range in form from a weak embodied cognition through to a more radical embodiment known as enactivism (e.g. Gallagher, 2017; Gallagher & Lindgren, 2015; Coello & Fischer, 2016; Shapiro & Stoltz, 2019). Enactivist origins can be traced to the work of Varela et al. (1991) arguing against the idea of cognition as mental problem solving, involving representations in the head, and rather emerging from processes distributed across brain, body and the environment. According to this view, cognition is grounded in our bodily movements (embodied action) and that adaptation to our environment has both emerged from and resulted into a cognitive system that is enacted through 'structural coupling' of organism and context (for an extended account, see Shapiro & Stolz, 2019).

While classical cognitive science directed its attention towards internal mechanisms (individualised focus), embodied knowing emphasises the intersubjective and socially situated dimensions of being and learning. Most crucially, the enactivist view of cognition lays emphasis on movement, with the idea that the sensory-motor apparatus structure our perceptions of the world (Noë, 2004). In other words, what 'enters' our field of perception literally depends on 'what we can do' in that field of action-perception, thus in terms of its pragmatic meaning. As it was already explored by John Dewey almost a century ago, the pragmatist approach to cognition underscores that perception is not a passive phenomenon triggered by an external sensory stimulus, but it is an active process, one that comes into being in sensorimotor coordination. "It is the movement that is primary, and the sensation which is secondary, the movement of body, head and eye muscles determining the quality of what is experienced (Dewey, 1896, p. 358). For example, a chair is perceived as 'apt' for sitting not because of the shape itself, but because of the characteristics of the human body, which is able to bend legs and knees at particular angles. By the same token, a chair may afford other functions, as other objects may afford the opportunity to sit down. In such interplay of bodily movement, perception and context lies great potential for making meanings and for developing linguistic creativity, e.g. a large rock on the beach may be named as a 'throne' for admiring the sea-view.

2.6 The 'Thinking' Body

In relation to learning, the most significant contribution of embodied cognition is a renovated understanding of language, not as the labelling activity of a dispassionate observer but as an expression of sensorimotor engagement of the organism-in-the-world. The well-known work of Lakoff and Johnson (1980) illustrated the role of basic bodily movements and gestures (e.g. stepping forward; moving back; pointing to) in the formation of abstract concepts expressed through metaphors. Their work literally illustrated the work of the body understood as a nexus of information, energy and materials flows in intra-actions. For example, we experience the body as a 'container' when we eat; or as a box into which we can 'bury' our thoughts and feelings when we feel unable to move, relate or communicate with others. Such dynamic exchanges come to our awareness in the form of 'image-schemas', which do not attempt to 'depict' an image of the world in the manner of a snapshot, but they capture the mind-body-state in its complex intra-actions at any one time.

From a scientific perspective, the creation of metaphors is central to the ability to conceptualise processes that may not be accessible to the field of human, everyday experience (e.g. at the micro or macro scale); most interestingly however, the embodied cognition perspective is also suggestive that image-schemas are situated 'in the environment' as opposed to being solely 'in the head'. Metaphors are thus considered as "tools for working with the experience" (Kirmayer, 1992, p. 335), and which emerge from experience in an ecological fashion (Gibson, 1979).

From an educational perspective, the enactivist position shines new light on the processing of information. While it is known that pictorial representations require less mental effort than verbal representations (Cox, 1999; Mayer, 2001), recent experiments with participants reading or listening to language describing sensorimotor events – either in action or in visual language – have shown the fast activation of cortical, motor hand areas with verbs related to hand actions (e.g. Tettamanti et al., 2005). In addition, such sensorimotor involvement is dependent on both task and linguistic context (Willems & Francken, 2012).

Returning to the image presented in Fig. 2.1, we discussed some of these ideas with a group of first year students enrolled in a Degree Course in Primary Education, as part of an introduction to an elective course on outdoor learning led by Donald Gray. Students contributed free responses to the pictorial representation, aided by the use of three stimulus questions[2]:

> *it looks like the body's function is to be a vessel in which knowledge can be uploaded*;
> *The body seems to be useless, and its only purpose is to sit and allow knowledge to be unnaturally absorbed.*

Such responses are interesting as they emphasise processes of transmission and assimilation, a vertical structure literally channelling information from point A to point B. It is notable the level of metaphorical language that is being used, with the body being 'like a vessel' and 'knowledge being either uploaded' in the manner of computers; or 'absorbed' in what may be the manner of a sponge. The use of metaphorical language directs attention to the mind-body split whereby the body is the carrier of abstract representations...

> *the body seems to be portrayed as a working tool...*;
> *The body seems to be almost part of a machine.*

[2]The three stimuli were set as follows: (a) What is the role of the body (if any)?; (b) What are the children learning? (c) Is it like that today (How is it different; How is it the same)?

But most interestingly, it also points to the vicarious triggering of sensorimotor perception associated with familiar technological structures in a manner that the students will have experienced in their everyday lives, and such vicarious experience is underscored by implicit value-positions. As indicated by Leavy (2015), students 'voices are revealing of how they respond affectively to the teacher-student split:

children learn whatever the teacher decides

Seemingly affirming a political and ethical stance, which arises from the particular configuration of the visual and technological structure in the picture.

Students' comments are of particular interest, as they spur us to think differently about the function of signs and tools in visual representations. Image-schemas do not exist in isolation but are constituent parts of participants and their relations – they are in fact the means of their relations. According to an enactivist view, thought does not pre-exist doing (or talking) but instead "the movement of thinking from thought to word is a developmental process. Thought is not expressed but completed in the word" (Vygotsky, 1987, p. 250, emphasis added). Secondly, according to the pragmatist branch of enactivism, not only thought is completed in the word, but also words do not stand by themselves, continuously becoming in the ongoing process of transaction (Dewey, 1896) of the organism through the environment. Thinking is not expressed – or externalised – by the gestures of the body, but it exists *through* it.

2.7 New Clothes for the Emperor…?

Taking an enactivist position on learning and teaching in biology, is the equivalent of a tectonic shift in the way we understand children's ideas and conceptions in biology, and our choices of methods for teaching and research. In the first instance, we are required to think of thinking 'as the moving body', and thus always 'in the making'. In this sense, as remarked by Pink (2012), embodied knowledge is not simply 'stored information' but it involves biological processes. Secondly, such a dynamic view of thinking challenges the relationship between researchers (or teachers) and pupils: understanding of mental conceptions can only take place through ongoing physical intra-actions. Thirdly, in order to support critical thinking in biology – as well as other disciplines – we need to be more acutely aware of the role that the body plays in the development of the relation – that is – how bodies relate through the activity of learning.

Roth (2017) provides an example of such principles translated in engineering teaching practice by showing how "the work-related (hand, body) movements that build and manipulate artefacts, or sensing (hand) movements deployed during an investigation, later function as symbolic movements" (p. 257). And in a similar vein, the field of mathematics education has taken an embodied cognition stance towards learning by recognising the link between abstract thinking and physical gestures (Duijzer et al., 2019). Communication is thus interwoven with doing, employed for the purpose of doing engineering design more so than being about the subject. This is so because, for example:

> when I describe a circle with my hand on a piece of paper . . . my body . . . comes into a state fully identical with the form of the circle outside my body, into a state of real *action* in the form of a circle" (Il'enkov, 1977, p. 69, original emphasis).

Similarly, recent contributions from the field of childhood studies (see Lenz-Taguchi, 2010) are shifting attention from interpersonal human relationships towards intra-active connections amongst all living organisms and the material environment, such as objects and artefacts, spaces and places. In alignment with a process view of biology and learning, the focus is on "how thinking and living in pedagogical practices is an entwined material-discursive business which will make us think about and perform our work differently" (p. 90).

To some readers, such dynamic articulation of knowing, doing and thinking may feel familiar, as there is a long tradition in science teaching of deploying kinaesthetic approaches to bring scientific knowledge alive in the classroom. Notable examples include the application of dramatic performances for modelling of scientific concepts (Abrahams & Braund, 2011); simulations of decision-making processes on environmental issues (Colucci-Gray et al., 2006) and embodied modelling of conceptual metaphors to generate understanding of common biological processes (Niebert & Gropengiesser, 2015). As Treagust and Duit (2015) underlined, the body of consolidated scientific knowledge is overflowing with metaphors; so it is not yet clear what may be the value added of embodied cognition to science education and how this approach can contribute to critical thinking. Traditionally, it is the skill of the teacher to create space through discussion "to allow learners to critique whatever analogous model or method is used so that its successes in promoting understanding and its limitations as a version of scientific reality are clear" (Braund, 2015, p. 115).

In order to overcome this impasse, it is helpful to return to Gallagher and Lindgren (2015) who distinguished 'sitting' metaphors – which are part of language, are taken for granted, and may help with understanding of texts; and 'live' metaphors, as ones that are brought into existence through action. Literally, through the process of 'acting' in the '*as if*' state. While the two metaphors may coincide, (e.g. see for example the metaphor of 'cycle') what changes is that sitting metaphors have long lost the original connection with the context in which they first originated. So, in biology for example, we recognise a number of sitting metaphors which accrue the body of knowledge but are derived from different strands of biological thought, connected to society at different historical points, from conservation genetics (e.g. bottleneck; genetic drift), to system theory (e.g. networks, nodes; for an extensive account see Larson, 2011). Hence, in the first instance, selecting a 'good' metaphor requires an implicit understanding of the larger linguistic ecology in which that metaphor is located. The 'choice' of metaphor will thus be telling of the social and cultural context of any speaker/learner, and revealing something of his or her own selves. Arguably however, engaging with the literary content of metaphors brings a new set of implications for research in biology education, calling for greater attention to the meaning of the word *per se*, and not simply for their functional use as 'indicators' of acquired scientific knowledge. In fact, if metaphors are not "merely shorthand for facts" (Larson, 2011, p. 128), the modelling of conceptual metaphors will not be simply an epistemic decision but it will be more correctly understood as a 'rhetorical act' inviting viewers *to act upon* the world *as if* it were configured in a specific way (Bono, 1990).

2.8 Enlivened Metaphors in 'Rhetorical Acts'

Following Andrews (2014), the rise of interest in democratic representation, internationally, and across different sectors of society, has been accompanied by the rise of interest in argumentation, that being as a function of social communication and as a domain of scientific education. However, understanding the rhetorical dimension of scientific learning is key for making sense of the role of argumentation in society, and primarily, sharpening understanding of "the power and limitations of communication in whichever modes [. . .] and a critical perspective of communication and action and how they work in relation to each other" (Andrews, 2014, p. 73). A contemporary view of rhetoric therefore rejects intellectualised models of communicating to acknowledge instead the wide range of thought and feeling that goes into communication.

In this view, we can attempt to provide a renovated understanding of language in biology and its connections with critical thinking. We will conclude with two examples. The first one is addressed to the field of research on argumentation. In a study focussed on students' argumentation explaining the links between sickle cell disease and malaria (Ageitos et al., 2019), we invited readers to adopt 'a double lens' to the reading of students' discussions related to genetics and evolution. Far from seeking to label all argumentative moves, we sought instead to uncover and make explicit the relational-discursive configurations underpinning students' understanding of causal events, often presented in the form of silent metaphors. While the problem of the relationship between sickle cell disease and malaria centres on the complex adaptive responses of organisms and environments, students in the study continued to focus on traits or causes that were specific to individuals, thus re-presenting the old issue of linear causality in genetics (Jiménez-Aleixandre, 2014). However, when analysed through the lenses of rhetorical analysis, we found that the 'silent' metaphor framing their dialogue was the 'mystery culprit' identified as the original carrier of the infection, as opposed to looking for systemic, multi-levelled interactions. Such metaphors were both used and 'enacted' by the students through the search for actors, places, times and specific actions, combining induction and deduction to justify the account of events there portrayed.[3] From a research perspective, such 'critical' reading of students' discussions involves a shift from approaching biology learning as a set of content to engaging with students' discussions in the manner of a text: whose voices are represented? Who is included? Who is excluded? Who has remained silent?

The second example is derived from recent work with future teachers in a University-based teacher education context at the first author's (Colucci-Gray) university. As part of a session outdoors that due to coronavirus restrictions took place in the University's Rugby pitch, students were asked to 'take a walk' around the space, collect any item they would find significant and arrange it into a piece of 'land art'. The session titled Sustainability in STEM was an introduction to inter-disciplinary learning and place-based education. There in the pitch we were standing at the foot of Arthur's Seat, the volcanic plug that is the main peak of hills in the city of Edinburgh. One of the chemistry students returned with several items of discarded plastic that she collected from under the trees at the margins of the pitch. She titled the display as a 'volcano of rubbish'! (Fig. 2.2)

A powerful metaphor, which contained in itself the imagery of the volcano overlooking the pitch along with the troubled history of fossil fuel extraction; the linear thinking connecting production with disposal and the gradual growth of awareness of the unintended consequences. As part of a Gestalt switch, the newly made 'art form' brought the problem 'from the margins to the centre'. From a cognitive point of view, a shift of *'reference-point'* occurred here, understood as a schematic procedure of imaginative nature mediated by the visual art form leading to a switch of target focus: from the green grass that had been presented to us upon arrival, to the rubbish that was hidden away (and yet ubiquitous feature of our urban lives). But the conceptual switch was also a rhetorical act, inviting to '*see and act upon the world'* in the manner emphasised by Bono (1990), and an ontological shift from the world 'out there' to a world in the making.

[3]Capra and Luisi (2014) effectively illustrate how even the classical Aristotelian syllogism is not a form of disembodied reasoning but it grows out of our bodily experience. For example, the physical experience of the body as container (e.g. of infected blood) or the body as 'carrier' (e.g. of an infection) is projected onto an abstract category (e.g. the idea of cell disease) thus giving legitimacy to the logical reasoning that follows. As the authors stated: "the structures and bodies and brains determine the concepts we can form and the reasoning we can engage in" (p. 273). From this, it follows that an education for critical thinking will pay greater attention to the experience of the body and its unconscious thought.

Fig. 2.2 'A volcano of rubbish' – land art

2.9 Conclusions

In this chapter, we reviewed some of the current debates on the use of visual representations, as they are part of conventional practice of both teaching and research in biology education. This chapter took a stance against intellectualist views of knowledge as being discordant with current contributions of processual biology and post-humanism proffering the need for a different stance on knowledge and learning: one that is rooted into the body as a prime locus of knowing. In this view, we argue that an education focussed on knowing about the world is no longer sufficient. As humans that are an integral part of the Earth's web of energy and material transformations, a critical stance on biology education is one that will enable us to 'read' the transformations of which we are a part. This requires a practical, experiential and fundamentally relational understanding of how the world works, but also an openness towards literary and aesthetic engagement with experience, as a source of new metaphors that brings together cognition with feelings and action in current times.

We argue that a new approach to critical thinking in biology is one that affords itself of greater epistemological sophistication, supported by the possibility of a 'moving/feeling/thinking' body in a field of space-relations. We have illustrated the importance of recognising the role of space – e.g. the outdoors – as a key target source for conceptual thinking, but also as a place where we can immediately assess the impact of our actions, both cognitively but primarily affectively. A feeling body is one that pays attention to what is deemed relevant or irrelevant and it is open to switching the target focus from the inside to the outside; from what is noticed to what is silent.

The voices of our students from the outdoor learning course keep resonating in our heads… what is education like today? Is it similar…? Different?

> *I think that nowadays there is an emphasis on learning data as opposed to knowledge or wisdom. Contextless data provides little opportunity for reflection or space for the learner to ask questions and be critical of their education and society.*

An education for critical thinking is thus an education, which moves from a vertical mode of thinking, whereby the reference points are the systems of symbolic abstractions, to a horizontal thinking mode. Here the primary goal is not the control and management of knowledge but the opportunity to put in relation and connect information and experiences from different contexts and different sources. More research is needed on learning and teaching processes seeking to integrate these different forms of thinking, for example, by combining linguistic analysis with practical modelling, ranging from illustration, design, making and dance. In such process, greater attention is paid to visual aesthetics. Not as a means to aid the transfer of scientific knowledge, but as a source of affective metaphorical thinking which can extend our abilities – as students, teachers, researchers – *to think the world* (for a fuller account of these different discourses on knowledge, and the connections between arts and science via STEAM, see Colucci-Gray et al., 2019). A biology education for critical consciousness is also concerned with attention to the world as we make it, directing attention towards solidarities with different social movements, such as those struggling for food sovereignty (https://viacampesina.org), climate justice (https://climatejusticeaction.net) or racial justice (https://blacklivesmatter.com), whereby biological knowledge is brought into direct contact with the lives of different people.

Resonating with Keller (2000, p. 148): "At my optimistic, I even imagine the possibility that new concepts can open innovative ground where scientists and lay persons can think and act together to develop policy that is both politically and scientifically realistic".

References

Abrahams, I., & Braund, M. (Eds.). (2011). *Performing science. Teaching chemistry, physics and biology through drama*. Continuum.

Ageitos, N., Puig, B., & Colucci-Gray, L. (2019). Examining reasoning practices and epistemic actions to explore students' understanding of genetics and evolution. *Science & Education, 28*(9), 1209–1233. https://doi.org/10.1007/s11191-019-00086-6

Andrews, R. (2014). *A theory of contemporary rhetoric*. Routledge.

Aronowitz, S. (1977). Mass culture and the eclipse of reason. The implications for pedagogy. *College English*, April, 768–774.

Bar, V. (1989). Children's views about the water cycle. *Science Education, 73*(4), 481–500.

Bono, J. J. (1990). Science, discourse, and literature: The role/rule of metaphor in science. In S. Peterfreund (Ed.), *Literature and science: Theory and practice*. Northeastern University Press.

Braund, M. (2015). Drama and learning science: An empty space? *British Educational Research Journal, 41*(1), 102–121.

Capra, F., & Luisi, P. L. (2014). *The systems view of life. A unifying vision*. Cambridge University Press.

Coello, Y., & Fischer, M. (Eds.). (2016). *Perceptual and emotional embodiment: Foundations of embodied cognition* (Vol. 1). Routledge.

Colucci-Gray, L., Camino, E., Barbiero, G., & Gray, D. (2006). From scientific literacy to sustainability literacy: An ecological framework for education. *Science Education, 90*, 227–252.

Colucci-Gray, L., Burnard, P., Gray, D. S., & Cooke, C. F. (2019). STEAM (Science, technology, engineering, and mathematics, with art). In P. Thomson (Ed.), *Oxford research encyclopedia of education* (pp. 1–26). Oxford University Press.

Cox, R. (1999). Representation construction, externalised cognition and individual differences. *Learning and Instruction, 9*, 343–363.

Dewey, J. (1896). The reflex arc concept in psychology. *Psychological Review, 3*(4), 357–370.

Diakidoy, A., Vosniadou, S., & Hawks, S. (1997). Conceptual change in astronomy: Models of the earth and of the day/night cycle in American-Indian children. *European Journal of Psychology of Education, 12*(2), 159–184.

Dove, J. (1998). Alternative conceptions about weather. *School Science Review, 79*(289), 65–69.

Duijzer, C., Van den Heuvel-Panhuizen, M., Veldhuis, M., et al. (2019). Embodied learning environments for graphing motion: A systematic literature review. *Educational Psychology Review, 31*, 597–629.

Facer, K. (2011). *Learning futures. Education, technology, social change*. Routledge.

Frède, V., et al. (2011). The acquisition of scientific knowledge: The influence of methods of questioning and analysis on the interpretation of children's conceptions of the earth. *Infant and Child Development, 20*(6), 432–448.

Freire, P. (1978). *Pedagogy in process: The letters to Guinea-Bissau*. Writers and Readers.

Gagliasso, E. (2001). *Verso un'epistemologia del mondo vivente*. Guerini Studio.

Gallagher, S. (2017). *Enactivist interventions : Rethinking the mind*. Oxford University Press.

Gallagher, S., & Lindgren, R. (2015). Enactive metaphors: Learning through full-body engagement. *Educational Psychology Review, 27*, 391–404.

Gibson, J. J. (1979). *The ecological approach to visual perception*. Houghton Mifflin.

Haraway, D. (2008). *When species meet*. University of Minnesota Press.

Haraway, D. (2016). *Staying with the trouble: Making Kin in the Chthulucene*. Duke University Press.

Hewson, P. W., & Hewson, M. G. A. (1984). The role of conceptual conflict in conceptual change and the design of science instruction. *Instructional Science, 13*, 1–13.

Il'enkov, E. V. (1977). *Dialectical logic: Essays on its history and theory*. Progress.

Jablonka, E., & Lamb, M. J. (2005). *Evolution in four dimensions. Genetic, epigenetic, behavioural and symbolic variations in the history of life*. MIT.

Jiménez-Aleixandre, M. P. (2014). Determinism and under-determination in genetics: Implications for students' engagement in argumentation and epistemic practices. *Science & Education, 23*(2), 465–484.

Keller, E. F. (2000). *The century of the gene*. Harvard University Press.

Keller, E. F. (2010). *The mirage of a space between nature and nurture*. Duke University Press.

Kirmayer, L. J. (1992). The Body's insistence on meaning: Metaphor as presentation and representation in illness experience. *Medical Anthropology Quarterly, 6*(4), 323–346.

Lakoff, G., & Johnson, M. (1980). *Metaphors we live by*. Chicago University Press.

Larson, B. (2011). *Metaphors for environmental sustainability. Redefining our relationship with nature*. Yale University Press.

Leavy, P. (2015). *Method meets art. Arts-based research practice* (2nd ed.). The Guilford Press.

Lenz-Taguchi, H. (2010). *Going beyond the theory/practice divide in early childhood education: Introducing an intra-active pedagogy*. Routledge.
Mayer, R. E. (2001). *Multimedia learning*. Cambridge University Press.
Nicholson, D., & Dupré, J. (2018). *Everything flows: Towards a processual philosophy of biology*. Oxford University Press.
Niebert, K., & Gropengiesser, H. (2015). Understanding starts in the mesocosm: Conceptual metaphor as a framework for external representations in science teaching. *International Journal of Science Education, 37*(5–6), 903–933.
Noë, A. (2004). *Action in perception*. MIT Press.
Ohlsson, S. (1992). Information processing explanations of insight and related phenomena. In M. Keane & K. Gilhooly (Eds.), *Advances in the psychology of thinking* (pp. 1–44). Harvester-Wheatsheaf.
Osberg, D., & Biesta, G. (2021). Beyond curriculum: Groundwork for a non-instrumental theory of education. *Educational Philosophy and Theory, 53*(1), 57–70.
Pierce, F. (2015). *The new wild. Why invasive species will be nature's salvation*. Iconbooks.
Pink, S. (2012). Principles for sensory ethnography: Perception, place, knowing, memory and imagination. In *Doing sensory ethnography*. Sage.
Roth, W.-M. (2017). The thinking body in/of multimodal engineering literacy. *Theory Into Practice, 56*(4), 255–262.
Russell, E. S. (1930). *The interpretation of development and heredity: A study in biological method*. Clarendon.
Russell, E. S. (1945). *The Directiveness of organic activities*. Cambridge University Press.
Saldaña, J. (1999). Playwriting with data: Ethnographic performance texts. *Youth Theatre Journal, 13*(1), 60–71.
Schnotz, W., Baadte, C., Müller, A., & Rasch, R. (2010). Creative thinking and problem solving with depictive and descriptive representations. In L. Verschaffel, E. De Corte, J. Elen, & T. de Jong (Eds.), *Use of external representations in reasoning and problem solving: Analysis and improvement* (pp. 11–36). Routledge.
Seo, J., Kim, S., Park, H. H., Choi, D. Y., & Nam, J.-M. (2019). Nano-bio-computing lipid nanotablet. *Science Advances, 5*(2), eaau2124.
Shapiro, L. (2011). *Embodied cognition*. Routledge.
Shapiro, L., & Stolz, S. A. (2019). Embodied cognition and its significance for education. *Theory and Research in Education, 17*(1), 19–39.
St. Pierre, E. (2014). A brief and personal history of post qualitative research toward "post inquiry". *Journal of Curriculum Theorizing, 30*(2), 2–19.
Tettamanti, M., Buccino, G., Saccuman, M. C., Gallese, V., Danna, M., Scifo, P., et al. (2005). Listening to action-related sentences activates fronto-parietal motor circuits. *Journal of Cognitive Neuroscience, 17*, 273–281.
The Public Domain Review. *A 19th century vision of the year 2000*. Available at: https://publicdomainreview.org/collection/a-19th-century-vision-of-the-year-2000. Accessed 29 Oct 2020.
Treagust, D. F., & Duit, R. (2015). On the significance of conceptual metaphors in teaching and learning science: Commentary on Lancor; Niebert and Gropengiesser; and Fuchs. *International Journal of Science Education, 37*(5–6), 958–965.
Varela, F., Thomson, E., & Rosch, E. (1991). *The embodied mind. Cognitive science and human experience*. The MIT Press.
Viennot, L. (1979). Spontaneous reasoning in elementary dynamics. *European Journal of Science Education, 1*(2), 205–221.
Vygotsky, L. S. (1987). *The collected works of L. S. Vygotsky* (Vol. 1: Problems of general psychology). Springer.
Vygotsky, L. S. (1994). The problem of the environment. In R. van der Veer & J. Valsiner (Eds.), *The Vygotsky reader* (pp. 338–354). Blackwell.

Wertheimer, M. (1924/1938). Gestalt theory. In W. D. Ellis (Ed.), *A source book of gestalt psychology* (pp. 1–11). London Kegan.
Wetzels, S. J., Kester, L., & van Merriënboer, J. G. (2010). Use of external representations in science prompting and reinforcing prior knowledge activation. In L. Verschaffel, E. De Corte, J. Elen, & T. de Jong (Eds.), *Use of external representations in reasoning and problem solving: Analysis and improvement* (pp. 225–241). Routledge.
Whitehead, A. N. (1925). *Science and the modern world*. Cambridge University Press.
Willems, R. M., & Francken, J. C. (2012). Embodied cognition: Taking the next step. *Frontiers in Psychology, 3*, 582. https://doi.org/10.3389/fpsyg.2012.00582

Chapter 3
Emotional Sense-Making and Critical Thinking in the Era of Post-truth: The Case of Climate Change

Elizabeth Hufnagel

3.1 Introduction

How instructors attend to the intersection of school science and people's identities, that span and cross school walls and boundaries (Aikenhead, 1996; Darner, 2019), have raised important questions about the permeability of classroom walls in science sense-making. In particular, uncertainty exists about how to teach science in an era when ideas that are substantiated and established in science communities are contested by policy makers, world leaders, and the general public. While this anti-science trend is not new, it has become more amplified and powerful in the current post-truth world in which social media use abounds. Critical thinking has been put forth as one tool "for confronting pseudoscience and credulity" (Jiménez-Aleixandre & Erduran, 2007, p. 8). Attending to critical thinking, though, invokes questions about how science teachers grapple with teaching science in ways to accurately represent scientific knowledge and practices while attending to students' identities. What comes to the fore, then, are the emotions of the teachers and their students, and how teachers navigate critical thinking with respect to emotions. As such, the discussion of what counts as truth, as science, and as critical thinking are embedded in emotional relationships with epistemologies and identities.

The aim of this chapter is to consider the intersection between what counts as truth and emotions and how unpacking these spaces makes salient assumptions about emotions within critical thinking. In particular, I bound this exploration to the case of climate change denial. I first briefly describe critical thinking theorizations that highlight the situated ways in which it is accomplished. In doing so, I draw

E. Hufnagel (✉)
University of Maine, Orono, ME, USA
e-mail: elizabeth.hufnagel@maine.edu

B. Puig, M. P. Jiménez-Aleixandre (eds.), *Critical Thinking in Biology and Environmental Education*, Contributions from Biology Education Research,
https://doi.org/10.1007/978-3-030-92006-7_3

from theorizations of epistemology in science education to describe how underpinnings of critical thinking link to anti-science stances in the public discourse on climate change and science. A thread throughout these spaces is identity, which is the reference point for emotions. As such, I describe emotional sense-making, highlighting the role of identities, to critical thinking. I then present a set of findings from a methods course on climate change education for pre-service secondary science teachers to underscore the intersection between the teachers' shared goal of critical thinking and how they orient to students who deny human-caused climate change.

3.2 Critical Thinking in Science Education

Drawing from fields of philosophy and psychology, critical thinking has been conceptualized in a variety of ways. In science education, critical thinking is associated with scientific argumentation as it largely involves the evaluation of claims or knowledge by examining evidence (Jiménez-Aleixandre & Puig, 2012). Critical thinking entails "developing criteria for evaluating" data, evidence, and experts (López-Facal & Jiménez-Aleixandre, 2009, p. 694). According to Jiménez-Aleixandre and Puig (2012), critical thinking extends beyond aspects of scientific argumentation – knowledge or claims evaluation and a disposition to seek reasons, including challenging authority – to include emancipatory features. One such emancipatory aspect is the questioning of information so as to not subscribe to group opinions as a default. Another feature is the critical analysis of discourse for inequalities or the justification of inequalities. Within this theorization, Jiménez-Aleixandre and Puig (2012) orient toward the role of discourse of critical thinking and in doing so, open the door to what is salient in the discourse when critical thinking is negotiated, practiced, and accomplished.

While an extensive overview of influences on critical thinking outside of emotions is beyond the scope of this chapter, the ways in which epistemology and the situatedness of critical thinking shape emotions within critical thinking is important. Critical thinking is closely linked to epistemic beliefs and practices (Greene & Yu, 2015; Jiménez-Aleixandre & Erduran, 2007). In particular, Jiménez-Aleixandre and Erduran (2007) point out the distinction between personal (or practical) epistemologies and those of scientists in scientific argumentation. While the demarcation is needed for robust epistemologies of science areas of inquiry, I orient to epistemological classifications as disciplinary, personal, and socially practiced (Kelly et al., 2012). The disciplinary perspective centers on the normative philosophical features of disciplinary structure and evidence embedded in specific science communities. The personal perspective focuses on "personal views of truth, rather than on disciplinary considerations of rationality, truth, and justification" (Kelly et al., 2012, p. 282). The crux of the social perspective is on the social practices in which knowledge is constructed *in situ*, which Jiménez-Aleixandre (2014) describes as enacted epistemic disciplinary practices. Across these different orientations to epistemology is one particular theme that is most salient to the nexus of critical thinking and emotions,

> epistemology is interpreted, not only in the traditional sense, concerning the origins, scope, nature, and limitations of knowledge, but as an interactional accomplishment among members who define for themselves what counts as knowledge in a particular context. . .This view suggests that knowledge be examined as it occurs in practical actions, rather than as measured by students' decontextualized views of epistemology, nature of science, and so forth. Thus, through interaction with the world and each other, members of communities come to define what counts as knowledge, evidence, explanation, and so forth, and embody an epistemology through such actions (Kelly et al., 2012, p. 288).

Although Kelly et al. (2012) did not consider science denial in their chapter and focused their theorizations on research within science learning settings, their suggestions about knowledge construction and group affiliation lend themselves to this discussion on critical thinking. Since knowledge is constructed through interactions with the world and others and that affiliation shapes what counts as knowledge, including evidence-based explanations, it is not much of a jump to understand how people have come to deny particular science explanations. Within communities (face-to-face and online), affiliation shapes what counts as knowledge—and what doesn't. Hence, epistemology is imbued with identities—singular and collective—which are frames of reference for making sense of the world.

An eye toward the three perspectives of epistemology (as disciplinary, personal, and socially practiced), then, suggests that critical thinking is dynamic. However, underlying many theories of critical thinking is the assumption that it is a static skill (e.g. Danvers, 2016; Pithers & Soden, 2000). Given that critical thinking spans settings, whether in science classrooms or outside of them, the act of thinking critically varies as it is influenced by context and the norms and routines of a given context (Pithers & Soden, 2000). In other words, critical thinking takes shapes interactionally in unique ways depending on the specific people, tools, objects, and knowledge involved and invoked. As such, critical thinking is not an essentialized skill but a social practice that is shaped by the discourse in which it is constructed and therefore shifts in its constitution across settings (Danvers, 2016). Furthermore, the accomplishment or enactment of critical thinking is not static not only due to the situatedness, but also the person involved. How people evaluate information is shaped by their identities and underlying epistemological orientations to what counts as scientific knowledge. Hence, it makes sense that teaching critical thinking is not prescriptive, as it is embedded within the context in which the learner is situated (Pithers & Soden, 2000), histories, personalities, and experiences of the learner (Jiménez-Aleixandre & Puig, 2012; Hofer, 2001) and the role of particular disciplinary knowledge and practices (Greene & Yu, 2015).

3.3 Anti-science Stances and Post-truth

Critical thinking has been put forth as one tool "for confronting pseudoscience and credulity" (Jiménez-Aleixandre & Erduran, 2007, p. 8). Anti-science stances have been taking shape for decades throughout the world, particularly with respect to evolution (e.g. Deniz et al., 2008; Eder et al., 2011; Glaze & Goldston, 2015; Skoog & Bilica, 2002). However, this stance has not been exclusive to evolution and now includes science explanations for climate change, vaccinations, and so forth. Anti-science stances are similar to science denial and post-truth, all of which comprise a nomenclature that captures ways people seek to refute or ignore aspects of scientific explanations. While the term post-truth is somewhat fluid in its definition, it involves a de-emphasis of facts and expertise while accentuating sensations of feeling or seeming true (Fischer, 2019). Underlying post-truth are epistemological and ontological questions about how people identify what counts as knowledge in relation to how it counts within science disciplines and areas of inquiry. Even with calls to attend to science learning as entangled in both relativism and objectivism, rather than the false binary of relativism or objectivism (Van Poeck, 2019), the dominant rhetoric has been that science facts are the best approach to addressing post-truth (Fischer, 2019).

Complicating this post-truth landscape is the decentralization of the publishing process, whereby digital resources and news sources can be accessed easily to align with one's ideologies. Hence, "every perspective can be catered to because, ultimately, consumers are a demographic to sell news to, profit from and whose views can be catered to. It is this new media landscape that has fostered a suitable environment for fake news to be believed" (Barton, 2019, p. 1028). Underlying this media landscape is that the meaning and interpretation of information is dependent on the user as every person interacts with information in relation to their own experiences (Barton, 2019). Compounding the access to anti-science stances is the role of algorithms that dictate what one interacts with in social media and other digital media settings (Boler & Davis, 2018), thus creating echo chambers. In Höttecke and Allchin's (2020) recent piece, they recommend that science education incorporate science media literacy as a prominent part of the nature of science in schools. In doing so, they recognize how the conclusions and practices of science are modified as they move from communities of scientists to the general public, mediated by internet resources. While the recognition of science communication in what counts as scientific knowledge is not new (Jiménez-Aleixandre, 2014), the broad reach of digital media is. Hence, the ways in which scientific information is taken up, framed, and shared intersects with science classroom learning, as students and teachers interact with science instruction in ways that are not divorced from their identities and sense making outside of the classroom. Given the permeability of classroom walls, students and teachers orient to science based on their identities, which are situated and imbued with learning and connections to ideas, events, and interactions (Avraamidou, 2020; Gee, 2000).

3.4 Emotions as Part of Critical Thinking

The deeply situated ways in which critically thinking is accomplished and the uptick in anti-science stances is then ripe for attending to emotions within critical thinking. Despite calls to attend to science topics as emotionally laden (Darner, 2019), particularly those that are considered socio-scientific issues or related to sustainability (Jiménez-Aleixandre & Puig, 2012; Sadler & Zeidler, 2005), the emotional dimension of post-truth has been limited. Specifically, the emphasis on emotions in post-truth centers on the ways sources of disinformation frame messages to appeal to emotion, while simultaneously positioning itself as an authority (Barton, 2019, p. 1027). The prevalence of emotions in media and politics does not represent a recent shift. Instead, the recognition of them as integral to politics and sense-making is recent (Boler & Davis, 2018; Hufnagel, 2019a).

In my work, I draw from theories of emotions in social psychology and neuroscience to challenge prevailing assumptions about emotions in educational research (see Hufnagel, 2019b for a full discussion). Emotions are evaluative mechanisms (Barrett, 2017) that make sense of the world. As such, they indicate personal relevance and a deep and urgent relationship to ideas, objects, and so forth more so than other forms of affect. Given this relational aspect, emotions are contextual, interactional, intertextual, and consequential in that they are embedded in and constructed in the discourse (Boiger & Mesquita, 2012; Hufnagel & Kelly, 2018) in which they are used, shaped, and shared (Hufnagel, 2020). As mechanisms that make sense of our world, emotions are neither discrete nor distinct from cognition (Barrett, 2017), including science learning (Jaber & Hammer, 2016a, b; Wickman, 2006). Yet, in the era of post-truth, the focus on emotions has been about *topics* that are deemed emotional to the exclusion of the *emotional relationship* between a person and idea, experience, or object.

3.4.1 Emotionality of Scientific Pursuits

The perception of objectivity and neutrality of science is pervasive and long-standing despite evidence to suggest otherwise (e.g. Broughton et al., 2013; Durnová, 2019a; Hufnagel, 2019a, b; Jaber et al., 2019; Lombardi et al., 2017; Sinatra et al., 2014a). Accounts of scientists' emotions within and about their disciplinary pursuits are well documented and illustrate the ways in which emotions permeate scientific work, such as attachment to particular theories (Mitroff, 1974), within scientific observations themselves (Fleck, 1979), and as part of their identities within their respective research communities (Osbeck et al., 2011). Due to the sociological nature of science (Barbalet, 2002, 2011), the actions and interactions of scientists with other scientists, tools, and texts are replete with emotions.

3.4.2 *Emotions, Identity, and Interactions with Science*

Despite these and other accounts of emotional sense-making within scientific practices, the false dichotomization of emotions from reason prevails (see Hufnagel, 2015, 2019a). As the field of neuroscience expands, new insights on the ways in which emotions come to bear are realized. One such current theorization is that the brain makes sense of sensory input and in doing so constructs meaning, emotional or not emotional (Barrett, 2017). Hence, emotions are not triggered but *in situ* mechanisms that make sense of our world – ideas, interactions, experiences, and so forth – in relation to what is most important to oneself. Thus, emotions signal conflict, enhancement, and maintenance of one's goals in a given moment (Hufnagel, 2015, 2019a). They indicate personal connections to goals emanating from our identities, regardless of the type of emotion experienced (Hufnagel, 2015, 2019b).

Identities, which are constructed in social interactions, are neither essentialized nor static (Avraamidou, 2014). Rather, identities are multi-faceted and contextual, as one person has many identities. In each interaction, particular identities are more relevant to a person's sense-making and interactional accomplishments. Personal goals are based in one's identity, which given its fluid nature, takes shapes differently across time and space. Additionally, emotional sense-making varies not just by person but by contexts that are imbued with sociohistorical and cultural routines, norms, and ways of being (Barrett, 2006; Barrett et al., 2009). As such, emotional sense-making offers a sense of how a person evaluates ideas and experiences in relation to their sense of self or identity. This fluidity of identity and context in relation to emotions is why one person's emotional sense-making can come to the fore in one situation but not another, despite similar experiences. This orientation aligns with the value of examining identity, particularly science identity, to make salient "the complexities of becoming a science person which are tied to political, structural, and societal problems" (Avraamidou, 2014, p. 325). Everyone processes or makes sense of information differently based on their identities, lived experiences, motivations, sense of self, and so forth (Danvers, 2016; Darner, 2019; Sinatra et al., 2014b). Since scientific sense-making is imbued with identity and context, emotions provide a focal point to clarify these complexities of sense-making by highlighting what one cares most about and why (Hufnagel, 2015).

3.5 Climate Change, Identity, and Facts

Scholarship on the denial of science, especially climate change, provides important considerations for the ways in which emotions intersect with critical thinking. For instance, when people make sense of climate change by denying it or its human causes, their arguments are not centered on the facts, but what the facts represent (Fischer, 2019). People who reject climate change do so because of what they perceive as the motives underlying scientists' work or policy makers who try to address climate change or both (Fischer, 2019). Climate change is imbued with fear for some people due to the perception that their rights will be infringed upon should the government take broad and aggressive action (Rosenau, 2012). In this way, the rhetoric of the urgency of climate change has been perceived as an assault on particular ideologies that oppose large-scale government interventions or perceived increases in regulations (Fischer, 2019). Hence, the sense-making entails how this phenomenon relates to their identity or sense of self through fear, anger, and other emotions. As such, the challenge of sufficiently and effectively confronting science denial will be insurmountable if emotional sense-making is not considered (Rosenau, 2012). Thus, it is not surprising that recommendations for communicating climate change involve knowing your audience to frame climate change in way that aligns with their ontological orientation (Center for Research on Environmental Decisions, 2009).

Emotions are integral to conflict and they emulate the deeply personal concerns people have. Identifying, acknowledging, and responding to the conflicts borne from identity – individual and collective, in the form of affiliation – is required. These personal conflicts or clashes with truth (of anthropocentric climate change) are emotional. As Rosenau (2012) states,

> Recognizing and defusing the social pressures underlying science denial are key in convincing people that it is even worth considering scientific ideas that seem contrary to those of their social identity. When science denial becomes entwined with group identity, the risk of social ostracism is probably costlier than scientific error (p. 568).

Therefore, approaches to addressing science denial ought to involve what constitutes truth, as "the debate about the facts is only a proxy for deeper and even more difficult sociopolitical questions" (Fischer, 2019, p. 135) in which truth is constructed and represented. Other scholars, suggest going further: shifting from a focus on what is truth to the rules of truth to understand the emotions in science denial (Durnová, 2019a). The rules of truth are that it is fact-based knowledge, objective and neutral, and constructed by scientists (experts). As such, a false bifurcation of emotions from reason and science is perpetuated and even qualifies the unscientific, as it "sustains the dichotomy between factual and emotional when science is used as a source of expertise" (Durnová, 2019a, p. 46). As such, the discussion of what counts as truth, as science, and as critical thinking requires attending to emotional sensemaking. Unpacking these spaces makes salient assumptions about emotions within critical thinking.

3.6 Emotions Within and About Critical Thinking

Given that emotions are evaluative sense-making mechanisms embedded in one's identity, it seems impossible that critical thinking could be unemotional when the crux of critical thinking is evaluating information. The act of evaluating information can be emotional and not exclusively because people do not wish to agree with a claim. For instance, Danvers (2016) reported that undergraduates who performed critical thinking did so in ways that felt like critical thinking. She elaborates, "critical thinking was always an affective experience of some kind, even if it seemed tempered or even neutral. These feelings were not simply emotional reactions to isolated performances of critical thinking. . .Students articulated the complex affects they felt in response to critical thinking's discourses and practices" (Danvers, 2016, p. 285). Danver's (2016) work demonstrates the ways in which critical thinking is an affective experience across a range of interactions and highlights why it is.

Within science education, shifting from the false bifurcation of rationality and irrationality to embracing emotions as part of sense-making, learning, and doing science (Jaber & Hammer, 2016a, b; Jaber et al., 2019; Wickman, 2006), including critical thinking and the denial of anthropocentric climate change (or the rejection of any well substantiated scientific claim), moves beyond a binary of right or wrong and rational or irrational to *why*. With this lens of making salient the why I present a condensed set of findings from an interactional ethnographic study (Castanheira et al., 2000; Hufnagel, 2019b) of how pre-service science teachers learn to teach climate change. For this study, I analyzed video recordings of every class meeting, all student written artifacts, and video recordings of sets of interviews of the seven participants in a secondary science methods course that centered on climate change. During my analysis a theme that became apparent was the intersection between the students' goals for teaching science and their perceptions of climate change. Within this theme was a sub-theme of the pre-service teachers grappling with their goal of teaching critical thinking and students' denial of human-caused climate change. Below I present findings from within this sub-theme to illuminate some of the complexities of supporting educators to teach climate change in a post-truth era.

3.6.1 Critical Thinking as a Goal of Pre-service Science Teachers

Throughout the course, the pre-service science teachers expressed a range of views about why people do not accept human-caused climate change as a robust scientific claim. In particular, the pre-service teachers often oriented to their concerns about teaching climate change, given the potentially contentious nature of the phenomenon, that may conflict with teaching critical thinking skills, a goal shared among all the participants. They grappled with how to teach a phenomenon that could cause conflict for their students. These expressed concerns were linked to their day-to-day interactions with friends and family members in face-to-face and digital (i.e. social media platforms, texting) settings and brought to bear in class discussions, written assignments, and interviews.

3.6.2 Intersection Between Critical Thinking and Emotions Emanating from Identity

Permeating the discourse in the course was the pre-service teachers' shared goal of teaching their students to be critical thinkers. Each teacher articulated it throughout the course in different ways. Carmen, for instance, when asked how she felt about teaching climate change during an interview, answered by addressing the role of critical thinking when she conveyed, "If students can walk away armed with evidence or the tools to find and vet evidence about climate change that they can then apply to their lives as informed citizens moving forward, I'd be satisfied." After being asked how she views her "role as a science teacher" during an interview, Violet shared, "I think that we need to educate students how to be critical consumers of information because there is so much media out there that is not telling the truth and even as an adult, you don't know which sources you can trust." These sentiments were shared across all seven of the participants and provided a reference point for how they oriented to prospective students' denial of climate change. It is worth noting that two students oriented to critical thinking as tightly linked to having sufficient scientific knowledge to perform critical thinking. Nadine, for example, shared that critical thinking was a goal of hers as a teacher during an interview, but it was couched in having enough "background" knowledge. She expressed that instructors should "provide students with all the information available and the background of understanding in climate science for the students who think critically to reach a reasonable conclusion."

3.6.3 Emotional Subtext of Climate Change Sense-Making and Identity

Over the course of the semester the pre-service teachers described complexities influencing their future students' critical thinking with respect to climate change. One such complexity centered on the intersection between their future students' emotions emanating from the intersection between their identities and views of climate change. This complexity took shape in various ways but all with an eye toward the entanglement of people's identities within their scientific sense-making of climate change. In doing so, what emerged was the pre-service teachers' prioritization of one of the emancipatory features of critical thinking: independence from a group (Jiménez-Aleixandre & Puig, 2012). For instance, Carmen talked about how people's views of a topic in science is wrapped up in their identity, as seen in the excerpt below:

> Often with climate change...people's view of the topic is wrapped up in how they see themselves or how they see the world at large. It's tangled up in their identity. It can be a scary thing. People often will have emotional responses when they learn things that then challenge the way that they see themselves or they see the world.

Carmen acknowledged that science ideas conflict with ways people make sense of the world and are wrought with emotion. This conflict, to Carmen, is emotional ("scary") as it challenges one's worldview and identity. Miranda shared a similar sentiment on a written assignment; that "a large resistance to the belief in climate change comes from the threat to people's current beliefs and mindsets." She later elaborated on her idea during a final interview, explaining how people make sense of climate change as "much more than just a debate of the actual science behind it" saying that it is linked to "politics," "the economy," and "religion and beliefs" suggesting "that you just can't think the scientific principles are going to be all that's needed to address it." Similar to Carmen's ideas, Miranda recognized that changing people's understandings of climate change involves an emotional process based in "threat." Shifting students away from denial toward critical thinking is neither easy nor straight-forward since they are, to use Carmen's term, "wrapped up" in their views of themselves and the world.

Matt also oriented to the role of identity, but in relation to group affiliation throughout the course. On numerous occasions he described people's membership to groups as defining oneself. While this took shape in numerous ways, it relayed the idea that social networks were based in what he described as shared "core values and core ideals." Furthermore, during class discussions he oriented to "group think" regardless of group (his definition of group being along a binary of accepting or denying human caused climate change) as "kind of sad. . .the loss of individuality I think surrounding it." He later elaborated that he has "a healthy level of skepticism for any group of any kind" in his final interview.

Some teachers drew parallels between climate change and religion. Violet shared this sentiment when she spoke of the ways in which "religion and money can just like take over a person's beliefs and what they choose to accept as knowledge and what they want to like block out" during her final interview. Violet invoked her experiences with members of her family when she likened climate change to religion. In doing so, she elaborated on her discussions with family members about climate change. Maya also connected climate change to religion but in a way that raised epistemology, as seen in the excerpt from a class discussion below:

> I went straight to like comparing it [denying climate change] to religion, as something that people hold very dear and true and believe very strongly but when you talk about like evidence or like proof of the existence of higher powers I don't think it necessarily has to be there. So by providing it [evidence] I don't know if it will like sway people from one end to the other.

In her explanation Maya referenced epistemic criteria—in this case that evidence can't prove the existence of a higher power. Interestingly, Maya also acknowledged the status of people in trying to talk with people about climate change. In particular she attended to what counts as an expert and the interaction that can ensue when she wrote on an assignment, "sometimes it's harder to try to educate or inform adults because so many people internalize "being educated" by a peer or colleague as a personal assault on their intelligence, beliefs." In these ways, Maya was orienting to ideas related to critical thinking: expertise and epistemology in ways to identify emotional tensions between critical thinking and identity.

3.7 Concluding Remarks

Throughout their experiences learning to teach climate change, the pre-service science teachers oriented to their shared goal of teaching students critical thinking in relation to the students' identities. Specifically, they recognized an important tension: teaching climate change is wrought with emotions in part due to the identity work at play. While the focus of the findings here was climate change education, the implications pertain to teaching science in an anti-science atmosphere.

In this chapter I explored the intersection between emotions and critical thinking in relation to anti-science stances, specifically in relation to anthropocentric climate change. I briefly described critical thinking to highlight the situated ways in which it is constituted in discourse and thus dynamic and situated. In doing so, I provided an overview of intersections with epistemology. I then elucidated the tensions within anti-science stances to make salient the intersection between critical thinking and emotions with respect to identities, as emotions indicate urgent personal relationships with ideas, experiences, and so forth. By highlighting particular aspects of emotional sense-making I suggested how emotional sense-making is part of scientific pursuits and people's interactions with scientific knowledge and practices, and therefore imbued within critical thinking itself.

As science education and other related fields such as science communication contend with the uptick in science denial or post-truth, scholars in educational research have suggested attending to emotions (Darner, 2019). Until discussions about climate change shift from a sole focus on the scientific data and/or facts to ontological, epistemological, and ideological features, fact-checking itself will not address the gap. Since emotions sit at the intersection of this divide, they offer a way to understand the relationships people have with climate change and other science issues in order to have discussions about people's fear, outrage, and other emotions that undergird their anti-science stances. Unpacking truth to include to the examination of what counts as truth and why (Durnová, 2019a) uncovers ways in which emotions are part of critical thinking and the constitution of knowledge. So long as emotions are marginalized, the status of divisiveness around what counts as truth will continue (Durnová, 2019a). "At a time when news organizations are capitalizing on "post-truth," we need a better understanding of emotion and affect that works beyond the simple opposition of rationality and emotionality that continues to overdetermine our political imaginary" (Boler & Davis, 2018, p. 84). Boler and Davis' plea is salient for science education as well.

Acknowledgements I thank the editors and Anica Miller-Rushing, Gabrielle Brodek, and Eliza Jacobs for their helpful feedback on earlier versions of this chapter.

References

Aikenhead, G. S. (1996). Science education: Border crossing into the subculture of science. *Studies in Science Education, 27*, 1–52.

Avraamidou, L. (2014). Studying science teacher identity: Current insights and future research directions. *Studies in Science Education, 50*, 145–179. https://doi.org/10.1080/03057267.2014.937171

Avraamidou, L. (2020). Science identity as a landscape of becoming: Rethinking recognition and emotions through an intersectionality lens. *Cultural Studies of Science Education, 15*, 323–345. https://doi.org/10.1007/s11422-019-09954-7

Barbalet, J. (2002). Science and emotions. In J. Barbalet (Ed.), *Emotions and sociology* (pp. 132–150). Blackwell Publishing.

Barbalet, J. (2011). Emotions beyond regulation: Backgrounded emotions in science and trust. *Emotion Review, 3*, 36–43. https://doi.org/10.1177/1754073910380968

Barrett, L. F. (2006). Solving the emotional paradox: Categorization and the experience of emotion. *Personality and Social Psychology Review, 10*, 20–46.

Barrett, L. F. (2017). *How emotions are made: The secret life of the brain*. Houghton Mifflin Harcourt.

Barrett, L. F., Gendron, M., & Huang, Y. M. (2009). Do discrete emotions exist? *Philosophical Psychology, 22*, 427–437. https://doi.org/10.1080/09515080903153634

Barton, C. C. (2019). Critical literacy in the post-truth media landscape. *Policy Futures in Education, 17*, 1024–1036. https://doi.org/10.1177/1478210319831569

Boiger, M., & Mesquita, B. (2012). The construction of emotion in interactions, relationships, and cultures. *Emotion Review, 4*, 221–229. https://doi.org/10.1177/1754073912439765

Boler, M., & Davis, E. (2018). The affective politics of the "post-truth" era: Feeling rules and networked subjects. *Emotion, Space and Society, 27*, 75–85. https://doi.org/10.1016/j.emospa.2018.03.002

Broughton, S. H., Sinatra, G. M., & Nussbaum, E. M. (2013). "Pluto has been a planet my whole life!" emotions, attitudes, and conceptual change in elementary students' learning about Pluto's reclassification. *Research in Science Education, 43*, 529–550. https://doi.org/10.1007/s11165-011-9274-x

Castanheira, M. L., Crawford, T., Dixon, C. N., & Green, J. L. (2000). Interactional ethnography: An approach to studying the social construction of literate practices. *Linguistics and Education, 11*, 353–400. https://doi.org/10.1016/S0898-5898(00)00032-2

Center for Research on Environmental Decisions. (2009). *The psychology of climate change communication: A guide for scientists, journalists, educators, political aides, and the interested public*. New York.

Danvers, E. C. (2016). Criticality's affective entanglements: Rethinking emotion and critical thinking in higher education. *Gender and Education, 28*, 282–297. https://doi.org/10.1080/09540253.2015.1115469

Darner, R. (2019). How can educators confront science denial? *Educational Researcher, 48*, 229–238. https://doi.org/10.3102/0013189X19849415

Deniz, H., Donnelly, L. A., & Yilmaz, I. (2008). Exploring the factors related to acceptance of evolutionary theory among Turkish preservice biology teachers: Toward a more informative conceptual ecology for biological evolution. *Journal of Research in Science Teaching, 45*, 420–443.

Durnová, A. (2019a). *Understanding emotions in post-factual politics: Negotiating truth*. Edward Elgar Publishing.

Eder, E., Turic, K., Milasowszky, N., Van Adzin, K., & Hergovich, A. (2011). The relationships between paranormal belief, creationism, intelligent design and evolution at secondary schools in Vienna (Austria). *Science and Education, 20*, 517–534. https://doi.org/10.1007/s11191-010-9327-y

Fischer, F. (2019). Knowledge politics and post-truth in climate denial: On the social construction of alternative facts. *Critical Policy Studies, 13*, 133–152. https://doi.org/10.1080/19460171.2019.1602067

Fleck, L. (1979). *Genesis and development of a scientific fact*. University of Chicago Press.

Gee, J. P. (2000). Identity as an analytic lens for research in education. *Review of Research in Education, 25*, 99–125.

Glaze, A. L., & Goldston, M. J. (2015). U.S. science teaching and learning of evolution: A critical review of the literature 2000–2014. *Science Education, 99*, 500–518. https://doi.org/10.1002/sce.21158

Greene, J. A., & Yu, S. B. (2015). Educating critical thinkers: The role of epistemic cognition. *Policy Insights From the Behavioral and Brain Sciences, 3*, 45–53. https://doi.org/10.1177/2372732215622223

Hofer, B. K. (2001). Personal epistemological research: Implications for learning and teaching. *Journal of Educational Psychology Review, 13*, 353–383.

Höttecke, D., & Allchin, D. (2020). Reconceptualizing nature-of-science education in the age of social media. *Science Education, 104*, 641–666. https://doi.org/10.1002/sce.21575

Hufnagel, E. (2015). Preservice elementary teachers' emotional connections and disconnections to climate change in a science course. *Journal of Research in Science Teaching, 52*, 1296–1324. https://doi.org/10.1002/tea.21245

Hufnagel, E. (2019a). The "subtext of everything": High school science teachers' views of emotions and their related teaching practices. *Canadian Journal of Science, Mathematics, and Technology Education, 19*, 430–445.

Hufnagel, E. (2019b). Emotional discourse as constructed in an environmental science course. In G. J. Kelly & J. L. Green (Eds.), *Theory and methods for sociocultural research in science and engineering education* (pp. 155–180). Routledge.

Hufnagel, E. (2020). Language, materiality, and emotions in science learning settings. In D. R. Gruber & L. C. Olman (Eds.), *Handbook of language and science* (pp. 253–261). Routledge.

Hufnagel, E., & Kelly, G. J. (2018). Examining emotional expressions in discourse: Methodological considerations. *Cultural Studies of Science Education, 13*, 905–924. https://doi.org/10.1007/s11422-017-9806-4

Jaber, L. Z., & Hammer, D. (2016a). Engaging in science: A feeling for the discipline. *Journal of the Learning Sciences, 25*, 156–202. https://doi.org/10.1080/10508406.2015.1088441

Jaber, L. Z., & Hammer, D. (2016b). Learning to feel like a scientist. *Science Education, 100*(189), 220. https://doi.org/10.1002/sce.21202

Jaber, L. Z., Hufnagel, E., & Radoff, J. (2019). "This is really frying my brain!": How affect supports inquiry in an online learning environment. *Research in Science Education*. https://doi.org/10.1007/s11165-019-09884-y

Jiménez-Aleixandre, M. P. (2014). Determinism and underdetermination in genetics: Implications for students' engagement in argumentation and epistemic practices. *Science and Education, 23*, 465–484. https://doi.org/10.1007/s11191-012-9561-6

Jiménez-Aleixandre, M. P., & Erduran, S. (2007). Argumentation in science education: An overview. In S. Erduran & M. P. Jimenez-Aleixandre (Eds.), *Argumentation in science education: Perspectives from classroom-based research* (pp. 3–27). Springer.

Jiménez-Aleixandre, M. P., & Puig, B. (2012). Argumentation, evidence evaluation and critical thinking. *Second International Handbook of Science Education, 2*, 1001–1015. https://doi.org/10.1007/978-1-4020-9041-7_66

Kelly, G. J., McDonald, S. P., & Wickman, P.-O. (2012). Science learning and epistemology. In *Second international handbook of science education* (pp. 281–291). Springer.

Lombardi, D., Bickel, E. S., Brandt, C. B., & Burg, C. (2017). Categorising students' evaluations of evidence and explanations about climate change. *International Journal of Global Warming, 12*, 313–330. https://doi.org/10.1504/IJGW.2017.084782

López-Facal, R., & Jiménez-Aleixandre, M. P. (2009). Identities, social representations and critical thinking. *Cultural Studies of Science Education, 4*, 689–695. https://doi.org/10.1007/s11422-008-9134-9

Mitroff, I. (1974). Norms and counter-norms in a select group of the Apollo moon scientists: A case study of the ambivalence of scientists. *American Sociological Review, 39*, 579–595.

Osbeck, L. M., Nersessian, N. J., Malone, K. R., & Newstetter, W. C. (2011). *Science as psychology: Sense-making and identity in social practice*. Cambridge University Press.

Pithers, R. T., & Soden, R. (2000). Critical thinking in education: A review. *Educational Research, 42*(3), 237–249. https://doi.org/10.1080/001318800440579

Rosenau, J. (2012). Science denial: A guide for scientists. *Trends in Microbiology, 20*, 567–569. https://doi.org/10.1016/j.tim.2012.10.002

Sadler, T. D., & Zeidler, D. L. (2005). Patterns of informal reasoning in the context of socioscientific decision making. *Journal of Research in Science Teaching, 42*, 112–138.

Sinatra, G. M., Broughton, S., & Lombardi, D. (2014a). Emotions in science education. In R. Pekrun & L. Linnenbrink-Garcia (Eds.), *International handbook of emotions in education* (pp. 415–436). Taylor & Francis.

Sinatra, G. M., Kienhues, D., & Hofer, B. K. (2014b). Addressing challenges to public understanding of science: Epistemic cognition, motivated reasoning, and conceptual change. *Educational Psychologist, 49*(2), 123–138. https://doi.org/10.1080/00461520.2014.916216

Skoog, G., & Bilica, K. (2002). The emphasis given to evolution in state science standards: A lever for change in evolution education? *Science Education, 86*, 445–462. https://doi.org/10.1002/sce.10014

Van Poeck, K. (2019). Environmental and sustainability education in a post-truth era. An exploration of epistemology and didactics beyond the objectivism-relativism dualism. *Environmental Education Research, 25*, 472–491. https://doi.org/10.1080/13504622.2018.1496404

Wickman, P.-O. (2006). *Aesthetic experience in science education: Learning and meaning-making as situated talk and action*. Lawrence Erlbaum Associates.

Chapter 4
Culturally Relevant Science Education and Critical Thinking in Indigenous People: Bridging the Gap Between Community and School Science

Alejandra García Franco, Lisber Ferrara Reyes, and Alma Adrianna Gómez Galindo

A secondary school in a community in the Chiapas Highlands. Students speak Tseltal and their teachers speak Spanish. Students belong to farming families and practice *milpa* as a subsistence crop. In the community there are also coffee plantations and collectives of knitter women who sell their handcrafts in larger cities. The community Yochib is one of the poorest towns in Mexico and subsistence depends mainly on agriculture. Yochib in Tseltal means' sump and this is literally what it is: a sump between two mountains where a river goes by.

What science is relevant in these communities and students in the second year of secondary school? How can they achieve critical thinking that is relevant for their lives?

4.1 Introduction

This chapter comes from the work we started nine years ago when we approached indigenous communities in southeastern Mexico (Montaña de Guerrero) as part of a large research team documenting traditional knowledge about cultivation techniques (*milpa*). The question back then was: what happens to this knowledge in school? How is this knowledge considered? The answer was that traditional/local knowledge

---

A. García Franco
Universidad Autónoma Metropolitana – Cuajimalpa, Mexico City, México
e-mail: agarcia@cua.uam.mx

L. Ferrara Reyes
Escuela Telesecundaria 714 David Gemayel Ruíz Estudillo, Yochib, Oxchuc, Chiapas, México

A. A. Gómez Galindo (✉)
Cinvestav Monterrey, Apodaca, Nuevo León, México
e-mail: agomez@cinvestav.mx

B. Puig, M. P. Jiménez-Aleixandre (eds.), *Critical Thinking in Biology and Environmental Education*, Contributions from Biology Education Research,
https://doi.org/10.1007/978-3-030-92006-7_4

was almost out of sight in school and, even if teachers were indigenous or farmers, the only relevant knowledge was the one prescribed in the school curriculum. Ever since, we have worked intermittently with teachers and students in indigenous and non-indigenous communities. We are three authors in this chapter: an educational researcher working in a large university in Mexico City; a science teacher in Yochib, an indigenous community in the Chiapas Highlands; an educational researcher in a public research institution in Monterrey in northern Mexico. We are not members of any indigenous group. Our basic preparation is in the natural sciences.

Our main concern comes from questioning which is the value of science education in rural schools and, particularly, in indigenous communities such as the one described in the epigraph. As non-indigenous researchers we are looking forward to constructing bridges between school science knowledge and indigenous/traditional knowledge. In this chapter we discuss how traditional knowledge could be valuable to promote critical thinking in the classrooms and why.

We recognize that global phenomena such as climate change and migrations are part of the current crisis and recognizable for everyone. However, these problems have different dimensions in particular communities (Gruenewald & Smith, 2007). For indigenous people global problems are accentuated because of the accelerated loss of indigenous languages and territories. For different reasons local knowledge that permits the recreation of life is being eroded (e.g. Maurer, 1977).

These problems and particular conditions of different communities are fundamental to think about the role of science education, how it can contribute to citizen preparation and provide students with basic tools to understand the world they inhabit and transform it in a sustainable way (Valladares, 2010). There is a global imperative to respect and conserve knowledge, language and practices of indigenous people. The relation between such knowledge and sustainability has been recognized in several studies (e.g. Oviedo et al., 2000; Boege Schmidt, 2008). In this sense, the school plays a relevant role. However, according to Harmin et al. (2016):

> In order to effectively and ethically engage with indigenous knowledge holders and address the complexity of sustainability problems in the context of socio-ecological systems, academic institutions are tasked with decolonizing approaches to knowledge creation and addressing ongoing privileging of some knowledge forms over others. (p. 2)

Our approach to science education comes from recognizing the need to establish dialogues between traditional/indigenous knowledge and school knowledge. In previous works we have explored the value of incorporating traditional knowledge in the science classroom, particularly indigenous knowledge about cultivating *milpa* in southeastern Mexico (Gómez Galindo et al., 2019). Milpa is a policrop, with maize, beans and squash as main components, that is fundamental to sustainability, food sovereignty and community organization.

We have argued for incorporating knowledge about milpa in school in order to achieve that what is learnt would be valuable to students' lives, their autonomy, community values, identity and, ultimately to promote social justice, where school gives something valuable to everyone. Establishing these dialogues between traditional/local and scientific/school knowledge has allowed the visualization of students' and teachers' knowledge about different topics and how they could be related to school knowledge (Gómez Galindo & García Franco, 2021). But it is clear that we need to expand our way of looking to incorporate not only knowledge but also ways of thinking that sustain diverse ways of life.

In this chapter we problematize, from a decolonial perspective, what we understand as school science where critical thinking and biology teaching are situated in specific communities. We present an analysis that allows the identification of relevant knowledge that can be approached from this critical stance. We recognize that there is a need to rethink what we understand as critical thinking and how we can promote it, so that science education would not be an artifice but rather that it provides tools to students and is relevant to their identity construction.

With this in mind, the question that guides our analysis is:

What is the contribution of considering students' traditional knowledge in the development of critical thinking in the biology classrooms?

In the second section of this chapter we present the theoretical perspective that allows reflecting about the role of traditional knowledge in developing critical thinking in situated practices and how the frameworks of interculturality and decolonialism can be related to social justice. In the third section we present our methodological approach and data gathered as part of a research trajectory. In the fourth we discuss our results, and in the last section we offer some concluding thoughts.

4.2 Theoretical Perspectives. Culturally Relevant Science Education and Critical Thinking

To think about a transformative education that allows students to construct meaning for school knowledge and use it to transform their circumstances, we need to extend the idea of critical thinking and recognize that school learning and instruction are not neutral and transparent processes (McLaren, 1997). To do so, we rely on the notion of critical pedagogy proposed by Henry Giroux y Paulo Freire.

Critical pedagogy construes education as a tool for *conscientization*,[1] to recognize one's own place in the world and critically interrogate it. From this stance it is indispensable that students and teachers voices are considered because these voices are agents of change and critical observation of reality.

According to Giroux (1997, 2003), schools are sites for struggle where teachers and students can take the spaces that school situations' offer in order to resist[2] the uniformity mandate. But this resistance is only possible if classroom knowledge

[1] Conscientization (concientização) is a concept proposed by Paulo Freire. It means 'critical awareness' or 'critical consciousness'.

[2] This term is derived from resistance. It is widely used in postcolonial and decolonial studies to describe the political position of native/indigenous people facing colonialism.

makes sense to students, is relevant for their lives, and allows them to have a voice. This implies retrieving wisdom, stories and cultural practices (Farrera, 2018). Curricular content and the way it is approached, as well as pedagogical practices should find resonance in students' vital experiences.

Freire recognizes that knowledge should promote emancipation (1970) and encourage students' critical stance. To do so, schools need to consider concrete problems that students and teachers face every day, so they are able to pose questions about topics that are relevant for them. This also promotes the recognition of their own knowledge as well as their communities' knowledge (Carreño, 2009/2010). Freire's critical pedagogy is particularly relevant for indigenous students who have seen their knowledge and language excluded from school.We should aim for students to participate in the classroom generating authorship processes where students head toward production and not reception of knowledge (Subero & Esteban-Guitart, 2020). Critical pedagogy signals the route to choose subjects and ways to work.

4.2.1 Intercultural Science Education from a Critical Perspective

Sociocultural perspectives are every time more frequent in science education (see Milne et al., 2015). Different critics to science education have shown its cultural, local and situated character and have modified the narrative of science as objective, neutral and universal knowledge (Carter, 2004; Gorbach & López Beltrán, 2008). Its coexistence with indigenous or local systems of knowledge has also been recognised. Moreover, the presence of students from different cultures and languages in the classrooms have made inevitable to question the different ways in which science and its normative culture are related with students that come from different cultures and talk different languages (e.g. Hutchinson, 2014).

Countries in Latin America have a pluricultural composition. In Mexico there are 64 different linguistic groups with more than 365 dialects. The relation between indigenous cultures and formal schooling has been complicated but, for the most part, education for indigenous people has been considered as a way to 'bridge the gap', assuming one culture as dominant (Ramírez Castañeda, 2006).

It was not until the last decade of the last century that intercultural education was considered an appropriate approach for education in Mexico. However, its implementation has been filled with lights and shadows for indigenous people. Interculturality can be considered as a result of indigenous people struggles and their demands for recognition. It can also be understood from globalisation and find it tied to power, capital and market. Walsh (2009) describes three perspectives for interculturality: relational, functional and critical. Relational interculturality recognizes the differences but does not question them, neither recognizes conflict. More dangerous is functional interculturality whose intention is to include socially excluded groups to the current system. The 'others' are recognized only to be co-opted, turning education into a domination strategy whose final objective is not the construction of more equitable and egalitarian societies but rather the control of ethnical conflicts and the conservation of social stability.

Critical interculturality recognizes that differences have been constructed in a colonial structure where native people occupy the lowest part of the social order. From this perspective, interculturality is a tool and a grassroot process that aims for the transformation of structures and institutions in order to construct different conditions for being, thinking, knowing, learning, feeling and being. Critical interculturality implies recognizing asymmetries that exist when school science knowledge and local knowledge come in contact.

The need of an intercultural science education has been put forward by authors such as Aikenhead (2002) and McKinley (2011) as an inescapable need when the culture of those who learn are far away from the dominant culture. One of the most common approaches from this perspective has been the incorporation of local knowledge in the classroom (Aikenhead, 2002), and the recognition that students have knowledge that could be related with the curricula in the design of teaching-learning activities (e.g. Santos Baptista & El-Hani, 2009). However, authors such as Carter (2004, 2008) warn about the danger when this type of interventions are undertaken without problematizing the role of indigenous knowledge, the reason it has been relegated in the classroom, and the relationship it keeps with dominant knowledge. This is why it is fundamental to look at these interventions from a decolonial stance.

Taking a decolonial stance would allow for an intercultural dialogue in the school that, according to epistemic pluralism (Olivé, 2009), incorporates, recognizes and values students' own knowledge and culture. This critical perspective allows us to reread texts that have been highly influential in the field of intercultural science education (Carter, 2004) and point out how some of these approaches have a functional perspective incorporating local/traditional knowledge in the classroom without problematizing or recognizing conflict and asymmetry.

In order to construct this intercultural scientific education, there is a need to consider real conditions in secondary classrooms where students speak a language different from Spanish and have relevant knowledge that could be related to school knowledge. We need to find ways to make intercultural dialogues possible and promote critical thinking that contributes to social justice.

4.2.2 Critical Thinking as Situated Practice

Different proposals have been developed to teach critical thinking in science in general, and in biology in particular. Some of them stress the development of scientific practices such as argumentation (Erduran & Jiménez-Aleixandre, 2007). In recent times there are proposals that target the development of arguments about socio-scientific issues that could be of interest to large segments of the population. Amongst them we find those associated to discrimination (Puig et al., 2012) and nutrition (Brocos & Jiménez-Aleixandre, 2020).

In science education critical thinking has a privileged place. In a sense it is recognized as 'good thinking'. It allows students to recognize weak arguments, inconsistent generalizations and non-trustable asseverations (Harrison, 2004). Critical thinking also implies a series of dispositions amongst which we can find: the search for intellectual rigour, conceptual clarity, and the desire to seriously consider other people's viewpoints (Ennis, 1996).

In general, critical thinking is presented as a cognitive attribute, but it could also be framed as a situated practice (Harrison, 2004). Understanding critical thinking as a situated practice could imply identifying its exercise not only as a way of thinking but also as ways of acting which could be related to community commitments (Varelas et al., 2012). Critical thinking as situated practice requires recognizing how thinking, doing, and living together can be incorporated in the classroom.

Before we started working in Yochib – the community where part of this work is situated-, teacher Lisber (second author of this paper) wrote a reflection on what she understood as critical thinking and how it could be promoted amongst the students. In a written reflection she points out:

> Critical thinking in indigenous communities is bred from a different perspective, it incorporates how learning comes from modelling and recognizing wisdom. You learn through practice and by using all your senses to learn. You are wise according to how you live and feel well. Their quality of life is measured according to how you feel, to the state of your heart. This expression has different connotations, *lexil xchanel* (good learning), *k'uuxubinel* (considering others' feelings), *ich'elta muk'* (respecting and considering others). [In the community] good thinking generates good living.[3]

The integration of "good thinking" and "good living" perspectives allows to amplify the initial theorization of critical thinking and support the interpretation of the information. As Rappaport mentions "theoretical frameworks that are not entirely traditional, but neither those of western academia, which brings innovative perspectives about subaltern history and culture" (in Vasco Uribe, 2002, p. 471).

This good thinking opens space for an intercultural science education that supposes recognizing the 'other' and affirming oneself (Moya, 2009, p. 28). In this case, 'the other' represents school science culture incorporating the nature of science, the story of the construction of ideas, and the development of scientific practices such as argumentation. In an indigenous school, affirming oneself leads to the recognition of a cultural identity: recognizing a common history, relations to the environment, worldviews, values, language, similarity of problems derived from exploitation, exclusion and social and economic marginality.

It is necessary to recognize how the very meaning of education is different in different communities. For Tsotsil people "education is conceived as a slow acquisition of the soul, which is analogous to a total conscience. Soul reaches maturity

[3]Good living is a translation from '*buen vivir*' which is a common characteristic of the worldview of different Latinamerican indigenous people.

through learning how to become a good cultivator of maize".[4] For indigenous people in Northern Canada, learning is a journey centred in participation: "The Eurocentric meaning of *to learn* becomes *coming to know* in most Indigenous contexts, a meaning that signifies a personal, participatory, holistic journey toward gaining wisdom-in-action" (Aikenhead & Elliott, 2010, p. 322, italics in the original).

We need to question: What does critical (scientific) thinking bring to good (community) thinking? How are they related? How does one achieve critical thinking as situated practice?

4.2.3 Students' Knowledge, Their Identity and Critical Thinking

Situated learning frameworks allow us to incorporate students' reflexive activity considering sociocultural context as inseparable of the activity itself (Lave, 2009). From this perspective, the development of critical thinking always reconstructs the identity of the learner. The concept of identity turns out particularly useful for theorizing the relationship between individuals and their social world. Identity as a theoretical methodological framework has been interpreted in science education in different ways. We consider that identity is negotiated in a continuous reconstruction in our interactions and is mediated through language (Pozzer & Jackson, 2015). Intersubjectivity and the development of funds of identity become relevant for such reconstruction (Subero & Esteban-Guitart, 2020). Funds of Identity are inserted in the theory of subjectivity from a cultural-historical perspective where it is understood as a complex system articulated through the learner's life story, through her experiences in the different and diverse contexts in where she lives. These funds of identity are understood as the "resources that are socially distributed, historically accumulated and culturally developed that are essential for self-understanding, self-expression and self-definition" (Subero & Esteban-Guitart, 2020, p. 220).

In the social negotiation of identity, different situations allow for the emergence of moral contingencies where members of a community ponder through instituted and interiorized values and recognize themselves in decision making. In modern societies, this decision making is associated with the possibilities of self-determination and the development of a life project.

González-Escallón (2017) points out:

> ... it is worthwhile to question the reason why self-determination of some individuals seems to produce no immediate effects that in other groups are taken for granted. An example is the case of homosexual persons whose life plan does not imply the benefits of heterosexual persons. Deep down, this implies that the right to self-determination is not guaranteed because they need to change their life plan if they want to have access to all the benefits. (pp. 171–172)

---

[4]Taken from: National Institute for Indigenous People https://www.gob.mx/inpi/es/articulos/etnografia-de-los-pueblos-tzotzil-batsil-winik-otik-y-tzeltal-winik-atel?idiom=es

In the same way, for indigenous students there is a dialectic relationship, established in a socially negotiated way, between their identity development, their self-determination capacity and the development of their life project. This is associated with the possibilities they have to value their viewpoints, take decisions, to plan, and develop their life projects.

4.3 Methodological Approach. A Research Trajectory

In order to explore the contribution of students' traditional knowledge in the development of critical thinking in the biology classrooms we have undertaken recurrent approaches to schools situated in indigenous communities in the Highlands and the Central Depression of Chiapas in southeastern Mexico. In these communities Tsotsil, Tseltal and Spanish are spoken. Field work in indigenous communities poses challenges associated to language use and to documentation and access time (such a situation has been theorized in the field of ethnography by different authors, see for example, Rappaport, 2007). We used a situated learning theory (Lave, 2009) to understand activity in context and an approach from critical theory (Johnson, 2008) which aims to identify structures in science education that can originate or perpetuate inequalities and subordination relations and, from there, revalue forms and work traditions in the classroom.

We have constructed research trajectories in which vinculation and collaborative work with teachers in the zone has been crucial to gather information and construct questions. In the part of the trajectory that we are presenting in this chapter, we undertook fieldwork in two stages.

In the first stage, we worked with a group of 18 students in secondary school (14–16 years) in the community of Yochib (Chiapas Highlands). The biology teacher of this group is the second author of this chapter. From september to december 2019 students worked on aspects related to milpa: describing diversity, cultivation techniques, animals and plants associated. The three authors met for four remote work sessions. In these meetings we planned activities, discussed results, and talked about results' interpretation. After activities were completed the teacher posed two questions for which students provided a written answer:

- How do you feel about learning about milpa in your school?
- Why do you think that your teacher is teaching about the milpa in the school?

Categories resulting from the analysis of these two questions are presented in Tables 4.1 and 4.2.

In the second stage, after identifying that the topic of fertilizers was relevant for the students, we envisioned a second stage in the research trajectory with the aim of exploring its potential in the development of critical thinking. This second stage started with a workshop for teachers who work in schools situated in indigenous communities. Two of the participant teachers invited us to develop activities related to fertilizers with their students. On february 2, 2020, we went to a secondary school

Table 4.1 Categories obtained from the analysis of students' answers to the question: how do you feel about learning about milpa in your school?

Category (code)	Description
Is life and food (LF)	Point out feelings of relevance and satisfaction because milpa and maize are life and nurture.
Learning to sow/harvest and preparing food (LS)	Point out they feel proud, good, happy. It is good and important to learn about how to sow/harvest and prepare food. They mention they should not use fertilizers to obtain a good crop and healthy food.
Have economic independence and food sovereignty (FS)	It is important, valuable and urgent. Women point out their future roles as responsible for a family and food providers. They recognize this is the moment to learn about milpa, to have their own milpa in their future without using fertilizers and achieve economic independence and food to prevent starvation.

Table 4.2 Categories obtained from the analysis to the question: Why do you think that your teacher is teaching about the milpa in school?

Category (code)	Description
Teacher as learner (T-L)	Students mention that the teacher wants to learn about milpa cultivation, and they will teach her.
Teacher as field examiner (T-FE)	Teacher wants to assess how students are working their milpa. They point out that she wants to verify they are not using fertilizers.
Teacher as school examiner (T-SE)	Teacher wants to revise that their school work is correct and that they know how to present their work about milpa in school.
Teacher as guardian of the future (T-FG)	Teacher knows that milpa is relevant for their future and that is why she brings it as a school topic.

in San Cristobal de las Casas with a group of 25 students. These students spoke Spanish and some of them Tsotsil. On February 3, 2020, we went to a secondary school in Aldama with a group of 27 students who spoke Tsotsil and Spanish (with different levels of command). Students worked with the material *Aprendiendo en la milpa* (Learning in the milpa) (García Franco & Gómez Galindo, 2020) that was presented both in Spanish and Tsotsil. They undertook an individual task where they elicited their knowledge about fertilizers, how they are used in milpa and what they thought of their usage. Here we present the analysis of questions posed by students about what they wanted to know about fertilizers (Table 4.3). After that they wrote, in couples, a letter to other families that will cultivate their milpa giving advice on the use of fertilizers.

To construct categories we transcribed all the activities, and used an hermeneutic circle (Weiss, 2017) as a tool to interpret our data. We undertook a first reading of the data with our initial question in mind. We identify common topics and also particular expressions that spark our attention: these are our first intuitions on meaningful relations. Then we start a systematic analysis where questions and categories are fine-tuned. We advance in the comprehension of words, paragraphs and metaphors, relating statements with the whole text, but also considering the implicit or explicit

Table 4.3 Questions that students pose about fertilizers arranged by category

About milpa and its growing	About fertilizers and the pollution
What can we do to make milpas grow well? Why is it that sometimes they do not grow well? Why do you sow beans near maize? What is the usage of maize leaves? What kind of bacteria are there in milpa?	Why do we use fertilizers? What is in the fertilizer that makes plants grow fast? What happens with the milpa if we always use fertilizers? Are the fertilizers that they use on milpa good? Why are fertilizers bad? Are there non-natural fertilizers that do not pollute the soil? Why is it that fertilizers damage plants? What happens when you are in close contact with fertilizers?
About the natural fertilizers and their production	**About alternatives to fertilizers**
What chemicals do fertilizers have? Are there chemicals in the natural fertilizers? What elements and materials are used to make fertilizers? How many different kinds of fertilizer are there? How is fertilizer produced? Which was the first country to use fertilizer? Who sends us fertilizer? Who created the fertilizer?	Why is an industrial fertilizer used knowing you can use a natural fertilizer? What can we do to stop using fertilizers? What is the difference in the growth time when using or not using fertilizer? Why aren't other things used? Instead of fertilizers?

relation of texts with the larger context trying to establish a pattern of meanings and generate analytical categories. Meaningfulness of a category is not necessarily related to how frequently it appears. In the results and discussion section we present examples of transcribed students' texts associated with the categories in order to confer reliability to our arguments. We have translated students' answers trying to convey students' meaning and writing style. We have assigned pseudonyms to each student.

4.4 Results and Discussion. Traditional Knowledge and Its Contribution to Reflections on Critical Thinking

As stated in the introduction, in previous works we have explored the subject of milpa and its relation to school science (Gómez Galindo et al., 2019). Milpa is a Mesoamerican policrop whose main components are maize, beans and squash. It has been recognized as an efficient agroecological system that profits from symbiotic relations among plants. Besides the main components of plants there are different milpas where you can find fruit trees, vegetables, chilies, as well as animals. Milpa has also been recognized as relevant to face climate change (Boege Schmidt, 2008) and is fundamental to communities social life and is in the centre of rituals and festivities (Carrillo Trueba, 2016).

Throughout the categorical analysis, we found different contributions from considering students' traditional knowledge in the development of critical thinking in the biology classrooms. We will discuss the aspects that we found more relevant in terms of their potential to promote critical thinking as situated practice.

4.4.1 Emotional Bond

In activities undertaken by students in the classroom it is evident that they have a relation to milpa beyond its cultivation. Categories in Table 4.1 indicate that the milpa is highly valued by students and has more than cognitive implications. Students feel happy learning about milpa, some of them state they feel proud. We consider this emotional bond is relevant because emotions can be tools to construct a life project and to generate links with school topics (Ahmed, 2015). Students also assert that milpa for them is life (not part of their life, but literally, life).

> This is what I like to learn about maize. When I learnt I felt very happy because we are living with the maize.
> Elena – LF

> Milpa is life, without milpa there is no life.
> Susana – LF

Another relevant category is the one in which students state that they like to learn about sowing/harvesting or to prepare food from the milpa. These kinds of answers also show an emotional bond and one that is related to motivation, to the recognition that what they are learning in school is valuable for their own interests and activities. This is also relevant in terms of students' identities and life-projects as will be discussed further.

> I like to learn about milpa, I like to learn how to sow in the milpa, how to sow maize, beans, that is why I like to learn about milpa, I like to learn how you can take the weeds out of the milpa (...) this is why I feel happy and proud to learn about the milpa because later it will be useful to sustain my family, milpa is important to me because it gives us maize, beans, everything comes from there.
> Susana – LS

4.4.2 Transforming Traditional School Roles

When the teacher questions students about the milpa, the traditional order in the classroom can be subverted, modifying students' identity who, in this case, turn into teachers or 'the ones who know'. There is a term in Tseltal, *sk'oplal jbiteswanejetik*[5], which means the one who knows, and which is fundamental to indigenous pedagogy.

[5] "The experience of those who teach".

The category ‘teacher as learner’ (T-L) shows us how bringing milpa into the classroom allows them to have an authorized voice that they can use to teach their teacher. This is unusual in school and perhaps even more in indigenous schools, where teachers are always the authority (Ramírez Castañeda, 2006). Students state that the teacher wants to learn from them.

> The teacher also wants to learn about maize to live. Also to cultivate her milpa.
> Elena – T-L

> Teacher wants to learn everything we do with the maize, maybe learn about the food.
> Gelda – T-L

It is worth noting that students consider this knowledge as their own knowledge, part of a larger system that we have called traditional knowledge. This larger system is also associated to their identity construction, for example, as a farmer that knows how to cultivate:

> My teacher wants to do all the questions. We are farmers and we know how to cultivate seeds of maize, beans, squash, chilies. We know how to cultivate. This teacher and the other teachers do not know how to cultivate milpa, but the people in Yochib know.
> Adelina – T-L

We understand traditional knowledge or indigenous knowledge as the knowledge that has been developed by people with ample histories of interaction with the natural environment and that originated independently from science in a particular cultural environment separate from occidental culture (Pérez Ruiz & Argueta Vilamar, 2011). We recognize that students in secondary school are in a process of appropriating traditional knowledge in their community. In this sense, students' knowledge can be partial and incomplete but, nonetheless, important.

The categories ‘teacher as field examiner’ (T-FE) and ‘teacher as school examiner’ (T-SE) show us how the examiner notion, attributed to the teacher prevails in school. However, in students' answers the teacher is recognized not only as supervising school work but also practices outside school and the respect to community consensus such as not using fertilizers. One aspect that teachers should take into account when bringing traditional knowledge to the fore is related to helping students recognize themselves in this formative process and visualize what it brings to school beyond teacher's supervision and her examiner character. The category ‘teacher as guardian of the future’ (T-FG) retrieves the interest that students assign to studying milpa in relation to their life project.

> My teacher knows that it is important for our future and that it could be useful for us somehow; we live it, we see it and it grows in front of us and because the teacher is right in teaching us about milpa because we could not forget our traditions and customs because those are in our blood, about how we are, where we were born...
> Sabrina – T-FG

4.4.3 Relevant Topics with Potential to Promote Critical Thinking

In the questions about how they felt about learning about milpa, when students talk about learning how to sow and prepare food they constantly refer to fertilizers. This term is generally associated with industrial fertilizers (when they talk about natural fertilizers they use the term manure). We found this recurrent mention worth noting because this topic could be suitable to develop critical thinking activities. Students' statements about fertilizers are, for example:

> Other knowledge in our milpa is that I can not use fertilizer because it is poison, it is not good for the milpa, it is better that we work with our own hands with the hoe, in this way the corn will grow very big.
>
> Vilma – LS

> If you give fertilizer it is possible that the plant dies, the teacher wants to know what is she going to do with maize. This teacher wants to know everything we do with maize, perhaps she wants to learn everything about the food.
>
> Gelda – T-L

In this part of the country the usage of industrial fertilizers has been the center of a debate. Not using fertilizer implies a political position that breaks the circle of dependence on industrial producers. Government and political parties have used industrial fertilizers as a way of control and the dependency on such fertilizer generates vicious circles, because the land impoverishes due to its continuous usage. Not using industrial fertilizers can be related to identity and food sovereignty. This is why we found it relevant that it appears in students' answers in a recurrent way, opening a clear window for resistance and for the development of critical thinking.

Students in the schools in San Cristobal de las Casas and Aldama posed different questions about fertilizers that let us recognize this is a topic with an important potential to develop students' critical thinking. In Table 4.3 we present the four identified categories and examples of the students' questions related to each one. In this case we identify interest in relation to different aspects of fertilizers.

Students establish differences between industrial and natural fertilizers. When writing a letter to a family about the usage of fertilizers, we can see an example of traditional knowledge associated to natural fertilizers in the description made in couples, by Rosa and Heidi:

> To have a good crop without using fertilizers, before you sow your milpa you have to prepare the terrain and once it is clean your terrain you mix the soil with *gallinaza*[6] or you can also put fruit rinds and peels and once they are rotten you give it in the feet of milpa. You can also make an organic compost with the leaves of the trees but they have to be dry, you put them together with cow manure and a bit of *pozol*[7] and leave it to settle for a while.

[6] *Gallinaza* is the manure of chicken mixed with soil, food residues and feathers.

[7] *Pozol* is a traditional drink in the Chiapas Highlands, made from fermented corn dough.

The differences between industrial fertilizers and natural fertilizers are questioned by students when they ask if industrial fertilizers could not contaminate or if the natural fertilizers have a lot of chemicals (Table 4.3). Listening to the questions made by students would allow us to generate an intercultural dialogue where traditional and scientific knowledge could come into play. Knowledge associated with plant nutrition, biochemistry needed to understand composition of fertilizers, water and mineral cycles, could have meaning if integrated from a dialogic perspective.

We recognize that it is necessary to design specific strategies to promote critical thinking and go beyond information delivery. Controversies are already present, for example: between milpas' good growing and fertilizers' polluting effects; dependency generated by acquiring fertilizer from the industry and communities' self-determination.

4.4.4 Critical Thinking and the Identity of Students in Yochib

When we ask students how they felt about studying milpa in school, in category 'have economic independence and food sovereignty' (FS) in Table 4.1, students in Yochib relate the topics of fertilizers and milpa cultivation with their own identities and their life projects. For these students the development of critical thinking associated with these subjects implies a larger commitment beyond the school institution. It implies learning about what is relevant and has a place in their lives. Critical thinking as situated practice, as we have pointed out, is not an abstract entity related only to thinking, it is thinking in action that will allow them to make decisions relevant for their lives:

> The knowledge I have and I am still learning, will be very useful in my future of what I am going to do later in life when I have kids, husband and my own family (...) I feel very happy and proud to learn about milpa in the school. I like to learn about milpa, I like to learn how milpa is cultivated, how you sow maize, beans. But if we do not have terrain, we do not have where to cultivate even if we like to cultivate, just like those who do not have terrain (...) they are the ones that steal from us, they steal our maize (...) until I go out of my house, that is why milpa for me is my life because I eat from there and if I don't know how to work I will starve with my kids.
>
> Susana – FS

> If we want the maize to maintain us then we have to cultivate our milpa because milpa is very important for us to take care of our house. Milpa is very important because it is food, we can make tortilla, pozol, etc.
>
> Adela – FS

For these students this knowledge is relevant, not only for the constitution of habits of mind, but for their life, for their future. This is why knowledge that comes into play goes beyond environmental and biological specific knowledge. Beyond knowledge that students can have about biodiversity in the milpa, this is relevant for their life and their future. We believe this is why it is important that students have spaces in the school to develop critical thinking understood as the possibility to interrogate their reality and, if possible, transform it.

4.5 Concluding Thoughts. A Life Project and Critical Thinking

We have discussed four different contributions of bringing traditional knowledge into the classroom and its relation to critical thinking as practice: emotional bond, transforming traditional school roles, relevant topics, and students' identity. We have tried to underscore that bringing traditional knowledge into the classroom, recognizing students' and community knowledge opens a door to develop critical thinking as a situated practice and as part of an intercultural dialogue.

We are aware that our dataset is limited and that we still need to develop collaborative teaching approaches and gather evidence of the ways in which critical thinking as situated practice emerges and thrives. Nevertheless, the findings of this study emphasize how critical thinking acquires relevance when topics in school are important for students, when knowledge is related to their own life projects and to their possibilities to construct a stance within the community, as well as taking decisions that have relevant consequences for life. We want to emphasize the contribution of this chapter in the reflection of what critical thinking could mean for the lives of students in communities with diverse worldviews.

Science education is not only a sociocultural activity but a sociopolitical one "where issues of authority, power, and hierarchy affect social relations, access to ideas, and positionings that learners of a particular socially constructed racial group, ethnolinguistic affiliation, class, gender, and so forth, must negotiate" (Varelas et al., 2012, p. 6). Indigenous people have traditionally seen how their knowledge is devalued in school. Therefore, it is relevant that students are able to discuss in class about relevant issues for their lives (such as the milpa and fertilizers), where their voice can subvert some of these hierarchical relations, as we have shown in this chapter.

Bringing knowledge related to milpa to the classroom is relevant to promote critical thinking as situated practice. To promote critical thinking in a variety of communities there is not a ready-made recipe. We need to rethink and reinvent, bring cases such as the one presented to the arena for a collective reflection, aiming for research to help us think about different ways of being and contributing. In this case we have discussed data from classroom experiences with indigenous students. However, this viewpoint can be applied in other contexts where inequalities have generated the exclusion of contents, ways of thinking and even specific groups of students (migrant communities, for example).

The development of critical thinking is related to knowledge, abilities, and dispositions that can strengthen what the students in our study call "good living", associated with individual identity development and the construction of life projects (Subero & Esteban-Guitart, 2020). The objective is the development of individual self-determination: "the disposition to exert the capacity to elect between alternatives and, with a moral north constructed or individually chosen, each individual can take the moral determination according to her values and practices in an individual way" (p. 169). However, these alternatives to choose are restricted by the possibilities that the community allows and that are considered acceptable. In this margin for election is where the school would be situated, trying to develop students' critical thinking in order to give them opportunities to identify the range of possibilities for action, and develop justifications for their elections. For this reason, classroom experiences as the one presented here, which supports the legitimization of alternative ways of knowledge are valuable.

Currently we are collaborating with some teachers to design activities that consider their students' knowledge, their interests about fertilizers and their ways of being and knowing the world with the objective of generating dialogue spaces that promote decision making and justification. These activities will, on one hand, let us recover the experience of teachers in their work with indigenous students and, on the other hand, open spaces for in-service preparation and collaboration in learning communities. Activities will be used by teachers in their classrooms and shared and analysed in groups.

We aim to develop critical thinking in school in a way that it is useful for students to make decisions about issues that are relevant for their lives, congruent with their communities ways of living and consider the power relations that have been historically established. Critical thinking should be useful to craft students' and communities' 'good living'.

This line of research, that aims to integrate traditional knowledge in the classrooms, promote critical thinking and develop decolonizing strategies is barely developed but highly relevant and challenging. Especially important would be that groups of teachers, such as the ones with whom we have collaborated, be consolidated, permitting that reflections and actions in the classroom come from the joint interests of a diversity of voices: academics, teachers, and particularly, students and their communities.

Acknowledgments We would like to thank students of the second grade in secondary school in Yochib, San Cristóbal de las Casas and Aldama, and the teachers who participated in the workshop in San Cristóbal de las Casas for letting us work besides them and learn from their experience.

To Yeison Arboleda for his support in data collection in some of the activities in Chiapas. To Rocío Balderas and Nallely Jiménez for organizing data and transcribing. To Yei Rentería for her support in organizing information.

To the members of the GRECI Seminar: Eurídice Sosa, Liliana Valladares and Luz Lazos for their opinions on a previous version of this document.

This work was made possible with the support of the project Cinvestav-SEP, Conacyt No. 217.

References

Ahmed, S. (2015). *La política cultural de las emociones*. Programa Universitario de Estudios de Género de la UNAM.

Aikenhead, G. (2002). Cross-cultural science teaching: Rekindling traditions for aboriginal students. *Canadian Journal of Science, Mathematics and Technology Education, 2*, 287–304.

Aikenhead, G., & Elliott, D. (2010). An emerged decolonizing science education *in Canada. Canadian Journal of Science, Mathematics and Technology Education, 10*(4), 321–338.

Boege Schmidt, E. (2008). *El patrimonio biocultural de los pueblos indígenas de México*. Instituto Nacional de Antropología e Historia. Comisión Nacional para el Desarrollo de los Pueblos Indígenas.
Brocos, P., & Jiménez Aleixandre, M. P. (2020). El impacto ambiental de la alimentación: argumentos de alumnado de Magisterio y Secundaria. *Enseñanza de las Ciencias, 38*(1), 127–145. https://doi.org/10.5565/rev/ensciencias.2802
Carreño, M. (2009/2010). Teoría y práctica de una educación liberadora: el pensamiento pedagógico de Paulo Freire. *Cuestiones Pedagógicas, 20*, 195–214.
Carrillo Trueba, C. (2016). Gente de maíz, maíz como gente. Variación y estabilidad en el infinito complejo-maíz. *Ciencias, 118–119*, 4–11.
Carter, L. (2004). Thinking differently about cultural diversity: Using postcolonial theory to (Re)-read science education. *Science Education, 88*, 819–836. https://doi.org/10.1002/sce.20000
Carter, L. (2008). Recovering traditional ecological knowledge (TEK): Is it always what it seems? *Transnational Curriculum Inquiry, 5*(1), 17–25. http://nitinat.library.ubc.ca/ojs/index.php/tci
Ennis, R. (1996). Critical thinking dispositions: Their nature and assessability. *Informal Logic, 18*(2 & 3), 165–182.
Erduran, S., & Jiménez-Aleixandre, M. P. (2007). *Argumentation in science education. Perspectives from classroom-based research*. Springer.
Farrera, I. (2018). *La biliteracidad y sus usos sociales en estudiantes tseltales de la telesecundaria 714. El caso de Yochib, Oxchuc, Chiapas*. Unpublished master's thesis, Universidad Autónoma de Chiapas.
Freire, P. (1970). *Pedagogía del oprimido*. Tierra Nueva, Siglo XXI.
García Franco, A., & Gómez Galindo, A. A. (2020). *Aprendiendo en la milpa*. Universidad Autónoma Metropolitana.
Giroux, H. (1997). *Los profesores como intelectuales. Hacia una pedagogía crítica del aprendizaje*. Paidós.
Giroux, H. (2003). *Pedagogía y política de la esperanza: teoría, cultura y enseñanza: Una antología crítica*. Amorrortu.
Gómez Galindo, A. A., & García, F. A. (2021). Multicultural and dialogic science education in indigenous schools in the Mayan highlands, México. In M. M. Atwater (Ed.), *International handbook of research on multicultural science education* (Springer international handbooks of education). Springer. https://doi.org/10.1007/978-3-030-37743-4_21-2
Gómez Galindo, A. A., García Franco, A., Gónzalez Galli, L., & Torres Frías, J. C. (2019). Cultural diversity and evolution: Looking for a dialogical teaching perspective. In U. Harms & M. Reiss (Eds.), *Evolution education re-considered* (pp. 227–247). Springer. https://doi.org/10.1007/978-3-030-14698-6_13
González-Escallón, J. D. (2017). Ciudadanía liberal, proyecto de vida y autodeterminación: los derechos individuales como camino o como límite de la construcción de la identidad. *Vniversitas, 135*, 65–192.
Gorbach, F., & López Beltrán, C. (2008). *Saberes locales: ensayos sobre historia de la ciencia en América Latina*. El Colegio de Michoacán.
Gruenewald, D., & Smith, G. (2007). *Place-based education in the global age*. Routledge.
Harmin, M., Barrett, M., & Hoessler, C. (2016). Stretching the boundaries of transformative sustainability learning: On the importance of decolonizing ways of knowing and relations with the more-than-human. *Environmental Education Research*, 1–12. https://doi.org/10.1080/13504622.2016.1263279
Harrison, N. (2004). Rethinking critical thinking: Indigenous students studying at university. *Teaching Education, 15*(4), 375–384.
Hutchison, C. B. (2014). Internationally inclusive science education: Addressing the needs of migrants and international students in the era of globalization. In M. M. Atwater, M. Russell, & M. B. Butler (Eds.), *Multicultural science education. Preparing teachers for equity and social justice* (pp. 137–158). Springer. https://doi.org/10.1007/978-94-007-7651-7_9

Johnson, D. P. (2008). *Contemporary sociological theory. An integrated multi-level approach.* Springer. https://doi.org/10.1007/978-0-387-76522-8

Lave, J. (2009). The practice of learning. In K. Illeris (Ed.). *Contemporary theories of learning, learning theorists…In their own words* (pp. 209–217). Routledge. https://doi.org/10.4324/9780203870426

Maurer, E. (1977). ¿Aprender o enseñar?: la educación en Takinwits, poblado tseltal de Chiapas (México). *Revista del Centro de Estudios Educativos (México), VII*(1), 84–103.

McKinley, E. (2011). Postcolonialism, indigenous students and science education. In S. Abell & N. Lederman (Eds.), *Handbook of research in science education* (pp. 199–226). Routledge.

McLaren, P. (1997). Prefacio: Teoría Crítica y significado. In H. Giroux (Ed.), *Los profesores como intelectuales. Hacia una pedagogía crítica del aprendizaje*. Paidós.

Milne, M. C., Tobin, K., & DeGennaro, D. (Eds.). (2015). *Sociocultural studies and implications for science education. The experiential and the virtual.* Springer. https://doi.org/10.1007/978-94-007-4240-6

Moya, R. (2009). La interculturalidad para todos en América Latina. In L. E. López (Ed.), *Interculturalidad, educación y ciudadanía. Perspectivas Latinoamericanas* (pp. 21–56). Plural Editores.

Olivé, L. (2009). Por una auténtica interculturalidad basada en el reconocimiento de la pluralidad epistemológica. In E. Sader (Ed.), *Pluralismo Epistemológico* (pp. 19–30). CLACSO.

Oviedo, G., Maffi, L., & Larsen, B. (2000). *Indigenous and traditional peoples of the world and ecoregion conservation: An integrated approach to conserving the world's biological and cultural diversity.* World Wild Fund.

Pérez Ruiz, M., & Argueta Villamar, A. (2011). Saberes indígenas y diálogo intercultural. *Cultura y Representaciones Sociales, 5*(10), 31–55.

Pozzer, L. L., & Jackson, P. A. (2015). Conceptualizing identity in science education research: Theoretical and methodological issues. In C. Milne, K. Tobin, & D. DeGennaro (Eds.), *Sociocultural studies and implications for science education. The experiential and the virtual* (pp. 213–230). Springer.

Puig, B., Bravo Torija, B., & Jiménez-Aleixandre, M. P. (2012). *Argumentation in the classroom: Two teaching sequences. Project Science Teacher Education Advanced Methods (S-TEAM).* Danú. http://www.rodausc.gal/wp-content/uploads/sites/2/2019/01/Argumentation-Two-teachingSequences-English.pdf

Ramírez Castañeda, E. (2006). *La educación indígena en México*. UNAM.

Rappaport, J. (2007). Más allá de la escritura: la epistemología de la etnografía en colaboración. *Revista Colombiana de Antropología, 43*, 197–229.

Santos Baptista, G. C., & El-Hani, C. N. (2009). The contribution of ethnobiology to the construction of a dialogue between ways of knowing: A case study in Brazilian public high school. *Science & Education, 18*, 503–520. https://doi.org/10.1007/s11191-008-9173-3

Subero, D., & Esteban-Guitart, M. (2020). Más allá del aprendizaje escolar: el rol de la subjetividad en el enfoque de los fondos de identidad. *Teoría de la educación, 31*(1), 213–236. https://doi.org/10.1420/teri20955

Valladares, L. (2010). La educación científica intercultural y el enfoque de las capacidades. *Revista CTS, 16*(6), 39–69.

Varelas, M., Martin, D., & Kane, J. (2012). Content learning and identity construction (CLIC): A framework to strengthen African American students' mathematics and science learning in urban elementary schools. *Human Development, 55*, 319–339.

Vasco Uribe, L. G. (2002). *Entre selva y páramo: viviendo y pensando la lucha india.* Instituto Colombiano de Antropología e Historia.

Walsh, C. (2009). Interculturalidad crítica y pedagogía de-colonial: Apuestas (des)de el in-surgir, re-existir y re-vivir. In P. Melgarejo (Comp.), *Educación Intercultural en América Latina: memorias, horizontes históricos y disyuntivas políticas* (pp. 25–42). Plaza y Valdés.

Weiss, E. (2017). Hermenéutica y descripción densa versus teoría fundamentada. *Revista Mexicana de Investigación Educativa, 22*(73), 637–654.

Part II
Research About Critical Thinking in Biology and Health Education

Chapter 5
The Role of Evidence Evaluation in Critical Thinking: Fostering Epistemic Vigilance

Ravit Golan Duncan, Veronica L. Cavera, and Clark A. Chinn

5.1 Introduction: Promoting Reasoning in Epistemically Unfriendly Contexts

The current times, with a global pandemic, have brought into focus the dangers of misinformation and the difficulties that lay individuals experience when trying to evaluate information (scientific and otherwise) on various media platforms. In fact, the problem of misinformation regarding COVID-19 is so severe that Tedros Adhanom, the Director-General of the World Health Organization claimed that "Fake news spreads faster and more easily than this virus, and is just as dangerous" (WHO, 2020). In this post-truth climate, it is more pertinent than ever to educate future citizens to be epistemically vigilant (Britt et al., 2019). By epistemically vigilant we mean evaluating and monitoring the credibility and trustworthiness of information while being aware of the potential of being misinformed (Sperber et al., 2010). Epistemic vigilance is vital to critical thinking.

We draw on Kuhn's (2018) definition of critical thinking as argumentation. That is, evaluating information in light of alternative claims and constructing an argument in support of one (or more claims) and that refutes alternative claims. A central component of such critical thinking is reasoning with and about evidence. In the U.S. context, the Framework for K-12 Science Education (National Research Council, 2012) argues that a common feature of science knowledge building across domains is "a commitment to data and evidence as the foundation for developing claims" (p. 26). Research in science education has accordingly attended to the role of scientific evidence in inquiry (e.g., McNeill & Berland, 2017; McNeill & Krajcik, 2012; Erduran & Dagher, 2014; Berland et al., 2016), argumentation, (e.g., Berland

R. G. Duncan (✉) · V. L. Cavera · C. A. Chinn
Rutgers University, New Brunswick, NJ, USA
e-mail: ravit.duncan@gse.rutgers.edu

B. Puig, M. P. Jiménez-Aleixandre (eds.), *Critical Thinking in Biology and Environmental Education*, Contributions from Biology Education Research,
https://doi.org/10.1007/978-3-030-92006-7_5

& McNeill, 2010; Driver et al., 2000; Jiménez-Aleixandre et al., 2000; Sandoval & Millwood, 2005), and modeling (e.g., Bamberger & Davis, 2013; Passmore & Svoboda, 2012).

However, there are several problems with the ways in which evidence is addressed in current science education. The evidence used in science instruction often does not reflect the complexity of evidence encountered in the real world. Rather, evidence in science instruction is usually highly simplified, consisting of straightforward experimental set-ups and observations, or simple descriptions of scientific data (Chinn & Malhotra, 2002). Moreover, a few pieces of evidence, sometimes just one or two, are treated as sufficient to develop and evaluate models (see critique by Samarapungavan, 2018). In contrast, even the abridged evidence that individuals encounter in the news or social media is more complex and varied in quality than the epistemically friendly evidence seen in classrooms. Students also have few opportunities to learn that evidence evaluation and synthesis are social epistemic practices that are governed by community norms (Longino, 1990, 2002). Thus, it is no surprise that adults reason poorly with authentic, and epistemically messy, evidence of the real world (Chinn & Malhotra, 2002), frequently basing judgments on low-quality evidence or single studies, and showing little appreciation for higher-quality, stronger evidence (Iordanou & Constantinou, 2014; Kelly & Takao, 2002; Nisbett & Ross, 1980).

If we want students to learn to think critically about the kinds evidence and claims they will encounter in the epistemically unfriendly world outside the classroom, we need to engage them with evidence that varies in quality and strength of support and reflects greater epistemic complexity. This raises the question of which aspects of evidence quality and strength are easier or harder for students' to grasp, especially when students are tackling a set of varied evidence.

In this chapter we attempt to address this question through analyses of secondary students' responses of an argumentation task involving a set of evidence that varied in quality and strength. Students' arguments reflected their critical thinking about the evidence. Our analyses of their arguments is grounded in the *Grasp of Evidence* framework (Duncan et al., 2018), which captures dimensions of reasoning with evidence.

5.2 Theoretical Framework: Grasp of Evidence

Our conception of grasp of evidence builds on work by Ford (2008), who discussed the grasp of scientific practice as an internalization of two critical, interrelated roles: *constructing* and *critiquing* claims. "Grasp" entails an understanding of both the elements of the practice and practical skill in engaging in the practice across various contexts. Such a grasp is socially constructed and negotiated within a community of practicing scientists (or a community of science learners in a classroom). The Grasp of Evidence (GoE) framework (Duncan et al., 2018) unpacks reasoning with evidence as an important aspect of grasp of practice.

Expert grasp of evidence involves extensive topic-specific knowledge of theories and methods in a field, such as highly technical methods of data analysis (Chinn & Duncan, 2018; Samarapungavan, 2018). Achieving an expert level of grasp of evidence surpasses what is feasible to accomplish in K-12 education. Therefore, GoE proposes a mid-level of domain specificity, a lay grasp, at which students understand many important dimensions of reasoning with evidence (e.g., understanding the importance of carefully controlled experiments) while also appreciating that working with evidence involves extensive connections with detailed topic knowledge (that they do not necessarily have). Despite lacking the advanced topic knowledge to attain an expert grasp, students can still understand—as competent outsiders (Feinstein et al., 2013)—many of the relevant aims, ideals, and processes that undergird scientific evidentiary practices.

What are these aspects of lay reasoning with evidence? The framework posits five *dimensions* encompassing both what students should understand about how experts work with evidence and how laypeople can use evidence reports themselves. These dimensions are complementary, because having a mid-level understanding of how scientists use evidence can help laypeople engage with scientific evidence in productive ways despite limited disciplinary knowledge. For this study we focused on four of the five dimensions, described below:

1. ***Evidence evaluation.*** The evaluation of the methodological quality of evidence is central to scientific practice and often a point of contention in arguments (e.g., Collins & Pinch, 2012; Galison, 1997; Mayo, 1996; Staley, 2004). Although full evaluation of methodology typically requires deep disciplinary knowledge, there are some mid-level methodological considerations that should be within the grasp of secondary students. For example, checking that experimental variables are appropriately controlled to rule out any confounding factors; or being aware of the limitations of some forms of data, such as self-reports, in terms of their potential bias.
2. ***Evidence interpretation.*** Evidence is interpreted in light of (often) competing claims and models. Different theoretical perspectives can lead to different interpretations of evidence (Thagard, 1992; Thomm et al., 2017), as well as considering the strength and diagnosticity of evidence in supporting or refuting alternative models (Chinn, Rinehart, & Buckland., 2014). Not all high-quality (i.e., methodologically sound) evidence is strong evidence; rather, strength is a property of model-evidence relationships. For example, stronger evidence might support more model components or core (rather than peripheral) model components.
3. ***Evidence integration.*** The dimension of evidence integration focuses on the challenges of coordinating large, diverse, and often conflicting bodies of evidence (Barzilai & Ka'adan, 2017; Thagard, 2012) with alternative models. It also relates to the idea of converging lines of evidence and why such occurrences provide strong support of models. Ideally, different kinds of evidence cohere—for example, when evidence using a particular method supports a model, there will be independent lines of evidence for the validity of that method.

4. ***Lay use of evidence***. Given bounded disciplinary knowledge, laypeople need to use practices for engaging with evidence that are informed by expert evidentiary practices but are distinct from them (Bromme & Goldman, 2014; Keren, 2018). This dimension details grasp of evidence for lay use. For example, although laypeople cannot evaluate the merit of conclusions from technically complex studies, they can seek corroborating accounts for these claims, consider the reactions of other scientists, and get a sense of where experts agree or disagree about them (Thomm et al., 2017).

These four dimensions are each unpacked in terms of the kinds of epistemic considerations entailed by them. We conceptualize these epistemic considerations in terms of the *AIR* model of epistemic cognition (Chinn et al., 2014). This model posits that epistemic cognition includes three central components:

A. ***Aims and values,*** which are the goals that individuals and communities set (aims), such as knowledge, and the importance of that knowledge (value) to those individuals and communities. For example, a lay person's epistemic aim may be to find out (gain knowledge) whether being sick with COVID-19 confers immunity and protects you from further infections.
B. ***Epistemic Ideals*** are the criteria used to evaluate the resulting epistemic products such as evidence, claims, or models. In relation to the COVID-19 immunity, the person may have an ideal of high source credibility—that is, information should be gathered from reliable news media who themselves obtained information from reliable experts.
C. ***Reliable epistemic processes*** are the diverse processes used to achieve epistemic aims, such as protocols for carrying out observations or conducting experiments, procedures for meta-analyses, social processes of critique, etc. In relation to the COVID-19 immunity, the person can search for information in credible news outlets, corroborate the information across multiple sources, and try to determine the extent of scientific consensus around these claims.

In the GoE framework, each dimensions is specified in terms of relevant aims & values, ideals, and reliable process associated with the dimension. For example, in the *evidence evaluation* dimension, the aim is to determine whether available evidence is of high quality and its conclusions trustworthy. High-quality evidence meets the epistemic ideals of being appropriate for addressing the research question and of ruling out potential confounds and alternative explanations. To meet these ideals, one needs to use reliable processes for creating good evidence, including designing studies using appropriate experimental design; creating valid, accurate, and reliable measures; using well-established procedures, and so on. Students can use such processes themselves when conducting investigations, or they can evaluate whether evidence from secondary sources, such as reports of scientific studies in the media, employed appropriate reliable processes. (See Table 5.1 for aims, ideals, and reliable processes for the other dimensions.)

In this chapter we discuss our analysis of students' arguments using the four dimensions of GoE shown in Table 5.1.

Table 5.1 Grasp of evidence framework dimensions

Dimension	Epistemic aims & value	Epistemic ideals (examples)	Reliable epistemic processes (examples)
Evidence evaluation	Determining if evidence is of high quality and conclusions are trustworthy	*Appropriateness* (i.e., the methods address the questions) *Determinacy/conclusiveness* (i.e., ruling out confounds and alternative explanations for the findings) *Source trustworthiness* (degree of expertise, track record, etc.)	Study uses proper processes of study design (e.g., appropriate samples and comparisons) Study uses valid and reliable measures and analysis Study provides evidence of relevant source expertise
Evidence interpretation	Determining model validity using strong evidence	*Relevance* (i.e., the evidence is related to the model) *Strength* (i.e., the evidence clearly supports core parts of the model) *Diagnosticity* (i.e., precise fit with the model, ruling out other models, providing severe tests)	Systematic coordination of alternative models with evidence. Model perspective taking (i.e., inferring what results different models predict and how different models explain the same evidence) Model-evidence fit evaluation (i.e., carefully analyzing which parts of models are supported)
Evidence integration	Determining model validity based on a large body of evidence	*Quantity* of evidence *Quality* of evidence *Variety* of evidence (i.e., multiple types/lines of evidence) *Consistency of support* (i.e., lack of contradictory evidence) *Coherence* within bodies of evidence	Reliable evidence selection methods are used to select all relevant evidence Quality checks ensure that only evidence of adequate quality is integrated Systematic integration methods such as meta-analyses, tallying methods, and so on, enable consistent and impartial weighting of evidence
Lay use of evidence	Determining the credibility of scientific claims in everyday communication	*Source credibility* (degree of expertise, lack of bias, etc.) *Acceptance of claims in scientific community* (degree of consensus) *Validation by knowledgeable others* (reports vetted by independent relevant experts)	Determine relevance of expertise trustworthy-ness of source (e.g., potential conflict of interest) Determine degree of consensus in the scientific community; identify points of expert agreement and disagreement. Determine whether publication and dissemination channels have mechanisms of vetting or filtering

5.3 Instructional Context- Model-Based Inquiry in Biology

The arguments used in our analysis are drawn from two model-based inquiry projects for life sciences students: The *Promoting Reasoning and Conceptual Change In Science* (PRACCIS) for middle school students (Rinehart et al., 2014, 2016); and the *Investigating Issues in Learning Progressions* (I^2LeaPs) in which we developed a genetics unit for high school students (Castro-Faix et al., 2020; Duncan et al., 2017). Both interventions engage students in the practice of modeling through the development, evaluation, and revision of models of biological phenomena. In our instructional activities and assessments, students typically weigh multiple models (either ones they create or ones we provide) in light of evidence sets. Evidence sets include three to seven pieces of evidence varying in form and equality, such as anecdotal accounts, videos of experiments or observations, and simplified reports of research studies. Students are tasked with appraising the models in light of all the evidence available and then writing an argument in support of the model they think is best.

To support students' modeling practices, especially their coordination of models and evidence, we developed a suite of scaffolds that includes the Model-Evidence-Link (MEL) matrix, evidence ratings, and shared epistemic criteria (Rinehart et al., 2014, 2016). The MEL matrix (Fig. 5.1) presents a table with three columns: the first lists the evidence, and the following two list each of the competing models. Students use five types of arrows to denote the relationship between each piece of evidence and each model.

In conjunction with denoting the relationship between evidence and models, students also rate the quality of each piece of evidence on a four-point scale (0–3).

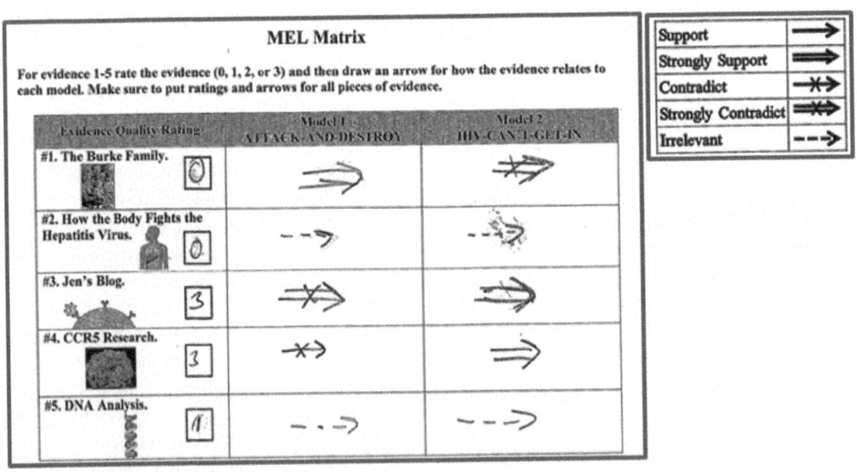

Fig. 5.1 Model-Evidence Link Matrix. Students use arrows to denote relationship between each piece of evidence and each model Students also rate evidence quality on a scale of 0–3 in the box next to each evidence thumbnail

A rating of 0 indicates that evidence is low-quality and untrustworthy (e.g., an anecdotal account), whereas evidence rated as 1 or 2 suggest that there are one or more methodological problems with the research design or its execution (e.g., lack of appropriate controls, small sample size, or possible confounds). Evidence quality pertains to the validity and reliability of the conclusion of the evidence, it does not imply any particular relationship between the evidence and the model. For instance, a high-quality study may be irrelevant to the models in question.

In addition to the MEL matrix, we also engaged students in the development of shared epistemic criteria for good models (Rinehart et al., 2016). Students first developed individual lists of criteria, and then the class a whole agreed on a shared community list of criteria for good models (Pluta et al., 2011). The class lists differed but overall included criteria for good models such as: "should be easy to understand and make sense", "supported by evidence" (fit with evidence), "show all the steps in a process", "explains why or how something happens", etc. These shared epistemic criteria were on public display in the classrooms and teachers drew students' attention to them as they engaged in modeling and argumentation tasks throughout the units.

5.3.1 Argumentation Tasks

We analyzed three different argumentation tasks. One of these tasks was part of a lesson in a five week unit in molecular genetics, for high school students, that focused on evaluating models that explain the genetically based resistance to HIV found in some individuals. Figure 5.2 displays the two models and one piece of evidence from the HIV task. The other two tasks, DEB and Asthma, were used as performance assessment task. The DEB task was about a genetically inherited skin disorder in which blisters are formed in the skin. While the disorder is real, the name "DEB" is fictitious. The DEB task was the performance assessment given at the end of the 5-week high school HIV unit. The Asthma task was a simpler performance assessment that was used in middle school (ages 11–13) as a post assessment after students had engaged in multiple model-based inquiry units. Table 5.2 summarizes the materials for each task.

5.4 Research Questions and Analysis

The argument tasks in these assessments entail reasoning about epistemic ideals and reliable processes involved in the four dimensions of reasoning with evidence (evaluation, interpretation, integration, and lay use). As described in Table 5.2, we engineered epistemic "messiness" into these tasks in terms of varied evidence quality, evidence strength and quantity, and the need to consider two models simultaneously when reasoning about model-evidence fit. This added epistemic

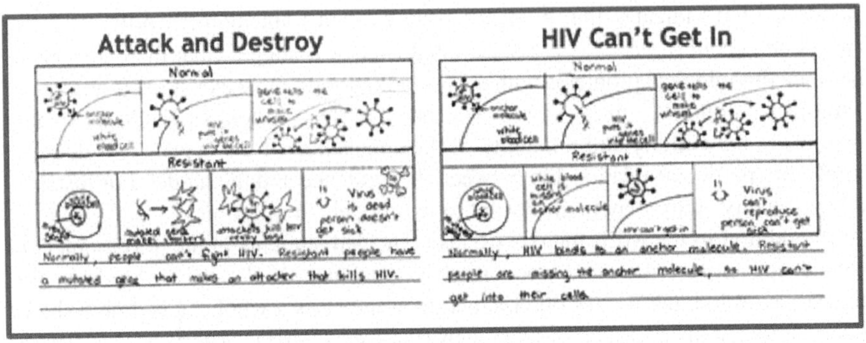

Evidence 2: A New Way to Fight Viruses?

Hepatitis B is a disease caused by a virus that attacks the liver. It can be fatal. Scientists have been testing a new way of treating people who are sick with the Hepatitis B virus. They are hoping that injecting sick people with interferon twice a week for a year will cure them of the virus.

What is interferon?

Interferon is a protein that is produced by the body's cells when attacked by bacteria or viruses. The interferon protein moves out of the cell and turns on the body's immune cells. Then immune cells destroy the invader.

Experiment:
People who are sick with Hepatitis B do not produce enough interferon to kill all the Hepatitis B viruses. So scientists think that injecting them with more interferon might turn on more immune cells to kill the Hepatitis B virus.

Method:
The scientists tested their treatment on 12 patients who were infected with Hepatitis B. They got injections of interferon every two weeks for a year.

Results:
Before the treatments, all 12 patients had high levels of the Hepatitis B virus.
After one year: 3 patients had no Hepatitis B virus in their blood.
5 more patients had very, very low levels of the Hepatitis B virus.
4 patients still had high levels of the Hepatitis B virus.

Fig. 5.2 Top box – Models for the HIV unit: normative model is "HIV can't get in" and an inaccurate model "Attack and Destroy". Bottom box – Evidence 2 regarding a new treatment for Hepatitis B involving interferon

messiness raises the question of whether students will be able to cope with it all. Will students notice problematic methodologies? Will they attend to relative strength of support? Will they attend to the amount and quality of the supporting or contradicting evidence? Moreover, will tasks that have more epistemic messiness (HIV) compared to ones with less messiness (DEB or Asthma) result in arguments that are richer in epistemic considerations related to GoE dimensions? By addressing these questions, we can gain insights about students' reasoning in epistemically messy contexts (in school and ultimately real world situations) and how to support it.

To address these questions we analyzed 20 written arguments of high school students from the HIV and DEB tasks (total of 40), and 20 written arguments of middle school students (ages 11–13) from the asthma task. The two high school

Table 5.2 Argumentation tasks: models and evidence

Genetic resistance to HIV	
Model A: Attack and destroy HIV resistant individuals have a mutated gene that codes for a new protein. This protein exits the cell and attacks the virus and kills it. (non-normative explanation)	Model B: HIV can't get in HIV resistant individuals have a mutated gene that codes for a non-functional protein, which is then missing from the membrane. Without this protein the virus cannot anchor and get into the cell.
Evidence	Features of evidence and relationship to models
E1: Burke family members suffer from AIDS	A family in which the father, mother, and one child are sick with AIDS. The first-born child is not infected (born before the parents were infected). Anecdotal account from a non-expert (father's opinion) of why the first born is not sick that implicates a protein that can attack the virus. Supports Model A
E2: Interferon	This evidence describes an experiment to treat hepatitis patients with interferon. The experiment was conducted by scientists but was not controlled. The sample size is small. The effect size is small. The phenomenon is different (Hepatitis vs. HIV). Supports Model A
E3: HIV resistant individuals are missing CCR5	This high-quality evidence describes an experiment showing that HIV-resistant individuals are missing the CCR5 protein (the anchor for HIV). Conducted by a scientists in a well-controlled and executed experiment. Supports Model B
E4: EM imaging of white blood cell and HIV	This high-quality evidence describes a well-controlled experiment conducted by scientists showing that resistant individuals do not have any HIV bound to their white blood cells (using an electron microscope image). Supports Model B
E5: DNA sequencing	Experiment comparing DNA sequence of HIV resistant and non-resistant individuals showing that they differ in one DNA section. Supports both Models (non-diagnostic)
Genetic disorder of the skin (DEB)	
Model A: Separatin Individuals with the DEB disorder have a gene for DEB which codes for a new protein called separatin. This new protein breaks down the proteins that hold together the skin layers. This causes the layers to separate. This causes blisters in the skin. (non-normative explanation)	Model B: Connectin Individuals with the DEB disorder have a mutated gene which codes for a non-functional connectin protein that is broken down by the cell. Normal connectin protein holds the skin layers together, without it the layers separate. This causes blisters in the skin.
Evidence	Features of evidence and relationship to models

(continued)

Table 5.2 (continued)

E1: DNA mutation	Shows the mutated sequence of DNA, noting that scientists think this mutation will affect the protein. Supports Model B
E2: Gaps in skin	Provides the results of an experiment conducted by scientists showing that affected individuals have blisters, but non-affected individuals do not. Method of experiment is not described. Supports both models (non-diagnostic)
E3: Image of skin layers	Labeled microscope image of the three skin layers (background information) Supports both models (non-diagnostic)
E4: Maria's treatment	Anecdotal account of a mother who bought medication from a website claiming the medication can break down separatin. After giving it to her son he had fewer blisters. Supports Model A
E5: Affected individuals missing a protein	Report of scientific study claiming that affected individuals (but not non-affected individuals) were missing a protein in their skin. Method of study not provided. Supports Model B
E6: Injecting connectin	Experiment conducted by scientists in which affected individuals were injected with connectin and most got better. Experiment was missing a control and sample size was not large. Supports Model B
Cause of asthma in brothers	
<u>Model A: Cockroach allergy</u> The boys are allergic to cockroaches. Cockroaches in their apartment are causing the asthma attacks.	<u>Model B: Inherited</u> Asthma runs in the family (the grandmother had asthma) and the boys inherited it from their parents. (non-normative explanation)
Evidence	Features of evidence and relationship to models
E1: Cockroach dust gets into lungs	Experiment conducted by scientists; individuals who spent a long time in a room with cockroach dust all had dust in their lungs. A control group did not. Sample size was moderate. Partially supports Model A
E2: Allergy study	Scientists studied the prevalence of allergies in a large sample of children. They found that over a third were allergic to cockroach dust. Partially supports Model A
E3: Grandma	Anecdotal account by neighbor claiming the boys sound like their sick grandmother and that genes can skip a generation. Supports Model B

classes (HIV and DEB tasks) were honors biology classes taught by the same high school biology teacher; the middle school class (asthma) was taught by a single middle school teacher. The schools, from which these classes were drawn, are from a suburban district with academically average performance in mathematics, language arts, and science. The district served 79% White, 9% Hispanic, 6% Black, 1% Asian, and the remaining students identify as mixed race. Approximately 34% of the students were classified as economically disadvantaged.

Our analysis of these arguments focused on which ideals and reliable processes students invoked in relation to the four evidence dimensions of GoE. In Table 5.3, we provide a coded example of an argument from DEB assessment task.

We were rather conservative with our coding. Students did not get credit for mentioning details of the evidence (e.g., that it was conducted by scientists, or had 25 patients, or the large impact of 80% improvement in patients). In order to get credit for noting source credibility students had to explicitly offer an appraisal, as in this example: "Evidence 4 also states the same thing, but is found using a controlled experiment by actual scientists". This statement was double coded as referencing evidence evaluation in terms of the *reliable epistemic process* of having controls and as referencing lay use of evidence in terms of the *ideal* of professional/credible source. In addition, it had to be clear that students were not simply repeating information provided as part of the evidence itself (e.g., sample size of 25 or noting that scientists conducted the study).

Table 5.3 Example of a student argument from the asthma task

The connectin protein explanation is better. This is so because the evidence is better quality and more reliable[1] for this theory. The connectin protein is clearly missing as proven by evidence 5 where scientists found that affected people were missing a protein. Then, as per evidence 6, when people were injected with connectin, 80% of the 25 patients did not produce as many blisters. Connectin is non-function because we see a genetic mutation in the gene that produces connectin in evidence 1. Finally, the lack of connectin explains the gap between the epidermis and dermis talked about in evidence 2. Without functional connectin proteins, the two parts separate and cause problems. Evidence 4, while it does support the separatin theory, has only one patient. This makes it unreliable[2].	1. This statement was classified as *evidence integration* in terms of GoE because it relates to the body of evidence supporting the preferred model. It was further coded as the *Ideal* of having better support due to the higher quality of the evidence. 2. These statements were classified as *evidence evaluation* in terms of GoE because they were specifically about a methodological issue with the evidence. This was further coded as related to R*eliable Epistemic Processes (REP)* because it references the small and unreliable sample size of one.

5.5 Results and Discussion

Students' arguments were fairly comprehensive. With the HIV arguments being longer on average (152 words average, ranging from 88 to 281), the DEB ones somewhat shorter (144 words average, from 50 to 207), and the Asthma arguments the shortest (134 words average, from 44 to 251). Most arguments across the three tasks included a claim supported by at least one piece of evidence.

There were differences in how many of these epistemic considerations (ideals and reliable processes) were evident in students arguments. Of the 20 HIV arguments, four failed to include epistemic considerations (in terms of our coding), six included 1–2 considerations, and ten included 3–5 considerations. Of the 20 DEB arguments, ten did not include any considerations, seven included 1–2 considerations, and three included 3–4 considerations. Of the 20 Asthma arguments (by middle school students), seven did not include any epistemic considerations, seven included one consideration, and six students included two. The HIV arguments included the most epistemic considerations. We believe that this is partly attributable to the greater complexity and epistemic messiness of the task and evidence in the HIV tasks. We discuss this point in more detail next.

Table 5.4 illustrates the ideals and reliable processes students noted in relation to the four dimensions of GoE across the three tasks. Numbers in the three rows represent how many arguments included the epistemic consideration noted in each row (total number of written arguments is 60).

Overall, for each set of considerations (i.e., lay use of ideals, evaluation ideals, evaluation reliable processes, interpretation ideals, integration ideals), there were more unique considerations in the HIV arguments. This was most evident for evaluation ideals of which there were three different types of considerations mentioned in HIV arguments but none in DEB or Asthma. We attribute this trend to the epistemic opportunities engineered into the task. The evidence set for HIV was more complex and epistemically messy; there was simply more to appraise in regard to epistemic ideals and reliable processes. The DEB and asthma evidence were less complex (and asthma had only three pieces of evidence) and thus afforded fewer issues to raise.

When problems did exist in the evidence, it is worth noting that students did notice and referenced them. For example, in the Asthma task evidence 3 was an anecdotal account and one student noted this. In fact, two additional students noted that the source of this evidence was not professional or credible, these students did not raise the issue of the account being a mere opinion (which would have been coded as anecdotal). Thus, about three students raised concerns about the quality of evidence 3 recognizing that it violated an ideal of source credibility (lay use of evidence).

However, given that there were multiple problems with the evidence (sample size, anecdotal account, lack of control) for DEB as well as HIV (there were fewer with the shorter asthma task), it is somewhat discouraging that students did not mention more epistemic considerations related to relevant ideals and reliable processes. In particular, evidence 6 in DEB and evidence 2 in HIV were both missing an

Table 5.4 Epistemic considerations in students arguments

Epistemic consideration categories	GoE dimension	Ideal or REP	Number of arguments		
			HIV	DEB	Asthma
Professional or credible source	Lay use	Ideal	5	1	4
Anecdotal, hearsay, or unproven idea	Lay use	Ideal	4	0	1
Non-professional or not-credible source	Lay use	Ideal	3	0	2
Totals and (number of unique considerations)			12 (3)	1 (1)	7 (3)
Accurate or reliable conclusion (result)	Evaluation	Ideal	2	0	0
Dated (old) evidence	Evaluation	Ideal	1	0	0
High quality evidence (unspecified)	Evaluation	Ideal	3	0	0
Totals and (number of unique considerations)			6 (3)	0 (0)	0 (0)
Adequate/missing controls	Evaluation	REP	2	0	0
Potential confounds	Evaluation	REP	0	2	0
Sample size (adequate or small)	Evaluation	REP	1	3	3
Evidence is detailed/shows process	Evaluation	REP	2	1	0
Empirical (experiment or series of experiments)	Evaluation	REP	1	0	1
Totals and (number of unique considerations)			6 (4)	6 (3)	4 (2)
Evidence contradicting favored model is weak	Interpretation	Ideal	1	0	0
Evidence is inconsistent with or contradicts other model	Interpretation	Ideal	2	2	0
Size of impact or effect (either large or small)	Interpretation	Ideal	1	1	0
Evidence is direct /conclusive/proves	Interpretation	Ideal	1	1	1
Evidence is irrelevant/indirect (different phenomenon)	Interpretation	Ideal	5	0	0
Totals and (number of unique considerations)			10 (5)	4 (3)	1 (1)
Multiple/more/most evidence supports model (amount)	Integration	Ideal	3	1	5
Explicit connection between two or more evidence pieces	Integration	Ideal	5	0	0
Stronger/better/higher quality support for preferred model OR insufficient support for other.	Integration.	Ideal	7	2	2
Totals and (number of unique considerations)			15 (3)	3 (2)	7 (2)

important control condition. Yet only two students wrote about this in their HIV arguments, and none mentioned this problem in their DEB arguments. Although even fourth graders can recognize confounds in simpler forced-choice tasks (Bullock et al., 2009), it may be more difficult to recognize confounds in these more complex tasks. Alternatively, some students may have found evidence 2 for HIV irrelevant (because it was about hepatitis and not HIV) and felt no need to reference the methodology of an irrelevant piece of evidence. Thus, students' attempts at developing a persuasive argument may have influenced which epistemic considerations they saw as worth mentioning.

Another noteworthy pattern is the distribution of considerations across the dimensions of the GoE framework. Most of the epistemic considerations related to evidence evaluation. This was also the only dimensions for which students referred to both ideals and reliable processes. We believe that students certainly used reliable processes for other dimensions. For example, students used tallying and weighting processes to decide which model has more and better supporting evidence; thus, they used reliable epistemic processes to integrate evidence (see Table 5.4). However, students did not explicitly mention the strategies they used for interpretation, integration, and lay use of evidence and so these considerations were not coded.

We next provide two examples of students' argument shown in Tables 5.5 and 5.6 below. We selected a-atypical arguments with five considerations (most arguments had fewer). However, these provide a good illustration of how these epistemic considerations are interleaved in an argument. The considerations in this example were all ideals and pertained to evidence evaluation (5), evidence interpretation (1 and 2), and lay use of evidence (3 and 4).

Table 5.5 Example of a student argument from the HIV task

The HIV-can't get-in model (specifically one where the CCR5 protein serves as an "anchor") is the superior model. In evidence 3, the CCR5 is shown to be lacking in HIV-resistant individuals. This shows that CCR5, rather than being a HIV-attacking protein, is a protein that facilitates HIV infection. One such way that a protein can facilitate infection is by acting as an "anchor" for the virus. <u>CCR5 was directly observed to do this in evidence 4</u>[1]; the HIV viruses were shown to be connected to the protein. Without the protein, as it was with HIV-resistant individuals, the virus could not be present on the white blood cells. There are pieces of evidence that seemingly contradict the model, most saliently evidence 2. While in that study viruses were directly attacked by immune cells activated by interferon proteins (which aligns with the attack-and-destroy model), <u>the evidence is not directly applicable. The hepatitis B virus, which was used in that study, is not the HIV virus, and it is not clear that the viruses are being inhibited by different mechanisms</u>[2]. Evidence 1 hints at the first model, but it's not worth considering- the "science" mentioned was <u>said by a nonprofessional</u>[3] (so all <u>he has to go on is hearsay</u>[4]), this was <u>during the 1980s, when HIV was little-known</u>[5]	1. This statement illustrates the interpretation ideal of "evidence is direct/conclusive". The idea that direct evidence provides stronger support. 2. These statements reflect a violation of the ideal above (i.e., evidence is irrelevant/indirect). The epistemic consideration notes that a different virus (phenomenon) may involve entirely different mechanisms and is thus irrelevant. 3. This statement reflects a violation of the ideal of credible source (i.e., non-professional/non-credible source) included in the lay use of evidence dimension. 4. This statement relates to the consideration that anecdotal evidence is problematic (lay use of evidence) and is essentially a violation of the trustworthy evidence ideal. 5. This statement reflects a violation of the ideal that evidence should be up to date and not old and is an important consideration in terms of evidence evaluation. Only this student mentioned it.

Table 5.6 Example of a student argument from the HIV task

Model 2 is the model I think to be better because there is better evidence supporting the model. Evidence 3 and 4 both are very good evidence they fit the criteria and give good information[1]. Evidence 3. Supports model 2 by showing by showing an experiment that people resistant to HIV do not have a certain protein that HIV locks onto so it can reproduce supporting the fact that HIV cannot get in. Evidence 4 also states the same thing, but is found using a controlled experiment [2] by actual scientists[3].	1. This statement appeals to the ideal of stronger/better support and explicitly notes multiple pieces of evidence taken together. This ideal relates to the evidence integration dimension of GoE. 2. Adequately controlling experiment is a reliable processes relevant to evidence evaluation. 3. The explicit pointing out that the study was done by actual scientists reflects a consideration of professional/credible source which is an ideal relevant to lay use of evidence.

The student appealed to the ideal of direct/conclusive support in order to highlight the strength of the model-evidence fit for the preferred model. Conversely, the student pointed out the weakness in the support evidence 2 (interferon) provided to the competing model, claiming that because it was a different virus the evidence was "not directly applicable" and thus resulted in a weaker model-evidence fit. Next, the student raised three epistemic concerns with evidence 1: the evidence is from a non-credible source, it is hearsay, and it is dated. This last epistemic consideration was only mentioned by this student and while true (the evidence was from the 1980s) this was not a potential weakness we actually considered or deliberately included as designers. That the student noticed and raised this as a problem (because HIV was "little known") suggests a fairly sophisticated understanding that during the early phases of inquiry knowledge is more tentative.

Our next example illustrates a variety of ideals and a reliable process from three of the four GoE framework dimensions: lay use of evidence (3), evidence evaluation (2), and evidence integration (1). We chose it specifically to illustrate the evidence integration dimension, which was not evidenced in the prior example. The student here appeals to the ideal that support by multiple and better evidence is stronger. Interestingly, the student notes that evidence 3 and 4 "fit the criteria and give good information"(1). We suspect the student meant that these evidence pieces fit the criteria for highly ranking evidence quality- the criteria used to determine the rating, from 0 to 3, of evidence. The explicit and specific focus on evidence quality and quantity as an important epistemic consideration when choosing a model is commendable as this consideration is particularly relevant to evaluating evidence in the epistemically messy real world. Later in the argument the student points out what contributes to the quality of the evidence, namely, that it is a "controlled experiment"(2) conducted by "actual scientists"(3). In this statement the student appeals to a reliable epistemic process (adequate controls) and source credibility, an ideal for lay use of evidence.

Overall, these students' arguments drew on multiple pieces of evidence, addressed the competing model and the evidence in support of it, and included multiple appeals to ideals and reliable epistemic processes across all four GoE dimensions (both arguments taken together). The task, with its two models and variety of evidence, afforded opportunities to notice and articulate these epistemic considerations rather impressively.

5.6 Conclusion and Implications

In this chapter we presented a theoretical framework, *Grasp of Evidence*, which we used to analyze students' epistemic considerations (ideals and reliable epistemic processes) relevant to the dimensions of evidence evaluation, evidence interpretation, evidence integration, and lay use of evidence. Our aim was to show that students can and do attend to issues of evidence quality and strength when these are available for them to comment on in the evidence. That is, students are responsive to engineered epistemic messiness.

However, this epistemic vigilance does seem to be more prevalent among the high school students compared to the middle school students. We do have to qualify this assertion as the evidence in the Asthma task that the middle school students completed afforded fewer methodological issues to comment on and was also an assessment task that lacked the MEL matrix and evidence rating scaffolds that were included in the HIV task. Thus, it is more likely that this result is a combination of multiple factors: development (age), fewer epistemic affordances (less messy), and a lack of scaffolds. The DEB task, which was a similar assessment completed by the high school students also lacked scaffolds and was somewhat less epistemically messy, and similarly led to fewer epistemic considerations per argument and overall (fewer unique considerations per dimension of GoE as seen in Table 5.4).

Also, worth noting is that not all epistemic messiness is equally salient to students. For example, the epistemic considerations raised most frequently were about source credibility (a lay use of evidence ideal) and having strong evidentiary support (an evidence integration ideal). In contrast, ideals and reliable processes related to evidence evaluation were mentioned less, despite the fact that problems with methodology (the evidence evaluation dimension) were specifically engineered into the evidence sets. This is somewhat surprising as these epistemic considerations are ones most relevant to the focus on experimental methods and control of variables that are relatively privileged in current science instruction.

A key instructional implication from this work pertains to making classroom environments more epistemically complex and messy. As we have noted (Duncan et al., 2018; Chinn et al., 2021) if we want students to develop competencies in reasoning about claims and evidence in the epistemically unfriendly real-world contexts outside of the classroom we need to engage them with epistemically messy evidence early and often. Our work here suggests that students are able to attend to these considerations when they are evident in instructional materials and assessments.

Acknowledgements The research presented herein was supported by National Science Foundation Awards #1053953 and #100863. Any opinions, findings, and conclusions or recommendations expressed in this material are those of the author(s) and do not necessarily reflect the views of the National Science Foundation. We also wish to acknowledge Aitan G. Duncan who kindly assisted with data analysis.

References

Bamberger, Y. M., & Davis, E. A. (2013). Middle-school science students' scientific modelling performances across content areas and within a learning progression. *International Journal of Science Education, 35*(2), 213–238.

Barzilai, S., & Ka'adan, I. (2017). Learning to integrate divergent information sources: The interplay of epistemic cognition and epistemic metacognition. *Metacognition and Learning, 12*(2), 193–232.

Berland, L. K., & McNeill, K. L. (2010). A learning progression for scientific argumentation: Understanding studentwork and designing supportive instructional contexts. *Science Education, 94*, 765–793. https://doi.org/10.1002/sce.20402

Berland, L. K., Schwarz, C. V., Krist, C., Kenyon, L., Lo, A. S., & Reiser, B. J. (2016). Epistemologies in practice: Making scientific practices meaningful for students. *Journal of Research in Science Teaching, 53*, 1082–1112. https://doi.org/10.1002/tea.21257

Britt, M. A., Rouet, J.-F., Blaum, D., & Millis, K. (2019). A reasoned approach to dealing with fake news. *Policy Insights from the Behavioral and Brain Sciences, 6*(1), 94–101.

Bromme, R., & Goldman, S. R. (2014). The public's bounded understanding of science. *Educational Psychologist, 49*(2), 59–69.

Bullock, M., Sodian, B., & Koerber, S. (2009). Doing experiments and understanding science: Development of scientific reasoning from childhood to adulthood. In W. Schneider & M. Bullock (Eds.), *Human development from early childhood to early adulthood: Findings from a 20 year longitudinal study* (pp. 173–197). Psychology Press.

Castro-Faix, M., Duncan, R. G., & Choi, J. (2020). Data-driven refinements of a genetics learning progression. *Journal of Research in Science Teaching*. Early view.

Chinn, C. A., & Duncan, R. (2018). What is the value of general knowledge of scientific reasoning? In K. Engelmann, F. Fischer, J. Osborne, & C. A. Chinn (Eds.), *Interplay of domain-specific and domain-general aspects of scientific reasoning and argumentation skills* (pp. 77–101). Routledge.

Chinn, C. A., & Malhotra, B. A. (2002). Epistemologically authentic inquiry in schools: A theoretical framework for evaluating inquiry tasks. *Science Education, 86*(2), 175–218.

Chinn, C. A., Rinehart, R. W., & Buckland, L. A. (2014). Epistemic cognition and evaluating information: Applying the AIR model of epistemic cognition. In D. Rapp & J. Braasch (Eds.), *Processing inaccurate information* (pp. 425–454). MIT Press.

Chinn, C. A., Barzilai, S., & Duncan, R. G. (2021). Education for a "Post-Truth" world: New directions for research and practice. *Educational Researcher, 50*(1), 51–60. https://doi.org/10.3102/0013189X20940683

Collins, H., & Pinch, T. (2012). Edible knowledge: The chemical transfer of memory. In *The golem: What you should know about science* (Canto classics) (pp. 5–26). Cambridge University Press.

Driver, R., Newton, P., & Osborne, J. (2000). Establishing the norms of scientific argumentation in classrooms. *Science Education, 84*(3), 287–312.

Duncan, R. G., Chinn, C. A., & Barzilai, S. (2018). Grasp of evidence: Problematizing and expanding the next generation science standards' conceptualization of evidence. *Journal of Research in Science Teaching, 55*(7), 907–937.

Duncan, R. G., Choi, J., Castro-Faix, M., & Cavera, V. L. (2017). A study of two instructional sequences informed by alternative learning progressions in genetics. *Science & Education, 26*(10), 1115–1141.

Erduran, S., & Dagher, Z. R. (2014). Reconceptualizing nature of science for science education. In *Reconceptualizing the nature of science for science education* (pp. 1–18). Springer.

Feinstein, N. W., Allen, S., & Jenkins, E. (2013). Outside the pipeline: Reimagining science education for nonscientists. *Science, 340*(6130), 314–317.

Ford, M. (2008). Grasp of practice' as a reasoning resource for inquiry and nature of science understanding. *Science & Education, 17*(2–3), 147–177.

Galison, P. (1997). *Image & logic: A material culture of microphysics*. University of Chicago Press.

Iordanou, K., & Constantinou, C. P. (2014). Developing pre-service teachers' evidence-based argumentation skills on socio-scientific issues. *Learning and Instruction, 34*, 42–57.
Jiménez-Aleixandre, M. P., Bugallo Rodríguez, A., & Duschl, R. A. (2000). "Doing the lesson" or "doing science": Argument in high school genetics. *Science Education, 84*(6), 757–792.
Kelly, G. J., & Takao, A. (2002). Epistemic levels in argument: An analysis of university oceanography students' use of evidence in writing. *Science Education, 86*(3), 314–342.
Keren, A. (2018). The public understanding of what? Laypersons' epistemic needs, the division of cognitive labor, and the demarcation of science. *Philosophy of Science, 85*(5), 781–792.
Kuhn, D. (2018). A role for reasoning in a dialogic approach to critical thinking. *Topoi, 37*(1), 121–128.
Longino, H. E. (1990). *Science as social knowledge: Values and objectivity in scientific inquiry*. Princeton University Press.
Longino, H. E. (2002). Science and the common good: Thoughts on Philip Kitcher's science, truth, and democracy. *Philosophy of Science, 69*(4), 560–568.
Mayo, D. G. (1996). *Error and the growth of experimental knowledge*. University of Chicago Press.
McNeill, K. L., & Berland, L. (2017). What is (or should be) scientific evidence use in k-12 classrooms? *Journal of Research in Science Teaching, 54*(5), 672–689.
McNeill, K. L., & Krajcik, J. (2012). *Supporting grade 5–8 students in constructing explanations in science: The claim, evidence and reasoning framework for talk and writing*. Pearson Allyn & Bacon.
National Research Council. (2012). *A framework for K-12 science education: Practices, crosscutting concepts, and core ideas*. The National Academies Press.
Nisbett, R. E., & Ross, L. (1980). *Human inference: Strategies and shortcomings of social judgment*.
Passmore, C. M., & Svoboda, J. (2012). Exploring opportunities for argumentation in modelling classrooms. *International Journal of Science Education, 34*(10), 1535–1554.
Pluta, W. J., Chinn, C. A., & Duncan, R. G. (2011). Learners' epistemic criteria for good scientific models. *Journal of Research in Science Teaching, 48*(5), 486–511.
Rinehart, R. W., Duncan, R. G., & Chinn, C. A. (2014). A scaffolding suite to support evidence-based modeling and argumentation. *Science Scope, 38*, 70–77.
Rinehart, R. W., Duncan, R. G., Chinn, C. A., Atkins, T., & DiBenedetti, J. (2016). Critical design decisions for successful model-based inquiry in science classrooms. *International Journal of Designs for Learning, 7*, 17–40.
Samarapungavan, A. (2018). Construing scientific evidence: The role of disciplinary knowledge in reasoning with and about evidence in scientific practice. In K. Engelmann, F. Fischer, J. Osborne, & C. A. Chinn (Eds.), *Scientific reasoning and argumentation: The roles of domain-specific and domain-general knowledge* (pp. 56–76). Routledge.
Sandoval, W. A., & Millwood, K. A. (2005). The quality of students' use of evidence in written scientific explanations. *Cognition and Instruction, 23*(1), 23–55.
Sperber, D., Clement, F., Heintz, C., Mascaro, O., Mercier, H., Origgi, G., & Wilson, D. (2010). Epistemic vigilance. *Mind & Language, 25*(4), 359–393.
Staley, K. W. (2004). Robust evidence and secure evidence claims. *Philosophy of Science, 71*(4), 467–488.
Thagard, P. (1992). *Conceptual revolutions*. University Press Princeton.
Thagard, P. (2012). *The cognitive science of science: Explanation, discovery, and conceptual change*. MIT Press.
Thomm, E., Barzilai, S., & Bromme, R. (2017). Why do experts disagree? The role of conflict topics and epistemic perspectives in conflict explanations. *Learning and Instruction, 52*, 15–26.
WHO. (2020). *Munich security conference* (WHO director-general speech). https://www.who.int/dg/speeches/detail/munich-security-conference

Chapter 6
Supporting Critical Thinking Through Engagement in Dialogic Argumentation: Taking Multiple Considerations into Account When Reasoning About Genetically Modified Food

Kalypso Iordanou

6.1 Introduction: Critical Thinking and Dialogic Argumentation

Being lockdown because of the COVID-19 pandemic (at the time of writing this chapter), the whole world is experiencing the multiple facets of a real-world problem. At the outset of the pandemic, the World Health Organization (WHO) organized a forum with the participation of almost 400 scientists of different disciplines to discuss preventive measures to help control the outbreak, including the scientific dimensions of the issue – diagnostics, vaccines, therapy – but also the social ones – ways to address fear, rumours, and stigma (Ghebreyesus & Swaminathan, 2020), showing that real life problems are complex, involving multiple dimensions. COVID-19 pandemic is an example of a socio-scientific issue (SSI), involving scientific and social dimensions, such as shortages of food, unemployment, education, environment, global economy, psychological effects of social isolation and racism, some of which are the results of the adopted preventive measures by countries to control the expansion of the virus. We are witnessing differences among countries in how they deal with the different dimensions of the issue of COVID-19. For example, some countries delayed or avoided escalating the measures to a lock down, to protect the country's economy for paralyzing, focusing on the economic dimension; while other governments took early and strong (with high impact on the economy) preventive measures to avoid the capacity degradation of the local health system and to save more human lives, prioritizing on the health dimension at a heavy expense on the economy. Different decisions might be the result of possible failure to consider all the diverse dimensions involved in a SSI

K. Iordanou (✉)
University of Central Lancashire, Larnaka, Cyprus
e-mail: kiordanou@uclan.ac.uk

B. Puig, M. P. Jiménez-Aleixandre (eds.), *Critical Thinking in Biology and Environmental Education*, Contributions from Biology Education Research,
https://doi.org/10.1007/978-3-030-92006-7_6

during decision-making. Another possible explanation of the different approaches observed among countries on how they handled the pandemic, at least at its outbreak, is how decision makers prioritize multiple dimensions of a SSI or the values they held.

It's not only politicians and scientists who are called to make decisions that affect citizens' lives about SSIs, they are individuals themselves who are called to make choices and take actions, such as following the instructions and recommendations issued by the public health authorities. Different decisions and actions lead to different consequences – including variations in the number of dimensions that would be affected and the intensity of the impact. Decisions, which fail to take into consideration the complexity of the issue and the different variables involved, may have detrimental consequences to the individuals and the societies. In challenging situations, such as the COVID-19 pandemic, politicians ask citizens to exhibit social responsibility. The President of the Greek parliament, Katerina Sakellaropoulou, asked citizens to put the social benefit above the individual one, to prioritize 'We' over 'I' (Ethnos, 2020). What could have prepared us to deal effectively with this situation and show the social responsibility expected? And what can prepare us to deal with such unforeseen situations in the future? The present chapter proposes that part of the answer is critical thinking. The answer may not sound novel, however a specific approach is proposed, along with new empirical evidence showing the effectiveness of the approach to achieve this objective. In particular, it is proposed that engagement in dialogic argumentation with real-world problems, such as socio-scientific issues, which are complex and multi-facets issues, can support individuals' critical thinking skills, particularly their ability to be aware of the complexity of the issue and the different dimensions it involves. To examine this hypothesis, primary school students were asked to engage in dialogic activities on the topic of Genetically Modified Food (GMF).[1] GMF constitutes another example of complex SSI, which shares some similarities to the features commented above for COVID-19. Unlike COVID-19 SSI, though, which is a new and pressing one, GMF is an on-going and contentious issue and its effects might be different from country to country, with one country, for example, experiencing to a greater extent its consequence on a particular dimension (e.g. environmental changes).

What is critical thinking? Critical thinking involves reflective thinking for deciding what to believe or do, according to Ennis' (2018) long-lasting definition, which it is adopted here. Critical thinking is considered as a key twenty-first century skill, which is fundamental for higher education and the workplace (Trilling & Fadel, 2009; Wagner, 2008). Kuhn (1999) proposed that critical thinking follows a developmental progression, along with meta-level thinking. Kuhn identified specific forms of meta-level thinking which are central to critical thinking, namely: (a) metacognitive, that is having control of one's own beliefs, (b) metastrategic, which involves the consistent application of standards of evaluation, and (c) epistemological, which describes one's view of knowledge and knowing.

[1] The data collection took place before the first cases of COVID-19 were reported.

The later form, the epistemological, has its roots in the early forms of the theory of mind understanding and follows a developmental progression (Iordanou, 2016a), embarking in childhood from a view of knowledge as an objective fact, conceptualizing later, in adolescence, knowledge as a subjective opinion and finally reaching an understanding of knowledge as a judgment amenable to evaluation, based on the criteria of argument and evidence. The later form of understanding, the evaluativist one, achieved usually in adulthood, is the form of understanding which supports critical thinking. Empirical data, show that an evaluativist epistemological understanding supports individuals' ability to consider multiple perspectives on an issue – being able to identify different dimensions of a SSI – (Baytelman et al., 2020), consider counterarguments (Zavala & Kuhn, 2017; Iordanou, 2016b; Iordanou et al., 2020; Shi, 2020), evaluate evidence (Iordanou et al., 2019b) and integrate divergent perspectives (Barzilai & Eshet-Alkalai, 2015; Bråten et al., 2011).

A fundamental component of critical thinking is individuals' ability to identify and take into consideration the multiple factors involved in a complex issue. In order to be able to decide effectively, individuals need first to be aware of the possible options. Without a consideration of the contribution of all possible factors to a problem, an effective solution can't be reached. Are individuals effective multiple factor thinkers? Several kinds of evidence suggest that most individuals think simplistically about complex issues. Individuals do not conceptualize multiple contributing causes as necessary to account for most phenomena, nor are they likely to take multiple considerations into account in making judgments (Kuhn & Iordanou, in press; Lytzerinou & Iordanou, 2020). Therefore, the question which appears pressing for educators is: How can we support individuals to be multiple factor thinkers? Being a multiple factor thinker involves both the skill to identify different dimensions of an issue – e.g. ethic, moral, social, scientific and economic dimensions – and the appreciation of divergent perspectives within a particular dimension – e.g. whether GMF is necessary to feed the entire world population. Explicit teaching (Kienhues et al., 2011) and engaging individuals in argumentative-based discussions (Fisher et al., 2017; Iordanou, 2010, 2016b; Shi, 2020; Zavala & Kuhn, 2017) appear promising in supporting students to acquire an awareness that multiple viewpoints may exist about an issue, focusing on a particular dimension. Yet, the question of how we can support students to acquire an understanding of the complexity of a problem, appreciating the diversity of dimensions involved in a particular issue, is still open.

Efforts to support directly the development of critical thinking skills yielded mix findings. Meta-analyses show that traditional undergraduate courses on critical thinking resulted in only minimal improvements in students' critical thinking skills, while subject-matter instruction involving student discussion on real life problems appears more promising (Abrami et al., 2015; Hitchcock, 2015). Still, there is a need for more empirical research examining which forms of interactions are more effective (El Soufi & See, 2019; Murphy et al., 2014). For example, research findings show that peers who engaged in discussions with peers holding an opposing view benefited more in developing their reasoning skills (Iordanou & Kuhn, 2020) and appreciating that there is an objective truth about a particular issue (Fisher et al., 2017) compared to a control group whose dialogs were confined to same-side peers.

The present study examines whether engagement in dialogic argumentation over an extended period of time can support individuals' critical thinking, particularly their ability to consider multiple dimensions of a particular SSI. Rather than conceptualizing critical thinking as an individual ability or skill, it is conceptualized as a dialogic practice (Kuhn, 2019). Consistent with Vygotsky's sociocultural theory, according to which skills developed at the social practice are internalised and are then exerted at the individual level, it is hypothesized that engagement and practice in dialogic argumentation with peers can support students' critical thinking. Students are engaged in an argumentative based curriculum, involving practice in dialogic argumentation and reflection (Iordanou, 2010; Iordanou & Constantinou, 2015; Iordanou et al., 2019a; Kuhn et al., 2016), about a SSI, namely Genetically Modified Food (GMF). Previous research has showed the benefits of engagement in dialogic argumentation for supporting students' skill to construct counterarguments (Kuhn et al., 2013; Iordanou, 2010), use evidence (Iordanou & Constantinou, 2015), acquire knowledge (Iordanou et al., 2019a), promote reading comprehension (Wilkinson et al., 2015) and advance individuals' epistemological understanding (Iordanou, 2010, 2016b; Zavala & Kuhn, 2017).

This study was part of a greater project that engaged primary school students in extensive application of the argumentative-based curriculum with SSIs for two years. The data presented here were collected in the second year of the implementation of the curriculum of a particular class and focus on students' ability to construct arguments of great diversity, taking into consideration the complexity of the issue and the multiple dimensions involved. Results of this intervention on promoting other aspects of students' argument skills over time – such as their ability to use evidence and the function that evidence serves, e.g. for supporting one's position or counterargue other's position – both on the intervention and a transfer topic, were reported in Iordanou and Kuhn (2020). Considering dialogic argumentation as a pathway of supporting critical thinking skills (Jiménez-Aleixandre & Puig, 2012; Kuhn, 2019; Murphy et al., 2014), the focus of this chapter is in examining dialogic argumentation as a vehicle for supporting multiple dimensions thinking, which is a fundamental component of critical thinking. This question was examined by asking primary school students to engage in an argumentative based curriculum. The topic of GMF was chosen, because it is a socio scientific issue which involves multiple dimensions which would make it ideal for supporting students' ability to take into consideration this range of dimensions when thinking about a particular issue. Engagement in meaningful discussion on this topic and making strong arguments for a particular position as superior to the alternative, adequate content knowledge was required. Building on the work of Iordanou et al. (2019a), we provided participants with multiple pieces of relevant information in the form of questions and answers, which remained available throughout the intervention so for them to make use of as they chose.

6.2 Modes of Inquiry

6.2.1 Participants

Nineteen primary school students participated in the study. They were 6th grade students, 12 years old, eleven of them girls, who attended a public school in Cyprus. Participants were mostly from middle-class families and within an average range of ability and academic achievement. Students participated as part of their science classes, taught by their science school teacher, who received training by the author.

6.2.2 Procedure

6.2.2.1 Initial Assessment

Participants' argument skills were assessed by an individual writing task. Participants were asked to write an essay taking a position on whether Genetically Modified Food (GMF) should be allowed or not in the European Union, in which their country is part of. The teacher provided a brief introduction about GMF. The prompt was to "Write the argument you would make to someone who didn't agree that your position is the better one."

6.2.2.2 Intervention

The intervention involved the implementation of a curriculum, which was based on practice in dialogic argumentation and reflection (Iordanou, 2010; Iordanou et al., 2019a; Kuhn et al., 2016). It took place over twelve 90-min class periods, twice per week, in the computer lab of the school. For motivational purposes, the participants were informed that the purpose of the activity was to prepare for a whole-class debate, where a winning team would be declared, while also learning about GMF. Participants were assigned to one of two teams – for or against allowing GMF in the European Union – based on the position they supported at the initial essay. Students who were undecided or neutral were asked to join the less popular team – for GMF – in order to have an equal number of students in each one.

Preparation In the first two sessions, participants assembled, randomly, into same-side groups of 5–6 each and were asked to generate reasons supporting their side's preferred position. First, participants were asked to write individually the reasons for supporting their position, and then, to discuss and reflect on those reasons in the context of their group. Once eliminating duplicates students recorded each reason on a separate card and ranked the reasons cards with respect to the reason's strength.

Paired Dialogic Electronic Argument with Opposing-View Pair Same-side pairs were formed in each team – for or against GMF –, who remained together throughout the next sessions, which involved dialogic argumentation. Each same-side pair engaged in an electronic dialog with a different pair from the opposing side in their classroom. The dialogs took place via an instant-messaging platform. Participants were asked to introduce themselves and the position they supported – either for or against GMF – at the beginning of the dialog. Then, they were instructed to collaborate with their partner to decide what they wished to communicate to the other pair they were to engage in dialog with. No explicit instruction with respect to argumentation was provided, besides the encouragement to focus on positions and not judging individuals.

Participants had available pieces of information that they could use if they wished. The information was provided in the form of a set of 24 "question and answer" (Q&A) cards. Each card contained a question on one side and a short answer to the question concealed on the other side. The set was balanced overall with respect to support of the two positions, including some neutral statements, statements for GMF and statements against GMF. Table 6.1 shows some examples of the evidence that were provided.

Table 6.1 Examples of questions and answers provided to students on the topic of genetically modified food

What are the major Genetically Modified crops?
Cotton, corn, and soybeans are the main genetically modified crops grown in the United States. Most of these are used to make ingredients for other foods, such as Corn syrup used as a sweetener in many foods and drinks, Corn starch used in soups and sauces, Soybean, corn, and canola oils used in snack foods, breads, salad dressings, and mayonnaise, Sugar from sugar beets. Other major GE crops include Apples, Papayas, Potatoes and Squash.
Do genetically modified crop increase yield?
A study by the Corn Producers Organization in 2006 showed that genetically modified corn produces 10 tonnes of corn per hectare while conventional corn yields 9 tons per hectare.
What are the effects of growing genetically modified plants on the environment?
The environment is benefiting as farmers increasingly adopt conservation tillage practices, build their weed management practices around more benign herbicides and replace insecticide use with insect resistant genetically modified crops. The reduction in pesticide spraying and the switch to no till cropping systems is continuing to reduce greenhouse gas emissions from agriculture.
How do the genetically modified plants affect the ecosystem?
Cultures of genetically modified plants affect various species of the ecosystem, such as weeds, which are undesirable for the crop. Weeds can reduce the harvest of a farmer, but they are important for the ecosystem and the environment, because they provide food and protection to insects, birds and other organisms. Reduction of weeds leads to reduction of beneficial insects such as bees and birds.
Why Australia abandoned a 10-year project on genetically modified pea?
Genetically modified peas under development in Australia were evaluated by tests normally applied to medicine. The peas created a dangerous immune response in mice which, if found in humans, might be life threatening. The 10-year pea project, costing over $2 million dollars (US), was abandoned.

At the beginning of each session, before the dialog began, pairs were presented with this information bank and were asked to choose three cards and access the answers. At each new session, the cards they had chosen in previous sessions remained available to them and three new ones could be chosen. By the end of these sessions, all participants had seen an identical set of 24 cards with questions and answers. Students were instructed to examine the new information and then engaged in electronic dialogs with other-side pairs. They were asked to try to convince the other pair that their position was the better one.

Reflective Analysis of Transcripts from Previous Argument Sessions After three dialogue sessions – 3rd, 4th and 5th – had been completed, reflective analysis was introduced in every other chat session, 6th, 8th and 10th (Table 6.2). In this activity, pairs were asked to reflect on the electronic transcript of the dialog they had completed in the previous session by using one of two reflection sheets (RS), the own-side RS and the other-side RS. First the own-side reflection sheet was introduced, which asked participants to reflect on the opposing side's position and on the strength of the counterarguments they had constructed to the opposing side's position, taking into consideration whether they had used evidence to support their position. Participants were also encouraged to construct a better counterargument to other side's argument, using evidence. Then the other-side reflection sheet was introduced which asked participants to reflect on the rebuttals they used to weaken counterarguments to their own position. Further, participants were asked to contemplate what a better rebuttal might have been. At the 10th session, participants were asked to complete both the own side and the other-side reflection sheets.

Review and Preparation for Showdown In Sessions 11 and 12, same-side groups reassembled and worked together to prepare for the whole-class showdown. The group reviewed and reflected on the reflection sheets they had completed in previous sessions to decide what arguments to use in the debate. An adult coach, the teacher and three research assistants, facilitated these discussions. Students after reflecting on the counterarguments and the rebuttals they had constructed, excluded duplicates and noted the arguments, counterarguments and rebuttals in different colored cards. In session 11, students had available the own-side reflection sheets that they had

Table 6.2 Activities involved in the curriculum, by session

Session	Activity
Sessions 1–2	Generating reasons supporting own side
Session 3–5	E-Chats (1, 2, 3)
Session 6	Reflective Activity: Own Side Reflection Sheet
Session 7	E-Chat 4
Session 8	Reflective Activity: Other side Reflection Sheet
Session 9	E-Chat 5
Session 10	Reflective Activity: Own and Other side Reflection sheets
Session 11	Preparation for Showdown
Session 12	Showdown

completed in the previous sessions and prepared a two-cards array of counterarguments that they could use to address possible arguments from the other sides. In session 12, using the completed other-side reflection sheets, students constructed an array of three-cards consisting of argument, counterargument and rebuttal. Those sets of cards provided a visual representation for students of the structure of an argument.

6.2.2.3 Final Assessment

One week following Session 12, all participants were asked to write a final essay on the discourse topic, of whether GMF should be allowed in the European Union. The prompt again was to "Write the argument you would make to someone who didn't agree that your position is the better one."

6.3 Results

6.3.1 Coding

Students' essays were segmented to arguments. Arguments were first coded based on their category to scientific, social or socio-scientific. Arguments were further coded based on their content to the following categories: Food Quality, Food Preservation, Economy, Health, Employability, Food availability, Poverty, BEnvironmen/Environment (Table 6.3).

In addition, arguments were coded as one-side or two-sided depending on whether they included both pro and con reasons or only one of the two (Kuhn & Udell, 2003).

6.3.2 Preliminary Analysis

At initial assessment, 5 students supported the use of GMF, 9 students were against the use of GMF, and 5 students were neutral or undecided. At the final assessment, 3 students were for GMF, 11 were against and 5 students were neutral or undecided. Regarding change of position, 14 students expressed the same position at initial and final assessment, while 5 students changed their position. Of those who changed their position, 2 students changed from for GMF to against GMF, 2 students changed from neutral or undecided position to for GMF and 1 student changed from a position supporting the use of GMF to a neutral position, from initial to final assessment.

Table 6.3 Categories of arguments

Categories of arguments	Examples
Food Quality	(+) Quality control of GMF (+) Enrichment of vitamin content of GMF (−) Reduced vitamin content in GMF (+) Improved physical characteristics of food (e.g. larger size, more attractive colours)
Food preservation	(+) Greater expiry time (+) Avoiding or minimizing the use of food preservatives which may pause a risk to health
Economy (Arguments related to the production, distribution and trade, as well as consumption of goods and services by different agents. It includes different perspectives: seller, buyer, companies, countries)	(+) Lower production cost (+) Genetically modified organisms are more resistant to pests and plan diseases, thus they require less or no spraying. Hence, farmers do not have to spend so much pesticides. (+) Consumers have to pay less to buy them.
Health (including human's and animals' health)	(+) Avoid effects to health from having to use less pesticides and insecticides compared to conventional plants. (+) Prevention of human diseases (by the enrichment of GMF with missing vitamins e.g. golden rice) (−) GMF are harmful for human health (−) GMF cause human diseases (−) GMF increase the number of deaths in humans (−) GMF are tested in animals (animal cruelty) (−) GMF cause increase in animal deaths, when animals are fed with these products
Employability	(+) Creating of new jobs in genetically modified organism production (−) GMF causes unemployment of farmers
Food Availability	(+) Increasing the quantity of food on the market (+) Finding food every season of the year (Reducing the seasonality on the availability of fruit and vegetable)
Poverty	(+) Increasing food production (which makes it available to a larger population) (+) Ability to feed a larger proportion of the population
Bioethical issues/ Environment	(−) Modifications in organisms' DNA (e.g. insects) (−) Genetically modified plants can have a negative effect on the (animal) food chain (+) Humans intervene in nature, as genetically modified organisms are not natural products

Note. (+) pro GMF, (−) con GMF

6.3.3 Number of Arguments

Before turning to the main analysis, examining the type of arguments used, a quantitative indicator was examined that, while hardly conclusive on its own, provide an initial, albeit superficial, indication of change in argument skill. This is change in the number of arguments in the essay. At initial assessment students produced 2.350 ($SD = 1.496$) arguments, while at the final assessment they produced twice as many arguments as they produced at initial assessment, $M = 4.60$ ($SD = 3.016$). Paired sample t-test showed that this difference was statistically significant, $t(19) = 3.028$, $p = .007$.

6.3.4 Categories of Arguments: Social, Scientific and Socio-scientific

Regarding the different categories of arguments produced – Social, Scientific and Socio-Scientific – while participants produced arguments of all of these categories at both initial and final assessment, the most prevalent category, as seen in Fig. 6.1, is social at both initial and final assessment. Participants produced about the same number of scientific and socio-scientific arguments at initial assessment, whereas at the final assessment they produced twice as many socio-scientific arguments compared to scientific arguments.

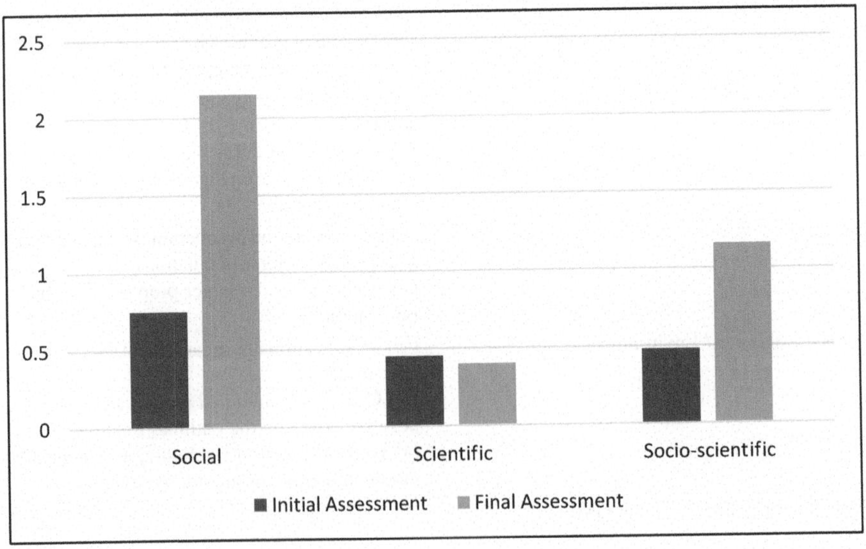

Fig. 6.1 Mean number of Social, Scientific, and Socio-Scientific arguments produced at initial and final assessments

To examine if there were changes in the categories of arguments in students' essays in terms of using social arguments, scientific arguments and socio-scientific arguments, from initial to final assessment paired sample t-test was used. Results showed that participants produced significantly more socio-scientific issues at the final assessment compared to the initial assessment, $t(19) = 2.896$, $p = .009$. In particular, they produced three times more socio-scientific arguments at the final assessment, ($M = 1.158$, $SD = 1.068$), than they did at initial assessment ($M = 0.47$, $SD = 0.612$).

There was also a significant change in the number of social arguments produced, from initial ($M = .750$, $SD = .910$) to final assessment ($M = 2.15$, $SD = 1.565$), $t(19) = 4.499$, $p < .001$. No significant change was observed in the number of scientific arguments produced, from $M = .450$ ($SD = .686$) scientific arguments at initial assessment to about the same number at the final assessment ($M = .400$, $SD = .754$).

6.3.5 Diversity of Arguments

To examine participants' ability to consider multiple perspectives of an issue, the diversity of arguments produced, that is, the number of different types of arguments produced was examined (i.e. Food Quality, Food Preservation, Economy, Health, Employability, Food Availability, Poverty, Bioethical issues/Environment). Comparing the number of different types of arguments produced at initial and final assessment, the essays at the final assessment included a greater diversity of arguments than the ones at initial assessment, $t(19) = 3.028$, $p = 0.007$. Participants' essays at initial assessment included $M = 2.350$ ($SD = 1.496$) different categories, whereas their respective essays at the final assessment included two times more diversity, with a mean of 4.600 different categories ($SD = 3.016$). Noteworthy, 84.2% of participants introduced at least one new type of argument at their final assessment, which they haven't reported at initial assessment.

As seen in Fig. 6.2, three new types of arguments were observed at the final assessment, which were not observed at the initial one. In particular, at the final assessment participants considered bioethical/environmental issues, food availability and food preservation, none of which were considered at initial assessment.

6.3.6 Sidedness of Essays

At initial assessment only 14.3% of the students produced two-sided arguments, whereas at the final assessment about half of the participants did so (47.4%). McNemar test showed that the change in the number of participants produced two-sided arguments from initial to final assessment was statistically significant, $p = .031$.

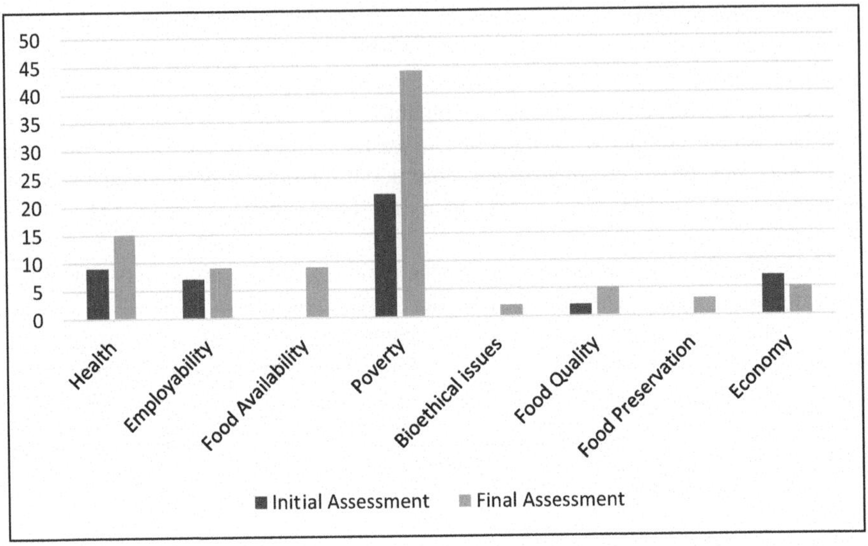

Fig. 6.2 Frequency of different types of arguments used at initial and final assessments by participants

6.3.7 *Examples of Students' Arguments at Initial and Final Assessment*

In this section, some illustrative examples are provided of the arguments provided during the initial and final assessment to exemplify the changes observed in students' argument skills. Students are identified using pseudonyms.

Example 6.1: Costas
Initial Assessment.

> *Some plants have to be genetically modified, because those that are not (genetically modified) may become contaminated while growing up.*

> Final Assessment

> *The use of genetically modified organisms should be allowed in Cyprus because GMF is more resistant to cold and have positive health effects. There are foods that can save lives if modified, such as golden rice. There are foods that prevent diseases. These foods are (genetically modified) soybean oil and soybean. Modified food lasts for more days. Also modified food is cheaper, bigger, and it is produced in larger quantities for consumers (to be accessed by more consumers). Non-genetically modified plants produce less quantity, therefore can feed less consumers. It (conventional food) is very expensive, so consumers have to spend a lot of money.*

In this example, at initial assessment only one dimension is considered, namely contamination while growing up, which is related to food quality. At the final assessment, besides food quality, other dimensions are now considered – food preservation, dealing with poverty and economy.

Example 6.2: Anne
Initial Assessment

I am undecided because on the one hand genetically modified food lasts longer, is bigger, so it is more worth buying. But on the other hand, genetically modified food can harm our body and be unhealthier. That's why I'm undecided, and I don't know which is best.

Final Assessment

I'm for Genetically Modified Food because there is more (availability). The use of genetically modified food is unhealthy but on the other hand it saves us from cancer, heart disease and (we can) add vitamins to some food. It also yields more crop and needs less pesticides. In addition, it is more durable, it is bigger and tastier. For example, (without genetically modification) corn's size would have been equal to our little finger, apple wouldn't have such a nice red colour, and tomato would have been tasteless and not juicy. They (scientists) will find out if the GMF includes something of danger for our health and they will correct it or remove it. For these reasons, I support the permission of Genetically Modified Food in my country.

In the second example, Anne at the initial assessment considered the food preservation, the economic and the health dimensions. She constructed arguments either for or against GMF for each dimension, for example for the health dimension the student supported that GMF has negative effects whereas for the economic dimension, she claimed that the GMF has a positive effect. At the final assessment, the student was able to see both pros and cons within a single dimension, constructing a two-sided argument. For example, for the health dimension, she acknowledged that GMF might by unhealthy, but at the same time, they can actually promote health, by adding vitamins that are missing in particular types of food. Then she continued her argument about health, counterarguing her initial argument – that GMF might not be healthy – by providing a counterargument – that if this is the case, scientists will identify and resolve the issue.

Example 6.3: Peter
Initial Assessment

I chose this view because some food after the (genetic) modification may have more vitamins. It is even more delicious than the regular one and consumers would want to buy it. Also, it is not bad and does not harm our health.

Final Assessment

I am neutral because both (options) have positives and negatives. GMF is cheaper and good value for money. But GMF causes health problems, such as cancer, but this is rare. If there were no GMF, there wouldn't be jobs in the factories and the economy would be affected negatively. Some people become allergic because of consuming GMF. I believe that both are good.

In the third example, Peter at initial assessment considers food quality, health and attractiveness of products leading most likely to increases in GMF's sales (economy), in his argument. He claims that GMF is more nutritious and tastes better than conventional food; these qualities make it more attractive to the consumers, leading most probably to an increased consumption ("consumers would want to buy it").

At the final assessment, Peter produced arguments regarding economy and health, but he also considered another dimension of the issue, namely employment, acknowledging the GMF industry has created new jobs. Interestingly, the student's thinking about a particular dimension has also changed. For the health dimension, although at initial assessment he didn't identify any risks, at the final assessment he identified the risks of some individuals being allergic to GMF and more importantly the possibility of developing cancer. Like Anne, though, in Example 6.1, the same student counterargued this argument, by claiming that this risk is rare. Those examples show that students have internalized the thinking first developed at the social level, of offering arguments and counterarguments, to the individual level.

Example 6.4: Loucas

Initial Assessment

I chose this answer because in this way farmers will not have to spray all the time, if for example a tomato is harder and makes it impossible for the various insects and worms to eat through it. But there should not be a lot of genetically modified food, because they harm our health.

Final Assessment

I believe that genetically modified food should be restricted because it is bad for health. When we (genetically) modify food, we eat something such as a tomato, but it is not (really) a tomato, it is something else. Also, if we modify a tomato the insects will not be able to pierce it, but gradually the insects will get stronger and they will become dangerous.

In the fourth example, Loucas, at initial assessment focused on producers' convenience who won't have to use chemicals to protect their products from insects and worms. At the final assessment, he extended his thought on the issue, identifying some cons on the issue which initially he considered as an advantage. At the final assessment, Loucas identified the risk of insects getting stronger, identifying bio-ethics risk on biodiversity.

Example 6.5: Mary

Initial Assessment

I chose this decision because if we hadn't modified some food, they would have been damaged (infected by pests), they would come in smaller sizes and they would last for a shorter time. We would have to spray them more, because without modification they wouldn't last long.

Final Assessment

GMF should not be banned, nor should it be allowed (in the EU) because modified fruits and vegetables are larger (in size) and last longer but on the other hand are not good for our health. Furthermore, genetically modified food is damaging for our health, but on the other hand, if it was actually damaging our health, it would have been published in the press that a particular type of food is not good for our health. Also genetically modified plants have greater production than the conventional ones and consumers have to pay less (are cheaper), but some individuals argue that if you eat large quantities of genetically modified food you will spend far more money for health care (due to illness caused by GMF) than (when eating) conventional ones.

Mary, as seen in Example 6.5, at initial assessment exemplifies the benefits of GMF, focusing on the dimensions of food perservation – GMF lasts longer –, food quality – GMF is bigger in size – and the environment – with cultivating genetically modified plants, spraing is minimised.

At the final assessment, Mary showed a two-sided thinking, identifying pros and cons. Interestingly, she provided an argument – GMF is of larger size and lasts longer –, then she counterargued her argument – GMF is damaging to health – and finally she provided a rebuttal to the counterargument she provided – "if it was actually damaging our health, it would have been published in the press". In addition, at the final assessment the student considered some new dimensions on the issue of GMF, namely availability of food – that genetically modified plants have greater production than the conventional ones – and the economic factor – GMF is cheaper than conventional food.

Notably, as seen in the examples above, participants took more dimensions into account when reasoning about GMF at the final assessment, after dense engagement in argumentation on this topic, compared to their performance at initial assessment.

6.4 Discussion

The aim of this chapter is to examine whether engagement in dialogic argumentation over an extended period of time on a real-life issue can support individuals' critical thinking, particularly their ability to consider multiple dimensions of a socio-scientific issue. Students engaged in a dialogic argumentative curriculum (Kuhn et al., 2008; Iordanou & Kuhn, 2020; Iordanou et al., 2019a), involving extensive discussions with peers on the SSI of whether Genetically Modified Food (GMF) should be allowed in their country. The results of the study show that after the intervention, participants produced longer and more socio-scientific arguments combining the social and the scientific dimensions of the issue and arguments of greater diversity – considering different aspects such as health, economy and the environment. While participants at initial assessment considered on average two different dimensions in their arguments, at the final assessment they considered twice as many dimensions as they did in the initial assessment. Interestingly, the majority of the participants – 16 out of 19 – considered a dimension at the final assessment that they didn't consider during the initial assessment, showing that peer-discussions gave them the opportunity to consider dimensions of the issue that they haven't considered before, when they were thinking on their own.

Besides constructing arguments of greater diversity, participants of the study exhibited gains in constructing two-sided arguments. Although at initial assessment only about 3 of the students produced two-sided arguments, at the final assessment, about half of them were able to do so. As seen in the qualitative analysis presented in the results section, students included counterarguments in their arguments either within a particular dimension or across dimensions. For example, Mary, in Example 6.5, provided an argument for GMF, regarding food preservation – claiming that GMF lasts longer – and then constructed a counterargument, on the dimension of health – claiming that GMF are not healthy. In other cases, such as Peter's and Anne's, students provided a counterargument within the same dimension that the argument was constructed. For example, Peter identified the risk of GMF causing health problems, but then he moved on to acknowledge that this possibility is rare.

The findings of this study provide support to our hypothesis that engagement and practice in dialogic argumentation with peers can support students' critical thinking. These results are in line with other empirical findings in the literature, both our own, using the same argumentative curriculum (Kuhn et al., 2008), and others', using similar methods (see Resnick et al., 2018, for a review), in exemplifying the benefits of engagement in dialogic argumentation in advancing students' quality of thinking. For example, the improvements observed in producing two-sided arguments are in line with the findings of other empirical studies which reported similar results after engagement in an argumentative-based intervention (Kuhn & Udell, 2003; Kuhn et al., 2008). Yet, our findings extend the literature in an important way, showing that engagement in dialogic argumentation can promote multiple dimensions thinking, and an important element of critical thinking. The finding of producing different types of arguments, considering multiple dimensions – such as health, economy, and the environment – show that engagement in dialogic argumentation supported participants to develop an appreciation of the complexity of the socio-scientific issue that they discussed. The opportunities provided in the context of the curriculum to articulate, explain, find relevant evidence, form arguments and counterarguments to convince peers, and reflect upon their own reasoning (Iordanou & Constantinou, 2015; Iordanou, 2022), may have supported students to think about the complexity of the problem and the multiple facets that it involves. These findings underline the necessity to incorporate dialogic argumentation with peers in the everyday learning process of science classes (Jiménez-Aleixandre et al., 2000; Resnick et al., 2018).

The current design does not enable us to identify the mechanism that supported the gains observed, ruling out the possibility that the improvements observed can be attributed to a particular element of the curriculum. For example, someone may argue that the new types of arguments observed at the final assessment are the result of learning new information provided during the intervention. Yet, empirical data from previous research makes this possibility less plausible. In particular, in the Iordanou (2010) study, an experimental group of middle-school students who engaged in the dialogic curriculum outperformed a control group of same age students who received the same pieces of information but didn't engage in the dialogic curriculum. The first group showed greater improvement in their argument skills, particularly their ability to produce counterarguments both in the intervention topic and in a new topic, in another domain, which had not been discussed (transfer topic). Similar results were observed in the Iordanou et al. (2019a) study, where we compared an experimental group who participated in the dialogic curriculum on the topic of alternative sources for producing electricity with a control group who engaged in the same curriculum but received the information in the form of a text instead of the question-and-answer format. The question-and-answer group outperformed the control group in their argument skills – their ability to construct evidence-based arguments and counterarguments – showing the superiority of the question-and-answer method over a traditional whole-text method in making available to students knowledge they need to engage in meaningful discourse about a topic they have little prior knowledge of, probably because the question-and-answer format supported students to appreciate the potential use of the information. Other experimental studies, focusing on other elements of the curriculum showed the importance of peer interaction and reflection. In Iordanou and Kuhn (2020) study, examining peer interaction itself, we found that the thinking of those who engaged in discourse with peers who held an opposing view benefited by hearing arguments favouring the opposing position expressed by individuals known to hold this position and outperformed peers who engaged in dialogs with same-side peers. Iordanou and Constantinou (2015) and Iordanou (2022) provided evidence for the contribution of engagement in reflective activities for promoting students' argument skills, either by examining students' dialogs during the intervention, using the microgenetic method (Iordanou & Constantinou, 2015) or by comparing a group of students who engaged in reflective activities in the context of the curriculum with a control group who engaged in peer dialogs only, without engaging in reflective activities (Iordanou, 2022).

This study shows that engagement in an argumentative-based curriculum on a real life socio scientific issue, such as the issue of genetically modified food, involving practice in argumentation with peers and reflective activities, can support students to develop an awareness of the complexity of socio-scientific issues. Whether the gains observed in the present study can have an impact on students' thinking but also their actions – critical thinking action – in real-world issues, such as the COVID-19, only future research can determine. Re-visiting the question we addressed in the introduction of "What could have prepared us to deal effectively with the situation of pandemic COVID-19 and show the social responsibility expected?", we believe that critical thinking could be part of the answer. Critical thinking could have prepared us, both by being more prepared long before the appearance of the pandemic – expecting a possible pandemic as scientific evidence suggested – and at the outset of the pandemic on how a country could have effectively dealt with this issue. The awful real-life experiment that humanity is currently going through with COVID-19 pandemic, shows that through engagement with a SSI, individuals experience the different aspects that it involves – such as scientific, social, environmental. The present study, which was pursued a couple of years earlier than the COVID-19 pandemic, shows that asking students to think of real life issues or possible scenarios, can support the development of multiple dimension thinking, an attribute of critical thinking which is necessary to deal with real life issues, for which we have little control on when and how they will arise. At the moment, supporting the development of students' critical thinking through engagement in argumentation on SSIs, appears as one of our best investment.

References

Abrami, P. C., Bernard, R. M., Borokhovski, E., Waddington, D. I., Wade, C. A., & Persson, T. (2015). Strategies for teaching students to think critically: A meta-analysis. *Review of Educational Research, 85*(2), 275–314.

Barzilai, S., & Eshet-Alkalai, Y. (2015). The role of epistemic perspectives in comprehension of multiple author viewpoints. *Learning and Instruction, 36*, 86–103.

Baytelman, A., Iordanou, K., & Constantinou, C. (2020). Epistemic beliefs and prior knowledge as predictors of the construction of different types of arguments on socio-scientific issues. *Journal of Research in Science Teaching. 57* (8), 1199–1227.

Bråten, I., Britt, M. A., Strømsø, H. I., & Rouet, J. F. (2011). The role of epistemic beliefs in the comprehension of multiple expository texts: Toward an integrated model. *Educational Psychologist, 46*(1), 48–70.

El Soufi, N., & See, B. H. (2019). Does explicit teaching of critical thinking improve critical thinking skills of English language learners in higher education? A critical review of causal evidence. *Studies in Educational Evaluation, 60*, 140–162.

Ennis, R. (2018). Critical thinking across the curriculum. *Topoi, 37*, 165–184.

Fisher, M., Knobe, J., Strickland, B., & Keil, F. C. (2017). The influence of social interaction on intuitions of objectivity and subjectivity. *Cognitive Science, 41*, 1119–1134.

Ghebreyesus, T. A., & Swaminathan, S. (2020). Scientists are sprinting to outpace the novel coronavirus. *The Lancet, 395*(10226), 762–764.

Hitchcock, D. (2015). The effectiveness of instruction in critical thinking. In M. Davies & R. Barnett (Eds.), *The Palgrave handbook of critical thinking in higher education*. Palgrave Macmillan.

Iordanou, K. (2010). Developing argument skills across scientific and social domains. *Journal of Cognition and Development., 11*(3), 293–327.

Iordanou, K. (2016a). From theory of mind to epistemic cognition. A lifespan perspective. *Frontline Learning Research, 4*(5), 106–119.

Iordanou, K. (2016b). Developing epistemological understanding through argumentation in scientific and social domains. *Zeitschrift für Pädagogische Psychologie., 30*(2-3), 109–119.

Iordanou, K. (2022). *Supporting strategic and meta-strategic development of argument skill: The role of reflection*. Metacognition and Learning. https://doi.org/10.1007/s11409-021-09289-1

Iordanou, K., & Constantinou, C. P. (2015). Supporting use of evidence in argumentation through practice in argumentation and reflection in the context of SOCRATES learning environment. *Science Education, 99*, 282–311.

Iordanou, K., Kendeou, P., & Michalinos, Z. (2020). Examining my-side bias during and after reading controversial historical accounts. *Metacognition & Learning, 15*(3), 319–342.

Iordanou, K., Kuhn, D., Matos, F., Shi, Y., & Hemberger, L. (2019a). Learning by arguing. *Learning and Instruction., 63*, 101–207.

Iordanou, K., Muis, K. R., & Kendeou, P. (2019b). Epistemic perspective and online epistemic processing of evidence: Developmental and domain differences. *The Journal of Experimental Education, 87*(4), 531–551.

Jiménez-Aleixandre, M. P., & Puig, B. (2012). Argumentation, evidence evaluation and critical thinking. In B. Fraser, K. Tobin, & C. McRobbie (Eds.), *Second international handbook of science education*. Springer.

Jiménez-Aleixandre, M. P., Bugallo Rodríguez, A., & Duschl, R. A. (2000). "Doing the lesson" or "doing science": Argument in high school genetics. *Science Education, 84*(6), 757–792.

Κατερίνα Σακελλαροπούλου: Να βάλουμε το «εμείς» πάνω από το «εγώ» [Katerina Sakellaropoulou: Putting 'we' over 'I'] (2020, March 24). *Όθνος [Ethnos]*. Retrieved from https://www.ethnos.gr/politiki/96135_katerina-sakellaropoyloy-na-baloyme-emeis-pano-apo-ego

Kalypso, I., & Deanna, K. (2020). Contemplating the opposition: Does a personal touch matter? *Discourse Processes, 57*(4), 343–359. https://doi.org/10.1080/0163853X.2019.1701918

Kienhues, D., Stadtler, M., & Bromme, R. (2011). Dealing with conflicting or consistent medical information on the web: When expert information breeds laypersons' doubts about experts. *Learning and Instruction, 21*, 193–204.

Kuhn, D. (1999). A developmental model of critical thinking. *Educational researcher, 28*(2), 16–46.

Kuhn, D. (2019). Critical thinking as discourse. *Human Development, 62*(3), 146–164.

Kuhn, D., & Iordanou, K. (in press). Epistemology as a core dimension of cognitive development. In D. Dunning & N. Ballantyne (Eds.), *Reason, bias, and inquiry: New perspectives from the crossroads of epistemology and psychology*. Oxford University Press.

Kuhn, D., & Udell, W. (2003). The development of argument skills. *Child Development, 74*, 1245–1260.

Kuhn, D., Hemberger, L., & Khait, V. (2016). *Argue with me: Argument as a path to developing students' thinking and writing* (2nd ed.). Routledge.

Kuhn, D., Goh, W., Iordanou, K., & Shaenfield, D. (2008). Arguing on the computer: A microgenetic study of developing argument skills in a computer-supported environment. *Child Development, 79*, 1311–1329.

Kuhn, D., Zillmer, N., Crowell, A., & Zavala, J. (2013). Developing norms of argumentation: Metacognitive, epistemological, and social dimensions of developing argumentive competence. *Cognition & Instruction, 31*, 456–496.

Lytzerinou, E., & Iordanou, K. (2020). Teachers' ability to construct arguments, but not their perceived self-efficacy of teaching, predicts their ability to evaluate arguments. *International Journal of Science Education. 42*(4), 617–634.

Murphy, P. K., Rowe, M. L., Ramani, G., & Silverman, R. (2014). Promoting critical-analytic thinking in children and adolescents at home and in school. *Educational Psychology Review, 26*(4), 561–578.

Resnick, L. B., Asterhan, C. S. C., Clarke, S., & Schantz, F. (2018). Next generation research in dialogic learning. In G. E. Hall, L. F. Quinn, & D. M. Gollnick (Eds.), *Wiley handbook of teaching and learning* (pp. 323–338). Wiley-Blackwell.

Shi, Y. (2020). Talk about evidence during argumentation. *Discourse Processes. 57*(9), 770–792.

Trilling, B., & Fadel, C. (2009). *21st century skills: Learning for life in our times*. Wiley.

Wagner, T. (2008). Even our "best" schools are failing to prepare students for 21st-century careers and citizenship. *Educational leadership, 66*(2), 20–25.

Wilkinson, I. A. G., Murphy, P. K., & Binici, S. (2015). Dialogue-intensive pedagogies for promoting reading comprehension: What we know, what we need to know. In L. B. Resnick, C. S. C. Asterhan, & S. N. Clarke (Eds.), *Socializing intelligence through academic talk and dialogue* (pp. 37–50). American Educational Research Association.

Zavala, J., & Kuhn, D. (2017). Solitary discourse is a productive activity. *Psychological Science, 28*, 578–586.

Chapter 7
Critical Thinking to Decide What to Believe and What to Do Regarding Vaccination in Schools. A Case Study with Primary Pre-service Teachers

Blanca Puig and Noa Ageitos

7.1 Introduction

Our world faces increasingly complex social, environmental and health problems marked by social tensions between individual rights, social aims, economic interests, and values (Reis, 2014). Socio-scientific issues (SSIs) affecting the health of our society require a citizenry that is well-equipped and capable of making critical decisions and taking action based on reason and critical thinking (CT), as the pandemic of Covid-19 demonstrates. The propagation of "fake news", the use of pseudo-therapies and the increasing number of opponents to vaccination have become worrying and are issues that should be addressed in teachers' training programmes and in biology lessons when looking at the topic of health. The introduction of modes of instruction capable of enacting and sustaining a culture of questioning in schools is essential nowadays (Peters et al., 2018). Teacher training "for" and "about" CT is necessary to face the challenges raised by postmodern thought in the teaching of health. This book considers CT reasonable and independent thinking (Jiménez-Aleixandre & Puig, Chap. 1) that entails a set of skills and dispositions, which can be developed through the practice, like other human practices (Facione et al., 2000).

School science practices and teacher training must move towards the use of new approaches, tools, and the design of learning environments that provide

B. Puig (✉)
Facultade Ciencias da Educación, Universidade de Santiago de Compostela, Santiago de Compostela, Spain
e-mail: blanca.puig@usc.es

N. Ageitos
IES Nº 1, Ribeira, Galicia, Spain
e-mail: noa.ageitos@edu.xunta.gal

B. Puig, M. P. Jiménez-Aleixandre (eds.), *Critical Thinking in Biology and Environmental Education*, Contributions from Biology Education Research,
https://doi.org/10.1007/978-3-030-92006-7_7

opportunities to practice CT without paying attention to subject matter, content, and procedural knowledge (Vincent-Lacrin et al., 2019). In the context of health teaching, this requires, among other issues, equipping future teachers and students with the kinds of knowledge and CT skills that will empower them to make health-enhancing decisions regarding their own health.

This chapter presents a case study research conducted with Primary pre-service teachers (PSTs) engaged in the practice of CT in a decision-making task on vaccination. The study was implemented while vaccination was mandatory for students in Early Childhood education in all state schools in our Spanish region, Galicia. In response to this regulation, a range of media spread information regarding this topic. Considering this context, and the importance of preparing teachers for dealing with this controversy in schools, a case study was designed. Although SSIs have been deeply explored, and there is evidence of their benefits for enhancing argumentation and knowledge application (Evagorou & Puig, 2017), in the area of teaching health issues there are no prior studies on how they could promote these competencies along with CT skills.

Our research aims to broaden empirical investigation on CT and health literacy in PST training. The main goal is to explore PSTs' positions regarding vaccination and their capacity to critically evaluate diverse premises supporting the anti-vaccination movement. The research questions are:

RQ1 – Which CT skills and what background knowledge are mobilized by a group of PSTs in a decision-making task on vaccination?

RQ2 – What are the connections between the practice of CT and the knowledge domain of immunization?

7.2 The Role of Critical Thinking and Health Literacy

We will first discuss CT on SSI instruction and then turn attention to the role of CT and health literacy as crucial components for addressing some of the challenges that arise in our times, such as the anti-vaccination movement.

7.2.1 Critical Thinking and SSIs Instruction

CT is one of the key competences included in the European Reference Framework (Hoskins & Deacon Crick, 2010). Curricula in most of OECD countries include, in one form or another, CT as an expected learning result (Elen et al., 2019). Nonetheless, teachers can find it unclear as to what CT means and what this will require in their daily teaching practice (Vincent-Lacrin et al., 2019). This is not surprising, since CT is a multifaceted notion that causes controversy concerning its conceptualization, evaluation, and instruction. The concept of CT has been used with a wide range of meanings in theoretical literature (e.g. Ennis, 1962; Facione, 1990). Views range from defining it merely as a commitment to evidence, to competencies related to the evaluation of evidence, the challenge of arguments based on authority, or the ability to criticize discourses that contribute to the reproduction of asymmetrical relations of power (Jiménez Aleixandre & Puig, 2012). This book builds upon this perspective and thus considers CT as a reasonable and independent mode of thinking entailing a set of components related to argumentation and social emancipation along with citizenship (Jiménez-Aleixandre & Puig, Chap. 1, this book). We draw from the revised proposal of characterization of these authors, which point to the importance of CT for action. SSI instruction, as health controversies, provides opportunities for dialogic practice and critical actions.

Both CT and argumentation overlap in their territories of engagement (Andrews, 2015). As Dewey (1910) suggests, CT is evaluative in nature; that is, it involves the active consideration and analysis of claims, suppositions, procedures and influences about whatever it is that the agent is thinking. One of the central features in argumentation is the development of epistemic criteria for knowledge evaluation (Jiménez-Aleixandre, 2007), which is a necessary skill to be a critical thinker. According to the experts of the Delphi panel of the American Philosophical Association (Facione, 1990), CT cannot be considered as a body of knowledge to be presented to students just as any other school subject. It should be embedded in the content programmes that are specific for each discipline and which rely on the events of everyday life as the foundation for developing one's CT. A key feature of CT is that it requires the mastery of context-specific knowledge in order to evaluate precise beliefs or claims (Anderman et al., 2012). These authors point out that even in environments that are appropriate to enhance CT and scientific reasoning, students frequently lack the required background knowledge to do so successfully. They suggest that the ability to reason effectively demands rich, interconnected, domain-specific knowledge, and that lack of content knowledge "makes the task of thinking critically challenging if not impossible". If this is correct, then how can we best prepare future teachers to succeed in this kind of teaching? Which teaching strategies and learning activities are adequate to engage students in the practice of CT along with the mobilization of knowledge?

There is a wealth of theoretical studies on CT, in contrast with little empirical research on which teaching strategies and learning environments better promote CT (Puig et al., 2019). According to Ennis (1989), there are four main approaches for the promotion of CT that emerged from the attempt to provide a framework that would help researchers and professionals: the *general approach*, the *infusion approach*, the *mixed approach* and the *immersion approach;* the last one being the most frequent in empirical studies on CT. The immersion approach means that CT is integrated into the subject matter without making its principles explicit to the students.

This research follows this approach, but expands it taking into account Brocos and Jiménez-Aleixandre (Chap. 12, this book) proposal, which distinguishes three dimensions or types of practices for teacher training in CT: (1) (instructors) teaching CT as content; (2) pre-service or in-service teachers' engagement in tasks intended for them to develop CT skills and dispositions; (3) transfer of practice: teachers' engagement in tasks intended to promote transfer of the CT skill to school pupils.

The task designed in this study deals with the second dimension and encourages participants to think critically, with attention to the following four aspects (Ten Dam & Volman, 2007): (a) students' development of their epistemological beliefs, (b) students' engagement in active learning, (c) interactions among students, to exchange their point of view (Paul, 1992), and (d) the use of real-life contexts (Brown, 1997). Health controversies are appropriate contexts for this purpose as they are topics that provide learners with a strong sense of a need to know where they are expected to make reasoned and independent decisions with a high impact on society. Prior investigations point to the need to promote personal decision-making on immunization to improve students' personal skills (Maguregi González et al., 2017). The decision about vaccination can be regarded as a SSI (Ratcliffe & Grace, 2003), in which CT and scientific knowledge play a crucial role.

7.2.2 *Critical Thinking and Health Literacy on the Topic of Vaccination*

Teacher education programs primarily focus on content knowledge and there has been a shift to focus more on practice (Grossman & McDonald, 2008) and on CT. Teaching SSIs through engaging learners in argumentation and CT goes beyond implementing a new biology curriculum. It requires dealing with some challenges presented by pseudoscientific thought, such as the anti-vaccination movement addressed in this chapter. Schools play a central role in preparing learners with the knowledge that will enable them to make health-enhancing decisions (Kickbusch, 2008). This chapter argues that promoting health literacy in primary education is crucial for this purpose. As a construct, health literacy encompasses the articulation of five core components: theoretical knowledge, practical knowledge, critical thinking, self-awareness, and citizenship (e.g. Paakkari & Paakkari, 2012; Nutbeam, 2000). Aligned with these authors we can agree that theoretical knowledge is not enough to make individuals take health-promoting actions. Learners need to comprehend health notions and processes, but they also must have CT skills to assess health messages, distinguish scientific facts from opinions, and make personal decisions on immunization, amongst other issues.

Despite the number of studies targeting PSTs training on SSIs continuing to expand, the efficacy of such designs for improving CT skills remains understudied. Furthermore, one of the problems identified is that PSTs often feel ill equipped to address the complexities and sensitive debates that derive from these issues (Evagorou & Puig, 2017). Added to all the difficulties above, pseudoscience has highlighted new challenges to health teaching. The increase in "fake news" has become a concerning issue (Peters, 2018). Research suggests that limited knowledge and misinformation about vaccines plays an essential role in public attitudes (Motta et al., 2018). In the case of vaccines, social media plays a large role in disseminating and sensationalizing vaccine objections (Lundström et al., 2012). This includes the question of safety of vaccines, for instance, concerning the alleged link to autism. This drives people to see only the risks of vaccines instead of their crucial role in public health care and the prevention of outbreak and global spread of diseases, as the COVID-19 pandemic is revealing.

Such objections are part of what has been called the "anti-vaccination movement" (Kata, 2010), which has had a demonstrable impact on vaccination policy, and individual and community health (Poland & Jacobson, 2001). The anti-vaccination movement is formed by people who reject the use of vaccines for different reasons. Some people have philosophical or religious objections; some see mandatory vaccination as interference by the government into what they believe should be a personal choice; some may believe that vaccine-preventable diseases do not present a serious health risk and others are concerned about the safety or efficacy of vaccines (WHO, 2019). We could also include in this group of opponents pseudoscience supporters, or people who feel aligned with them, and others that mistrust conventional medical practices and/or lack scientific knowledge. As Faasse et al. (2016) point out, it is important to note that this characterization of individuals as "pro-vaccine" or "anti vaccine" could lead to an oversimplification. Some people are either totally supportive or totally critical of vaccines. However, for the purposes of this study, we will consider that there is a continuum within "anti-vaccination" positions, between those who are totally opposed to vaccination and those who accept the value of vaccines, but think that their potential dangers pose real concerns for them.

Anti-vaccination messages are more frequent on the Internet than in other media, increasing the likelihood that vaccination decisions may be based on misleading information (Kata, 2010). Prior studies on attitudes to vaccination (Zingg & Siegrist, 2012) show that people with a higher level of general knowledge are more likely to vaccinate compared with people having a lower level of general knowledge. It seems likely that not only general, but also specific knowledge, influence vaccination decisions. These researchers found substantial correlations between knowledge and peoples' willingness to vaccinate their children or themselves, which points to the need to increase peoples' knowledge on this topic. However, our view is that the knowledge domain needs to be articulated with CT. As Paakkari and Paakkari (2012) suggest, CT relates to the ability to distinguish the conditions that promote health from those that do the contrary. Students face circumstances where they are required to assess whether a given source is reliable and can be regarded as scientific information. It is therefore essential to explore learning experiences and ways of teaching which support pre-service teachers in addressing these challenges.

Despite the increase of anti-vaccination movements over the world (Duggan & Gott, 2002), there are still few studies on this topic in Biology education. Maguregi González et al. (2017) examined PSTs' arguments and models of immunization in a teaching sequence about the vaccination controversy. Their results show PSTs' difficulties for applying this model to diverse contexts. Another study on this issue was carried out by Lundström et al. (2012), who examined which aspects a group of teenagers considered when making decisions about vaccination. The secondary effects of vaccines are considered as an important aspect for most of them, whether supporters or opponents to vaccines.

This chapter explores CT and arguments surrounding vaccination. We seek to explore the relationship between knowledge on vaccination and CT skills by means of case study research with PSTs working on the analysis of diverse premises about the anti-vaccination movement. The purpose of the study is to identify these connections and how they affect PSTs' positions on vaccination.

7.3 Methodology

The study employed a qualitative methods research approach (Denzin & Lincoln, 2005) with a single-case study design (Gerring, 2007). Discourse analysis methods were followed with the aim to explore how a group of PSTs put into practice CT in articulation with the application of knowledge on immunization in the context of assessing the anti-vaccination movement. Further, content analysis (Lincoln, 1995) was applied for the examination of the research questions using inductive and deductive procedures detailed in the data analysis section.

7.3.1 Participants, Instructional Context and Design

The study was carried out with a group of 39 PSTs during their third year in a public University in the northwest of Spain, Galicia. The students were enrolled in a science education methods course, which focuses on Biology instruction in Primary education. Content related with the human body, health education and controversies are part of this subject, among others, and they are introduced through practices of argumentation and CT. Author 1, with more than eight years of experience in this subject, was the teacher of the group. The students participated in argumentation and modelling activities throughout the entire course, prior to the implementation of the task addressed in this chapter.

The task, about vaccination and the anti-vaccination movement, was designed to engage PSTs in argumentation and CT. Previous to the application of this activity, participants completed an individual survey that included three open-ended questions to assess their prior knowledge of vaccination and to gauge their perception regarding how to introduce this topic in future teaching. Table 7.1 presents a summary of the sequence of activities carried out with the students. The design draws from a prior one developed and previously implemented by the authors with a group of secondary PSTs (Ageitos & Puig, 2016).

The task analyzed was developed in a three-part session totaling 90 min (see Table 7.1). Before the implementation, the teacher (author 1) and researcher (author 2) agreed to adjust time allotted for questions included in the activity based on students' needs and success and to maintain the time required for the class discussion and exit slip to suitably close the session.

Table 7.1 A summary of the tasks about vaccination and the anti-vaccination movement

Session	Task	Description of the task
1	What do you know about vaccines and vaccination?	An individual survey on vaccination composed by three open-ended questions
2	Lía's vaccines (part 1)	Identification of vaccines included in Lía's vaccination card (an hypothetical child living in our region) and comparison with the official vaccination calendar in our region.
	Lía's vaccines (part 2)	Students' communication of their prior knowledge about the anti-vaccination movement.
	Lía's vaccines (part 3)	Evaluation of five premises supporting the anti-vaccination movement.

In the first part, a hypothetical situation about vaccination in our context was presented. In particular, the case of a child called Lía whose parents are willing to enrol her in Early Childhood Education; however, they had decided to stop vaccinating her. Participants were provided with Lia's vaccination card and the official vaccination calendar in our region. They had to compare both in order to form a conclusion regarding Lía's vaccines. Afterwards, two pieces of news from 2019 reporting the introduction of mandatory vaccination in early education in all state schools within our region were provided. Students were asked to argue in favour or against this health regulation and to decide whether the girl should be allowed to enrol or not under those circumstances.

In the second part, the anti-vaccination movement was introduced. Firstly, students were asked about their previous knowledge about it and whether they had previously heard about it or not. Secondly, five premises that support the anti-vaccination movement were presented to students. They make reference to: (1) *individual free will;* (2) *dangerous secondary effects*; (3) *alleged reduced effectiveness*; (4) *they cause autism or allergies*; (5) *they introduce artificial substances to our bodies*.

Participants were asked if they agreed or not with each premise and to provide reasons to justify their answers.

The third part of the activity presented Lía's parents' arguments for stopping vaccination. Students were required to express their opinions on this decision and to argue their view.

The task ended by asking PSTs to reflect on their knowledge regarding vaccination and about their own views regarding health instruction in compulsory education.

7.3.2 Data Collection and Analysis

The first author implemented the vaccination activity and the second one attended the session as an observer taking field notes without interfering in the progress of the tasks. Individual reports (N=39) were collected and then numbered to guarantee

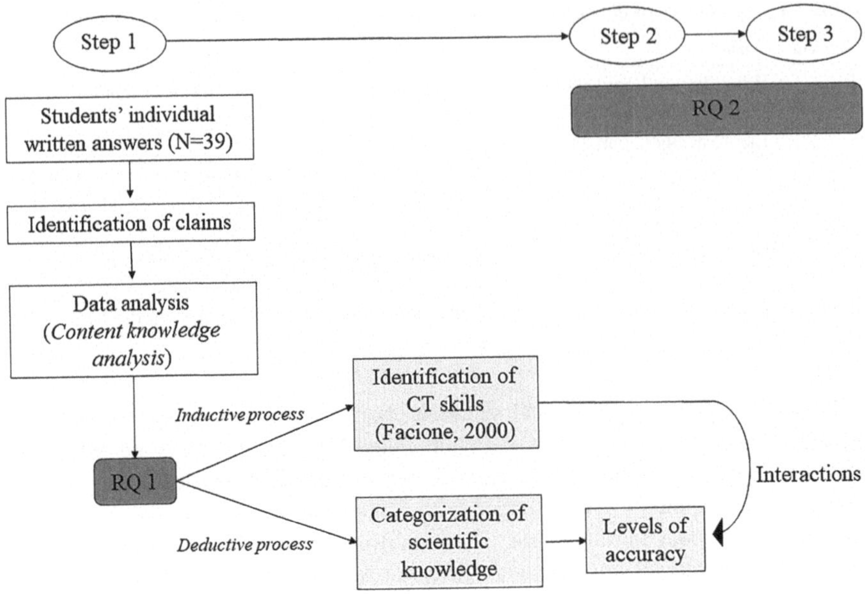

Fig. 7.1 Data analysis process

their anonymity for analysis. Figure 7.1 shows the process of analysis followed to address the two research questions. Content analysis methods (Lincoln, 1995) were applied for the examination of individual written statements. Both authors coded all the data together and a third person was involved in the analysis as a way of triangulating the findings.

The process of analysis of RQ1, about CT skills and scientific knowledge mobilized by PSTs, involves Step 1 (see Fig. 7.1). The identification of CT skills follows an inductive process, which adapts the definition of some categories of CT skills provided in Facione et al.'s (1990) proposal: (1) *interpretation*; (2) *evaluation*; (3) *explanation*, (4) *self-regulation*. The characterization of these skills, summarized below, was adjusted for the purposes of the research, in accordance with the content of the task.

1. *Interpretation*: to comprehend and express the meaning or significance of the premise being addressed.
2. *Evaluation*: to assess claims, particularly the credibility of statements supporting anti-vaccination movement.
3. *Explanation*: to reason about the premises with arguments.
4. *Self-regulation*: to self-consciously monitor one's cognitive activities and the elements used in those activities and the results educed. In order for students to be self-regulated they need to be aware of their own thought process, and be motivated to actively participate in their own learning process (Zimmerman, 2001). In the context of this task, this skill refers to the self-consciously process that learners use to manage and organize their thoughts and knowledge on the

topic of vaccination and make explicit when assessing the anti-vaccination premises.

It needs to be highlighted that more than one category could appear and overlap in an individual written response, or even within the same claim.

Concerning the examination of content knowledge, a deductive and interactive process was followed. We revised all individual answers and developed a coding scheme to capture the main ideas showed implicitly or explicitly in the data. First, we distributed all the responses in two main groups: (1) *not references to content knowledge*; (2) *references to content knowledge*. Four categories of knowledge related with immunization through vaccination emerged in the examination of responses included in the second group: *social benefits, vaccines function*; *vaccines effectiveness*; *vaccines composition*.

The analysis of RQ2, about the connections between CT skills and knowledge domain, was carried out in two steps (Steps 2 and 3) displayed in Fig. 7.1. Firstly, three levels of knowledge domain in individual responses were established (Step 2) through an iterative process followed by both authors:

- *Level 3 – Scientifically adequate*: appropriate use of scientific knowledge about immunization.
- *Level 2 – Partially scientifically adequate*: partial use of scientific knowledge about immunization. Only some scientific notions are used adequately.
- *Level 1 – Not scientifically adequate*: responses that do not incorporate appropriate scientific notions about vaccination.

Secondly, the connections between the level of knowledge domain and the type of CT skills mobilized were examined in Step 3.

7.4 Results

7.4.1 Critical Thinking and Background Knowledge on Vaccination

We begin by addressing RQ1, about the CT skills and scientific knowledge mobilized by PSTs while assessing five premises related to the anti-vaccination movement.

The results are developed in two parts. First, we focus on the CT skills by presenting the quantitative results obtained in each premise and discuss the categories with examples of the written reports. Secondly, we develop the results related to the scientific knowledge used in each premise.

7.4.1.1 Critical Thinking Skills in the Evaluation of Anti-vaccination Premises

Table 7.2 shows the frequencies of each CT skill identified in the PSTs' responses to each of the five premises.

Explanation was the skill most frequently displayed; it appeared 46 times. It was mobilized in the five premises, being the only one along with evaluation. In contrast, *evaluation* was the less frequent skill articulated by the students as Table 7.2 shows.

Participants were able to mobilize the four categories of CT skills in the five premises, except *self-regulation* in premise 1 and *interpretation* in premise 4, that did not appear in these responses. Premise 5 is the one in which the frequency of CT skills is higher (33 skills) and premise 4 the lowest (20 skills).

Next, we discuss our coding of CT skills by providing quotations of individual written responses. The examples chosen are the ones that appear more frequently related to the premised presented.

1. *Explanation*: responses related to the practice of reasoning, where students provide their own arguments in relation to the addressed premises. This skill is the most common in premise 1 (18 out of 27). Examples are provided below.

> I believe that getting vaccinated should not be voluntary as it is playing with the possibility of children getting the disease, and I also think that many parents do not have the scientific knowledge necessary to make such decisions and that they affect a person's health (S17, premise 1).

This response is coded as *explanation* since it shows how the student reasons his position regarding premise 1 (individual free will), providing two arguments rejecting vaccination as voluntary: one about the drawbacks of not getting vaccinated, and the other about the need of scientific knowledge in order to make decisions that affect human health.

> I consider that they have secondary effects because a disease is being introduced in the body (S5, premise 2).

Table 7.2 CT skills mobilized by PSTs in the responses of the five anti-vaccination premises. In bold, the most frequent category is highlighted. N=39

CT skills/ premises	P1. Individual free will	P2. Dangerous secondary effects	P3. Alleged reduced effectiveness	P4. Cause autism or allergies	P5. Introduction of artificial substances	N_T
Explanation	18	8	6	7	7	**46**
Self-regulation	0	4	10	6	10	30
Interpretation	3	9	3	0	13	28
Evaluation	6	6	4	7	3	26
Total skills	27	27	23	20	33	130
No skills	*6*	*13*	*15*	*15*	*7*	56

This student agrees with premise 2 (dangerous secondary effects) and reasons his position based on his own view about the composition of vaccines. The argument provided is that vaccines introduce diseases in the body.

2. *Self-regulation*: students self-consciously express explicitly their cognitive skills or knowledge domain on vaccination, particularly in relation to the content of the premise assessed. It is the second most frequent skill, with 30 appearances.

> I have not enough information on vaccines, their effects and all of this, so I do not consider myself able to comment (S3, premise 3).

> I do not have many notions of what exactly is introduced through the vaccine, as to say that they are totally artificial substances (S20, premise 5).

These responses are included in this category since they express their own limitations on the knowledge required to assess the content of the premises addressed. Both students are aware about their reduced notions and little information on the topic, and as consequence they decide not to provide their opinion regarding the premise presented.

3. *Interpretation*: during interpretation individuals engage in grasping and disclosing the meaning of the premise being assessed. This skill is enacted 28 times in the responses to four of the premises, being absent in the responses about premise 4 (vaccines cause autism or allergies). It appears more frequently in responses about premises 2 (dangerous secondary effects) and 5 (vaccines introduce artificial substances to our bodies).

> What is injected in the vaccines is a minimal amount of the agent that causes the disease so that the body creates the antibodies and is prepared for the possible arrival of that agent, so it would not be artificial (S11, premise 5).

This response shows how the student interprets the significance of premise 5, unpacking it's meaning and explaining the composition of vaccines and how they work. It shows a pro-vaccination position.

> We don't get vaccinated as often as it does to make us less effective (S6, premise 3).

The student interprets the meaning of premise 3, particularly what he understands by "reduced effectiveness". He associates this issue to the repetition of vaccines, what might be confusion between this notion and the idea of antibiotics resistance.

4. *Evaluation*: participants evaluate the content of the premises. In other words, they assess the credibility of the statements provided.

> I agree that vaccines have dangerous side effects, but I do not consider this a valid argument because certain medicines that we provide throughout life also have them and we do not question them (S25, premise 2).

The student rejects the credibility of premise 2 (dangerous secondary effects) as a valid argument for opposing vaccination. He (or she) appeals to the side effects of other medicines as data against this premise.

> I disagree, that was a conclusion from a doctor who came up with the experiment but was proven that the data was falsified. On the other hand, there are more than numerous scientific reports that prove otherwise. Autism cannot be got (S28, premise 4).

This response shows another example of *evaluation* in which the student opposes to premise 4 (cause autism or allergies), questioning its credibility; particularly, the study behind it, that used falsified data. He also points to the existence of many scientific studies that prove that autism is not caused by vaccination.

7.4.1.2 Background Knowledge on Vaccination

This section focuses on the scientific knowledge used by PSTs when assessing the five premises. Table 7.3 presents an overall picture of the frequencies of responses per premise. Each category is illustrated with selected examples, taking into consideration its higher frequency of appearance in relation to each premise.

1. *Vaccines function*: statements that point to the important role and contribution of vaccines to human health. The knowledge used is related to the need to prevent the appearance of a disease and create immunity against it (direct protection) and reducing the transmission of a disease, so the spread of an infection.

   It is the most frequent category, as Table 7.3 shows; it appears 32 times in the responses about all the premises. All students' responses reveal a pro-vaccination stance, except one. An example is:

   > I think they have side effects because we are introducing a disease directly, but I think it is best to have them rather than later as we would not have the antibodies needed to fight it [a disease] (S33, premise 2).

Table 7.3 Scientific knowledge used by participants when assessing the five anti-vaccination premises

Knowledge/ premises	P1 Individual free will	P2 Dangerous secondary effects	P3 Alleged reduced effectiveness	P4 Cause autism or allergies	P5 Introduction of artificial substances	Total
Vaccines function	4	11	3	1	13	32
Vaccines effectiveness	4	10	8	9	–	31
Vaccines composition	–	2	–	–	18	20
Social benefits	7	1	3	–	–	11

This student makes reference to the role of vaccines in the production of antibodies to "fight" a disease and considers that this is achieved by the introduction of the disease by vaccines.

2. *Vaccines effectiveness*: responses that point to the protective benefits of vaccines, and some factors that may affect their effectiveness. It is the second most frequent category, identified 31 times in the responses about all the premises, except premise 5.

> Vaccination is necessary to prevent a disease and is not effective if some people is vaccinated and other are not (S1, premise 1).

This student recognises that vaccines are a protection against diseases and also acknowledges a factor that affects the effectiveness of immunization (herd immunity).

> I disagree. It is true that some vaccines over time stop acting and need to be renewed, but this does not mean that they have reduced effectiveness, they protect us (S13, premise 3).

In this example the student points to the fact that some vaccines required more than one dose to produce a continued protection against a disease.

3. *Vaccines composition*: references to the ingredients or constituents of vaccines. This category appears, as expected, with a high frequency (18 times) in relation to premise 5 (vaccines introduce artificial substances in our bodies). However, it also appeared 2 times when assessing premise 2 (dangerous secondary effects).

The knowledge mobilized is related to the infectious agent introduced to the body through the vaccine. Five of the 19 students identified viruses as infectious agents and six mentioned microorganisms, agents or viruses and bacteria. Some of them introduced not only terms such as viruses or microorganisms, but also were able to build an explanation related to attenuated viruses. Three students mentioned that vaccines are made of particles responsible for the diseases, such as virus or bacteria, or that vaccines introduce directly the disease.

> Vaccines are cultures that are made from viruses or bacteria that are intended to protect people who will be vaccinated but weakened (S32, premise 5).

> I think they do have side effects since they are getting a disease in the body, but I think it's better to have them then than later since we wouldn't have the antibodies needed to fight it (S5, premise 2).

4. Social Benefits: references that acknowledge the social benefits of vaccination. This category was identified in the responses about three premises (1, 2 and 3) eleven times.

> I agree that not only is it an aspect that influences the person but also the society; it should be mandatory for everyone (S6, premise 1).

This student recognizes the importance of vaccination, not only individually, for the person that is being vaccinated and then protected against a disease, but also for the society. Based on this, he is in favour of compulsory vaccination.

7.4.2 Connections Between CT Skills and Knowledge Domain on Immunization

This section reports on RQ2, concerning the relationships between CT skills articulated by participants and their knowledge domain on immunization. We aim to investigate whether a higher performance in CT skills is related to a higher level of scientific knowledge domain in vaccination and/or vice versa. Table 7.4 shows the distribution of individual statements in three levels of knowledge domain adequacy (scientifically adequate; partially scientifically adequate; not scientifically adequate) along with the intersections with CT skills.

As Table 7.4 shows, most responses are accurate or partially accurate regarding the scientific knowledge used when assessing the five premises.

In the case of "social benefits," all statements were coded as adequate. In the rest of the categories, most responses were coded as scientifically adequate, however several responses were identified as partially adequate and one as not adequate regarding "vaccines composition" and "vaccines effectiveness". Vaccination is considered to be the inoculation of the disease and resistance to vaccines is related mistakenly to their lack of effectiveness.

Concerning connections between the scientific knowledge domain and CT skills, there seems to be a correspondence between a higher level of knowledge domain and a higher mobilization of CT skills. Students were able to put into practice diverse CT skills mobilizing scientific knowledge (81 times compared to the 69 that skills did not appear connected to scientific knowledge), although this rarely occurs in the opposite way. Examples of students using knowledge without being involved in the practice of CT skills are scarce (sixteen times).

Explanation and interpretation appear to be content-domain specific, whereas *evaluation* is not so frequently related to the content domain (approximately half of the times). Self-regulation is not content-domain specific as expected, since this category includes responses that express limited knowledge to assess the premises. Examples that show the connections between the scientific knowledge and CT skills enacted by participants may be:

> I don't agree because I can't find a relationship between them. One affects the immune system and the other is genetic (S29, premise 5).

The student is explaining why he disagrees with the premise by using his/her knowledge on the origin of allergies and autism. The skill of explanation is frequently related to the use of knowledge.

> I think that they aren't artificial substances, I think that it is the same (but in a fewer quantity) of what we are trying to eradicate (S8, premise 5).

Table 7.4 Interactions between CT skills and categories of scientific knowledge

	Social benefits			Vaccines function			Vaccines effectiveness			Vaccines composition				No scientific knowledge
Scientific knowledge adequacy	Adequate	Partially adequate	Not adequate	Adequate	Partially adequate	Not adequate	Adequate	Partially adequate	Not adequate	Adequate	Partially adequate	Not adequate	Total	
Explanation	8	–	–	13	2	–	9	2	–	2	1	1	38	18
Interpretation	1	–	–	8	1	–	3	2	1	9	3	–	28	4
Evaluation	–	–	–	4	–	–	8	–	–	3	–	–	15	17
Self-regulation	–	–	–		–	–		–	–		–	–	0	30
No CK skill	2	–	–	6	–	–	7	–	–		1	–	16	

The example shows the connection between knowledge about the composition of vaccines and the skill of interpretation. This student is pointing to the composition of vaccines, which can consist of a lower proportion of the agent that causes the disease.

Examples illustrating the lack of connections between CT and knowledge mobilization are provided below:

> I agree, as I know close cases that after vaccination a spot on the skin appeared (S9, premise 4).

> I don't agree, there are scientific reports that show evidence of its effectiveness (S28, premise 3).

Student 9 is involved in explanation, although he does not use any scientific notion; he reasons his position on his own experience. Student 28 is involved in evaluation; however, no scientific knowledge is used to support his argument. Evaluation is the skill in which students have more difficulties in mobilizing scientific knowledge and is the less frequent CT skill. This might point to the difficulties that students have engaging in this skill.

7.5 Discussion and Educational Implications

The research presented in this chapter was designed with the goal of exploring the practice of CT by a group of PSTs in the context of dealing with a health controversy. We aimed to advance the empirical research on CT, with an emphasis on the CT skills and interactions with the relevant knowledge domain when addressing premises supporting the anti-vaccination movement. Despite CT literature being extensive and there being a consensus regarding its central role in higher education, the number of studies depicting teacher training "for" and "about" CT in Biology education are still scarce.

The findings of this study indicate that PSTs engaged in the practice of CT were able to articulate diverse skills when they were asked to assess five anti-vaccination premises. However, not all CT skills appeared with the same frequency. The most frequent skill identified in the responses about all the premises assessed was *explanation*. Participants were able to reason their answers to the five premises, providing arguments supporting their positions in favour or not. The fact that explanation was the predominant skill in premise 1, *individual free will*, might be caused to its openness, which encourages arguments. While we think that CT skills cannot be exercised without some knowledge of the subject matter under consideration, this does not mean that CT always produce a well-reasoned argument. Students should feel free to express their own thoughts in biology classrooms, but they should be encouraged to support their views reasonably and to exercise informed critique. Drawing from Halpern's work (1998), we also think that CT requires awareness of one's own knowledge. When engaging in CT students need to monitor their thinking process. CT requires for instance insight into what one knows and the extent and importance of what one doesn't know in order to assess anti-vaccination premises and their implications. The results of this study have also highlighted that the CT skill of *self-regulation* was the second most frequent and emerged when students acknowledged their lack of scientific knowledge regarding the topic to give an opinion. Similar results were reported by Maguregi González et al. (2017), in an investigation about PSTs engaged in modelling and argumentation in the context of explaining immunization, in which participants acknowledge their lack of information on the topic.

In our study, part of the students' responses indicated the need to consult an expert in order to make a decision, and in other cases they stated that some people should listen to experts to decide whether to vaccinate their children or not. This result is consistent with Navarro Alonso et al.'s (2001) research, which shows that Spanish families identify paediatricians as the experts to be consulted to make decisions related to vaccination. However, a recent study in our country (Picchio et al., 2019) with public paediatric professionals reveals that one in four participants in the study showed doubts about some vaccines included in the official immunization schedule.

Regarding the scientific knowledge mobilized when dealing with the anti-vaccination premises, four categories were identified in the five premises. It seems that the content of the premise oriented or influenced the knowledge being used by the participants. The majority of students used appropriate scientific knowledge on immunization in their answers and few participants were not able to use appropriate scientific notions about vaccination. This means that most students have conceptual tools to critically evaluate the anti-vaccination premises.

Addressing the interactions between CT skills and use of scientific knowledge, the most frequent one corresponds to interaction between the two most frequent categories: vaccines function and explanation. The CT skill of *evaluation* closely related to the use evidence to support claims showed the least frequent interactions with the use of scientific knowledge. This result highlights the difficulties that students face when engaging in this skill.

We are aware that these results have limitations, as the nature of the premises provided might have influenced the type of CT skills that we identified in students' discourse and also the knowledge used.

Atwell and Salmon (2014) argue that labelling a person as pro-vaccine or anti-vaccine is oversimplification. Although it is not the focus in this chapter, the results stress that most participants are in the pro-vaccination spectrum, and no students are close to the anti-vaccination spectrum.

Previous studies such as Walker et al. (2002) suggest that scientific literacy is not significantly correlated to the degree of believing in pseudoscientific claims. Our results suggest that CT skills and scientific knowledge may help citizens to make decisions on pseudoscientific topics, since there seems to be a correspondence between a better performance in CT and a higher level of knowledge domain. What is a remarkable result is the fact that *self-regulation* is a skill frequently enacted, showing students' self-awareness of their knowledge limitations for critical assessment and evaluation of the topic. As Motta et al. (2018) point out, people with overestimation in their knowledge on a topic, such as the causes of autism, might show an opposition to mandatory vaccination. Furthermore, recognizing one's own bias on a topic is also important for CT development. Biology instruction designed to help students to be critical thinkers on SSIs should attend to these issues related with metacognitive skills. Halpern (1998) suggested a model for teaching CT that consists in four components, being the metacognitive component to assess thinking important for this purpose. An implication of this study is the possibility to tackle SSIs by promoting students' engagement in the practice of CT skills along with the application of content knowledge, making the skills performed explicit. Challenging anti-vaccination premises and pseudoscientific beliefs is not an easy task. Some resources and tools should be made accessible to students in order to assist them in engaging CT in Biology classrooms. This requires teaching activities that can equip teachers and students with the tools for more critical analysis of health topics in the media. Research on SSIs show that whilst Science teachers commonly use media information to highlight the relevance of a particular topic, they lack the skills required to teach how to critically analyse this information (Kachan et al., 2006). We agree with Johnson and Pigliucci (2004), who raise the point that we cannot assume that by improving only scientific knowledge, CT and scepticism will develop. However, we argue that the promotion of both scientific knowledge and CT skills in unison may help to increase levels of scepticism within the context of health controversies, health literacy and critical actions.

Acknowledgements This study was supported by The Spanish Ministry of Science, Education and Universities, partly funded by the European Regional Development Fund (ERDF, grant code: PGC2018-096581-B-C22; and European Union (Erasmus+) project P2D, code, 5016-D5ZD-64400. We gratefully acknowledge the participants.

References

Ageitos, N., & Puig, B. (2016). ¿Debería ser obligatoria la vacunación? *Alambique, Didáctica de las Ciencias Experimentales, 86*, 78–79.

Anderman, E. M., Sinatra, G. M., & Gray, D. L. (2012). The challenges of teaching and learning about science in the twenty-first century: Exploring the abilities and constraints of adolescent learners. *Studies in Science Education, 48*(1), 89–117. https://doi.org/10.1080/03057267.2012.655038

Andrews, R. (2015). Critical thinking and/or argumentation in higher education. In M. Davies & R. Barnett (Eds.), *The Palgrave handbook of critical thinking in higher education* (pp. 49–63). Palgrave MacMilan.

Atwell, J. E., & Salmon, D. A. (2014). Pertussis resurgence and vaccine uptake: Implications for reducing vaccine hesitancy. *Pediatrics, 134*, 602–604. https://doi.org/10.1542/peds.2014-1883

Brown, A. (1997). Transforming schools into communities of thinking and learning about serious matters. *American Psychologist, 52*, 399–413. https://doi.org/10.1037/0003-066X.52.4.399

Denzin, N. K., & Lincoln, Y. S. (2005). Introduction: the discipline and practice of qualitative research. In N. K. Denzin & Y. S. Lincoln (Eds.), *The Sage handbook of qualitative research* (3rd ed., pp. 1–32). Sage.

Dewey, J. (1910). *How we think*. D.C. Heath and Company.

Duggan, S., & Gott, R. (2002). What sort of science education do we really need? *International Journal of Science Education, 24*(7), 661–679. https://doi.org/10.1080/09500690110110133

Elen, J., Jiang, L., Huyghe, S., Evers, M., Verburgh, A., … Palaigeorgiou, G. (2019). In C. Dominguez & R. Payan-Carreira (Eds.), *Promoting critical thinking in European higher education institutions: Towards an educational protocol*. UTAD.

Ennis, R. H. (1962). A concept of critical thinking. *Harvard Educational Review, 32*(1), 81–111.

Ennis, R. H. (1989). Critical thinking and subject specificity: Clarification and needed research. *Educational Researcher, 18*(3), 4–10. https://doi.org/10.3102/0013189X018003004

Evagorou, M., & Puig, B. (2017). Engaging elementary school pre-service teachers in modeling a socioscientific issue as a way to help them appreciate the social aspects of science. *International Journal of Education in Mathematics, Science and Technology*. https://doi.org/10.18404/ijemst.%2099074

Faasse, K., Chatman, C. J., & Martin, L. R. (2016). A comparison of language use in pro- and anti-vaccination comments in response to a high profile Facebook post. *Vaccine, 34*(47), 5808–5814. https://doi.org/10.1016/j.vaccine.2016.09.029

Facione, P. A. (1990). *Critical thinking: A statement of expert consensus for purposes of educational assessment and Instruction. research findings and recommendations*. The California Academic Press.

Facione, P., Facione, N. C., & Giancarlo, C. A. (2000). The disposition toward critical thinking: Its character, measurement, and relationship to critical thinking skill. *Informal Logic, 1*. https://doi.org/10.22329/il.v20i1.2254

Gerring, J. (2007). *Case study research: Principles and practices*. Cambridge University Press.

Grossman, P., & McDonald, M. (2008). Back to the future: Directions for research in teaching and teacher education. *American Educational Research Journal, 45*, 184–205. https://doi.org/10.3102/0002831207312906

Halpern, D. F. (1998). Teaching critical thinking for transfer across domains. Dispositions, skills, structure training, and metacognitive monitoring. *American Psychologist, 53*(4), 449–455. https://doi.org/10.1037/0003-066X.53.4.449

Hoskins, B., & Deacon Crick, R. (2010). Competences for learning to learn and active citizenship: Different currencies or two sides of the same coin? *European Journal of Education, 45*(1), 121–137. https://doi.org/10.1111/j.1465-3435.2009.01419.x

Jiménez-Aleixandre, M. P. (2007). Designing argumentation learning environments. In S. Erduran & M. P. Jiménez-Aleixandre (Eds.), *Argumentation in science education. Perspectives from classroom-based research* (pp. 91–117). Springer.

Jiménez Aleixandre, M. P., & Puig, B. (2012). Argumentation, evidence evaluation and critical thinking. In B. Fraser, K. Tobin, & C. McRobbie (Eds.), *Second international handbook for science education* (Vol. 2, pp. 1001–1017). Springer.

Johnson, M., & Pigliucci, M. (2004). Is knowledge of science associated with higher skepticism of pseudoscientific claims? *The American Biology Teacher, 66*(8), 536–548.

Kachan, M. R., Guilbert, S. M., & Bisanz, G. L. (2006). Do teachers ask students to read news in secondary science? Evidence from the Canadian context. *Science Education Policy, 90*(3), 496–521. https://doi.org/10.1002/sce.20113

Kata, A. (2010). A postmodern Pandora's box: anti-vaccination misinformation on the Internet. *Vaccine, 28*(7), 1709–1716. https://doi.org/10.1016/j.vaccine.2009.12.022

Kickbusch, I. (2008). Health literacy: An essential skill for the twenty-first century. *Health Education, 108*(2), 101–104. https://doi.org/10.1108/09654280810855559

Lincoln, Y. S. (1995). Emerging criteria for quality in qualitative and interpretive research. *Qualitative Inquiry, 1*(3), 275–289. https://doi.org/10.1177/107780049500100301

Lundström, M., Ekborg, M., & Ideland, M. (2012). To vaccinate or not to vaccinate: how teenagers justified their decision. *Cultural Studies of Science Education, 7*, 193–221. https://doi.org/10.1007/s11422-012-9384-4

Maguregi González, G., Uskola Ibarlueza, A., & Burgoa Etxaburu, M. B. (2017). Modelización, argumentación y transferencia de conocimiento sobre el sistema inmunológico a partir de una controversia sobre vacunación en futuros docentes. *Enseñanza de las Ciencias, 35*(2), 29–50. https://doi.org/10.5565/rev/ensciencias.2237

Motta, M., Callaghan, T., & Sylvester, S. (2018). Knowing less but presuming more: Dunning-Kruger effects and the endorsement of anti-vaccine policy attitudes. *Social Science & Medicine, 211*, 274–281. https://doi.org/10.1016/j.socscimed.2018.06.032

Navarro Alonso, J. A., Bernal González, P. J., & Niguez Carbonell, J. C. (2001). Analysis of factors influencing vaccine uptake: Perspective from Spain. *Vaccine, 20*(Suppl 1), 13–15. https://doi.org/10.1016/S0264-410X(01)00300-0

Nutbeam, D. (2000). Health literacy as a public health goal: a challenge for contemporary health education and communication strategies into the 21st century. *Health Promotion International, 15*(3), 259–267. https://doi.org/10.1093/heapro/15.3.259

Paakkari, L., & Paakkari, O. (2012). Health literacy as a learning outcome in schools. *Health Education, 112*(2), 133–152. https://doi.org/10.1108/09654281211203411

Paul, R. (1992). Critical thinking: what, why and how. *New Directions for Community Colleges, 77*, 3–24. https://doi.org/10.1002/cc.36819927703

Peters, M. A. (2018). Education in post-truth world. In M. A. Peters, S. Rider, M. Hyvönen, & T. Besley (Eds.), *Post-truth, fake-news. Viral modernity & higher education* (pp. 77–88). Springer.

Peters, M. A., Rider, S., Hyvönen, M., & Besley, T. (2018). *Post-truth, fake-news. Viral modernity & higher education*. Springer.

Picchio, C. A., Carrasco, M. G., Sagué-Vilavella, M., & Rius, C. (2019). Knowledge, attitudes and beliefs about vaccination in primary healthcare workers involved in the administration of systematic childhood vaccines, Barcelona, 2016/17. *Euro Surveillance, 24*(6). https://doi.org/10.2807/1560-7917.ES.2019.24.6.1800117

Poland, G. A., & Jacobson, R. M. (2001). Understanding those who do not understand: A brief review of the anti-vaccine movement. *Vaccine, 19*(17–19), 2440–2445. https://doi.org/10.1016/S0264-410X(00)00469-2

Puig, B., Blanco-Anaya, P., Mosquera Bargiela, I., & Crujeiras, B. (*2019*). A systematic review on critical thinking intervention studies in higher education across professional fields. *Studies in Higher Education, 860–869*. https://doi.org/10.1080/03075079.2019.1586333

Ratcliffe, M., & Grace, M. (2003). *Science education and citizenship: Teaching socioscientific issues*. Open University Press.

Reis, P. (*2014*). Promoting students' collective socio-scientific activism: Teachers' perspectives. In J. L. Bencze & S. Alsop (Eds.), *Activist Science and Tecnology Education* (pp. 547–574). Springer.

Ten Dam, G., & Volman, M. (2007). Educating for adulthood or for citizenship: Social competence as an educational goal. *European Journal of Education, 42*(2), 281–298. https://doi.org/10.1111/j.1465-3435.2007.00295.x

World Health Organization (WHO). (2019). *Inmunization-stress related responses.* A manual for program managers and health professionals to prevent, identify and respond to stress-related responses following immunization. World Health Organization; 2019. Accesible at: http://apps.who.int/iris

Vincent-Lacrin, S., González-Sancho, C., Bouckaert, M., de Luca, F., Fernández-Barrera, M., Jacotin, G., Urgel, J., & Vidal, Q. (2019). Fostering students' creativity and critical thinking: What it means in school, *Educational research and innovation*, OED Publishing. https://doi.org/10.1787/62212c37-en

Walker, W. R., Hoekstra, S. J., & Vogl, R. J. (2002). Science education is no guarantee of skepticism. *Skeptic, 9*(3), 24–28.

Zimmerman, B. J. (2001). Theories of self-regulated learning and academic achievement: An overview and analysis. In B. J. Zimmerman & D. H. Schunk (Eds.), *Self-regulated learning and academic achievement: Theoretical perspectives* (pp. 1–37). Lawrence Erlbaum.

Zingg, A., & Siegrist, M. (2012). Measuring people's knowledge about vaccination: Developing a one-dimensional scale. *Vaccine, 30*, 3771–3777. https://doi.org/10.1016/j.vaccine.2012.03.014

Chapter 8
Students' Thinking Strategies and the Role of Argument as a Shared Thinking Tool

Marida Ergazaki

8.1 Introduction

Science education research is concerned with argumentation for quite a long time. A very rich and still growing body of studies (e.g. Christenson et al., 2014; Driver et al., 2000; Sandoval et al., 2019; Zohar & Nemet, 2002) attempts to shed light on students' discourse in various scientific and socio-scientific contexts and inform the design of learning environments and assessment tools. This high interest of science education researchers in argumentation seems to reflect the significance of argumentation's systematic introduction in science classes (Siegel, 1995). It has been suggested that the use of argumentation in teaching and learning science can support teachers and students in pursuing more effectively important educational goals, such as understanding science content, getting familiar with the scientific culture, acquiring scientific literacy, training for active citizenship and developing critical thinking dispositions and abilities (Jimenez & Puig, 2012; Siegel, 1995).

According to Vygotsky (1978), learning is performed first on the inter-personal and then on the intra-personal level. When students interact in argumentative discussions that take place either in small peer-groups or in the whole class, they have learning opportunities on the inter-personal level; in other words, they have opportunities for joint construction of knowledge, which later on can be internalized and thus become personal. Argumentation is also considered as a key scientific practice (Jimenez-Aleixandre & Crujeiras, 2017; Kelly & Licona, 2018). Scientists depend on it for the construction and evaluation of their hypotheses or theories (Kuhn, 1993; Knorr-Cetina, 1999). So, an introduction to the use of argument can be

M. Ergazaki (✉)
University of Patras, Patras, Greece
e-mail: ergazaki@upatras.gr

B. Puig, M. P. Jiménez-Aleixandre (eds.), *Critical Thinking in Biology and Environmental Education*, Contributions from Biology Education Research,
https://doi.org/10.1007/978-3-030-92006-7_8

considered as an introduction to the scientific culture as well (Jimenez-Aleixandre & Erduran, 2008). Moreover, building a sound background of scientific knowledge and getting familiar with the scientific culture in argument-based science classes that highlight socio-scientific issues too, can enhance students' scientific literacy and prepare them for becoming active citizens.

It seems that all these issues have something to do with critical thinking. Being a scientific literate citizen with strong insights from scientific culture and knowledge requires, and possibly enhances at the same time, one's disposition and ability to behave as a critical thinker. So, if argumentation can really support science education goals like the above, it can support critical thinking as well. Argumentation is based on justification and thus it involves commitment to evidence and indeed strong one. Rationality and reflection lie at its core. Critical thinking on the other hand, has been described as reasonable reflective thinking for deciding what to believe or what to do (Ennis, 1987, 2018). Critical thinkers are disposed to search for evidence and employ rational criteria for evaluating it (Siegel, 1989). Therefore, rationality and reflection that are intrinsic to argumentation are also intrinsic to critical thinking (Jimenez-Aleixandre & Erduran, 2008).

Although the relationship between argumentation and critical thinking appears to be easily traced, it is certainly worth exploring, even if briefly. Argument may be viewed from quite different perspectives. A comprehensive review can be found in Jimenez-Aleixandre and Erduran (2008). The most popular view among science education researchers is the one suggested by Toulmin (1958). According to it, argument is a reasoning strand with different structural elements, such as claims (statements about something), data (claim-supporting evidence), warrants (explanations for the shift from data to claims), backings (warrant-supporting statements), qualifiers (statements about the conditions under which claims can be valid), and finally rebuttals (oppositions to any other structural element of the argument). This model of argument has been proved very useful both as an analytical and as a teaching tool (Erduran et al., 2004; Erduran, 2008). However, in the study presented here, argument is considered as any justified claim or any justified concession, opposition or challenge, as explained in the methods' section.

Critical thinking can represent a theoretical notion, a set of measurable skills, or an educational goal (Kuhn, 1991, 2019). As argumentation, critical thinking can be viewed from several perspectives. Jimenez-Aleixandre and Puig (2012) suggest a holistic, science education relevant view of critical thinking, which emphasizes its argumentation-related components and combines them with components related to social emancipation (Jimenez-Aleixandre & Puig, Chap. 1, this book).

In sum, critical thinking and argumentation appear to be related either because they share core elements like rationality and reflection, or because argumentation seems to be part of critical thinking. But what if argumentation and critical thinking were not just related to each other? What if they could be conceptualized as one thing? In fact, according to Deanna Kuhn (2019), they can. Critical thinking can be viewed as a dialogic practice, which is both social and individual. People are involved in critical thinking while interacting with others, and as they do it more and more they can internalize it and start performing it individually; i.e. in the presence of an *inner* other and not necessarily *real* others who take part in actual argumentative discourse. This notion of critical thinking is consistent with Vygotsky's ideas (1978) about the social and individual dimensions of knowledge construction and thinking processes. Moreover, it implies that argumentation in interactional contexts is a pathway to both inter- and intra-critical thinking (Boyd, 2019; Kuhn et al., 2016; Makhene, 2017).

This chapter is concerned with the use of argument as a shared thinking tool for the construction of knowledge. More specifically, we are interested in the thinking strategies that biology students may employ when exploring the procedure of making and using a DNA library. Our focus is particularly set on what kind these strategies may be and on whether they are underlined by the use of argument as a shared thinking tool that allows for a critical examination of the emerging issues. Thus, the research questions addressed here are:

1. What kind of thinking strategies may be employed by biology students involved in making decisions and articulating experimental proposals for creating and using a DNA library?
2. What is the role of argumentation within these strategies? Do students use argument as a shared thinking tool and how?

8.2 Methods

To shed light on these research questions, we draw on a qualitative case study that is concerned with the argumentative discourse of thirty biology students interacting in three- or four-member groups, in order to build shared knowledge about making and using DNA libraries. In this context, students were involved in thinking as if they would really have to make and use the DNA library of a medically useful plant. They were asked to make decisions about experimental tools, suggest experimental procedures, predict experimental outcomes, articulate and test hypotheses that might explain not necessarily expected experimental observations (Ergazaki & Zogza, 2005a, b, c,; Ergazaki et al., 2007). This chapter is particularly concerned with the discussions of one of these groups on two different tasks.

8.2.1 Tasks

The first task had to do with peers' decision-making about the vector they would use for making the DNA library of the plant. In fact, peers were required to choose between a plasmid with a unique recognition site for the restriction enzyme they were supposedly using, and another plasmid with three recognition sites. The second task required the articulation of an experimental procedure, which would allow peers to locate a specific gene in the DNA library. More specifically, students were supposed to have already (a) selected their plasmid vector, (b) used a specific restriction enzyme to digest both plasmids and plant DNA, (c) mixed the 'open' plasmids with the pieces of the plant DNA, (d) attempted to transform bacterial cells with the products of the previous step, and finally (e) cultivated the bacteria to eventually form the colonies of the library. So, they were asked to come up with an experimental procedure for locating a specific gene in the DNA library in the light of two crucial pieces of information: (a) the plasmid that was supposedly used for the construction of the DNA library had two genes for resistance to the antibiotics kanamycin and ampicillin respectively, and (b) the restriction enzyme cut in the second gene. Students were provided with scaffolding questions that prompted them to (a) start by getting to know their search scope (i.e. the bacterial colonies and their differences in terms of the plasmids they might have taken in), and (b) try to narrow it down before proceeding with the quest of their target gene (i.e. get rid of the colonies that could not carry the target gene in any way).

8.2.2 Participants

The peer-group that carried out both of these discussions consisted of three female, second-year biology students of the University of Patras, who volunteered to take part in the study and were keen on collaborative work. All three were diligent students with at least average scores at the exams they had taken up to that point. Moreover, they showed interest in the molecular biology course that hosted the study, since they were active in the course's typical lectures and labs that had already taken place.

8.2.3 Data Analysis

The discussions were tape-recorded, transcribed and analyzed in NVivo, a computer program for the analysis of qualitative data. First they were coded with regard to peers' arguments. More specifically, the coding scheme used for the argumentative operations (Pontecorvo & Girardet, 1993; Resnick et al., 1993) allows for capturing not just the individual, but also the social aspect of argument construction within a peer-group. Since the focus is set on the use of argumentation as a *shared* thinking tool that could support critical examination of the issues that emerge while exploring the tasks, the coding scheme includes typical argumentative operations (claims and justifications), as well as non-typical ones (oppositions, concessions and challenges) which are equally important. For instance, a challenge posed to a peer's unjustified claim by another peer, may very well lead to the missing justification and thus to the construction of an argument. Similarly, an opposition to one's justified claim by another one may prompt the use of more elaborated justifications and thus enhance the argument's value as a thinking tool. Moreover, coding for the arguments' epistemic content (Ergazaki et al., 2007; Jimenez-Aleixandre et al., 2000; Mason,

1996; Pontecorvo & Girardet, 1993) allows for tracing what may count to peers as a good justification, and thus highlighting further whether peers' arguments can really give them a critical view to the issues that emerge while exploring the tasks. For instance, epistemic operations like evaluating consequences or appealing to knowledge seem to indicate a more critical approach than those of appealing to authority or to mere opinion. The inter-rater reliability of the argumentative and epistemic operations' coding was checked by having a second researcher also re-code part of the discussions.

Finally, the thinking strategies peers employed in making the decision required in the first task or in coming up with the experimental procedure required in the second, were re-constructed through the overall analysis of the discussions. In fact, considering argument as any justified claim, concession or opposition, makes it possible to trace all the partial arguments that shape students' overall strategies for coping with the two tasks, and thus to adequately describe these strategies.

8.3 Results

In this section we present the results of our analysis with regard to students' thinking strategies and use of arguments when involved (1) in deciding about the vector they would use for making a DNA library, and (2) in articulating an experimental procedure that would allow peers to locate a specific gene in this library.

8.3.1 *Thinking Strategies and Argumentation in Students' Decision-Making Discussion*

Students jointly made the right decision about the vector they should use for supposedly constructing the DNA library of the plant. More specifically, they selected the plasmid that had a unique recognition site for the restriction enzyme they were supposedly using (option A), instead of another plasmid that had three (option B). Their decision was made by evaluating both options through two different reasoning strands: (a) a reasoning strand that supported option A, and (b) a much more elaborated reasoning strand that rejected option B which actually seemed quite appealing to them at some point. The former strand evaluated the consequences of using the one-site plasmid as the vector that would be loaded with the plant DNA, whereas the latter strand evaluated the consequences of using the three-site plasmid for the same purpose. In other words, students started with each of the two available plasmids ('bottom') and explored whether their use would be appropriate for the production of recombinant plasmids carrying pieces of the plant DNA, i.e. appropriate for the required goal ('up'). So, students' thinking strategy was actually based on a 'bottom-up' approach regarding all given options. In other words, it was a strategy of deciding which option to use by evaluating the consequences of each with regard to the goal that needs to be achieved. Although used in a domain-specific task, a strategy like this cannot actually be considered as domain-specific. Apparently, it may also apply to decision-making tasks in everyday life situations, as it will be discussed later.

Shifting to argumentation in particular, it should be noted that argument appeared to have a central role as a shared thinking tool in the peer-group's discussion. In fact, students' thinking strategy was underlied by the use of argument, as shown by the analysis of their discourse. In this, the three students are presented in short as S1, S2 and S3.

- S1: *'The one-site plasmid is a good choice* (claim); *'It can just open at the recognition site by the restriction enzyme that cuts there, take in a piece of the plant DNA and then close. It does what we need it to do'* (justification). S1 argued for the one-site plasmid (option A) by evaluating the consequences of its use with regard to the experimental goal; she did this by appealing to background knowledge about how recombination may occur.
- S2: *'Isn't it a better idea to choose a plasmid that could probably carry more than one pieces of the plant DNA at the same time?'* (claim); *'One recognition site means a unique insertion; three recognition sites means a triple insertion' (justification).* S2 challenged the previous argument by highlighting the favorable consequences of the three-site plasmid (option B) through an analogy.
- S1: *'After all, we do need many copies of each piece, don't we? The three-site plasmid can give these copies to us'.* S1 reflected on the argument for the three-site plasmid and conceded to it by providing a more explicit evaluation of the consequences of its use with regard to the experimental goal.
- S3: *'The three-site plasmid is not a good idea' (opposition); 'It can be cut in three different pieces and it won't be easy for these to re-connect to each other the same way as before. They may be arranged in different ways and result in a non-functional vector'* (justification). S3 argued against the three-site plasmid by challenging the supposedly favorable consequences of its use. She did this by appealing to the idea of contingency on the molecular level.
- S1: *'Let's take the safe way. A three-site plasmid does not necessarily mean triple insertion. It may mean a disaster as well'.* S1 conceded to S3's argument against the three-site plasmid. She suggested avoiding risking the whole experiment just for the possibility of a triple insertion. S2, however, insisted on the argument for the three-site plasmid.
- S3: *'What if a piece does not re-connect at all? And what if this piece is the ori for instance? We'll get up with nothing at all. No replication will be possible'* (justification). S3 attempted to highlight further the risk that the use of the three-site plasmid may pose to the whole experiment. She appealed to contingency by giving a specific example. The peer-group reached consensus against the three-site plasmid, thanks to S3's concrete example, which made S2 to reconsider her own argument. S1 reminded her initial argument for the one-site plasmid in particular, which, in the light of the three-site plasmid's rejection, got accepted by all three peers.

In sum, students were engaged in argumentative discourse. They employed justifications to support their claims for or against each of the options. Moreover, they came up with challenges, oppositions and concessions, which points to using argument as a shared thinking tool. Finally, students drew upon several epistemic operations such as evaluating consequences or appealing to goals, analogy, contingency, and background knowledge, in order to build or critically reflect upon the arguments that shaped their 'bottom-up' thinking strategy.

8.3.2 Thinking Strategies and Argumentation in Students' Discussion on Articulating an Experimental Procedure

Students did not actually complete the articulation of a procedure for locating a target gene somewhere within the bacterial colonies that they had supposedly created earlier. However, they did make significant progress by arguing for or against several ideas as discussed below. First, they got to know their search scope by trying to describe the different colonies in terms of (a) whether they were formed by non-transformed or transformed bacteria, and (b) whether the transformed bacteria carried non-recombinant or recombinant plasmids. Second, they tried to narrow down their search scope (a) by using the idea of antibiotic resistance or sensitivity to eliminate the colonies of non-transformed bacteria, and (b) by inventing ways to get the target colonies alive. After using the task's scaffolding (see the methods section above) for narrowing down their search scope to the colonies of bacteria that were transformed with recombinant plasmids in particular, the students focused on locating the target gene, not by searching for it but by looking for the protein it codes for. Finally, when they realized that such an indirect search cannot be effective, they attempted to come up with the details of a direct search for the gene, although not quite successfully.

In sum, students' thinking strategy included four different parts: (a) getting to know the search scope, (b) narrowing it down, (c) exploring the possibility of an indirect search for the target gene, and (d) shifting to the idea of a direct one (the first two parts are encouraged by the scaffolding questions of the task). Once more, students' thinking strategy does not seem to be domain-specific. On the contrary, it may also apply to decision-making tasks in everyday life situations, as it will be discussed later.

To highlight the role of arguments as peers' shared thinking tool, we draw on an overview of their discussion about this task.

- S3: *'We have also bacteria without plasmid'* (claim); *'Transformation does not necessarily happen all the times. It is random'* (justification). S3 provided a claim about the presence of non-transformed bacteria in the culture and, after being challenged by S2, she justified her claim by appealing to background knowledge about contingency as an inherent characteristic of the process of bacterial transformation.

- S2: *'The colonies of bacteria that did not take in plasmids, are sensitive to both antibiotics* (claim), *since 'no plasmid' means 'no resistance"* (justification). S2 argued for the sensitivity of the non-transformed bacteria to both antibiotics by appealing to the task's data.
- S2: *'The colonies of bacteria that took in plasmids, are resistant to kanamycin and sensitive to ampicillin'* (claim); 'Th*e gene for ampicillin resistance gets destroyed by the restriction enzyme'* (justification). S2 appealed to the task's data, this time in order to argue for the sensitivity of the transformed bacteria to ampicillin.
- S3: *'But are all the plasmids that were used for bacterial transformation, recombinant? Is this the only possible case?'* S3 challenged the last argument by recognizing the underlying assumption that all transformed bacteria had taken in *recombinant* plasmids. After the recognition and challenge of peers' problematic assumption, both S2 and S1 seemed to follow S3 according to whom the idea of contingency applies not only to bacterial transformation but to plasmid recombination as well. This critical notion about the kind of plasmids that the transformed bacteria have taken in was used later on.
- S3: *'We have colonies of bacteria that haven't taken in any kind of plasmid* (claim), *and thus are sensitive to both antibiotics* (justification*); we can eliminate them by adding kanamycin'* (claim). S3's argument about how to eliminate the non-transformed bacteria started the group's attempt to narrow down their search scope.
- S1: '*We have also colonies of bacteria that took in a recombinant plasmid, as well as colonies that took in a non-recombinant plasmid* (claim); *some plasmids did not become recombinant; recombination is not a certain thing* (justification)'. S1 continued narrowing down the search scope. She did this by going back to the transformed bacteria and using the previously discussed idea of contingent recombination in order to identify their content.
- S1:*'We can add ampicillin* (claim), *because both of the types of transformed colonies that we have, are resistant to kanamycin* (justification)'. S1 completed her contribution by arguing for an experimental handling that would eliminate some of the transformed bacteria.
- S3: *'Yes, but this not convenient for us (claim); ampicillin will kill the colonies with recombinant plasmids and then we will never be able to locate our gene'* (justification). S3 challenged S1's argument (or at least its claim) by evaluating the consequences of the suggested handling. Both S2 and S1 conceded to this argument and the group realized that they needed to come up with a smart handling that would allow them to identify the colonies with recombinant plasmids and still be able to have them alive.
- S3: *'We have to use ampicillin and also to keep the colonies with the recombinant plasmids alive and these two things are not possible at the same time* (justification); *so we need to copy the colonies somewhere else* (claim). S3 suggested such a handling by taking into account both the data they have and the goal they need to achieve.

- S3: *'We can take a sample from each colony, cultivate them in a new Petri dish and add ampicillin there* (claim); *the original Petri will be intact and the colonies we are interested in will be alive* (justification); *when we see which ones die in the new Petri, we can go back to the original and spot them there* (justification)'. S3 completed her contribution with another argument in order to fulfill S2's request for an elaboration of the details.

Having narrowed down their search scope only to the colonies that were transformed with recombinant plasmids, students were now able to focus on locating the target gene. They started with exploring the idea of an 'indirect' search and then proceeded with the idea of a 'direct' one.

- S2: *'We should find the gene by looking for the protein it codes for'* (claim). S2 reframed the task by suggesting an indirect, protein-based search instead of the direct gene-search that was probably more expected.
- S1: '*In theory this is possible* (claim), *because gene and protein go together* (justification)'. S1, although worrying about practical details, built on her peer's claim by drawing on background knowledge about the 'gene-protein' relation. So, the task was shifted on the level of a protein search.
- S3: *'It would be convenient to isolate* [from the bacterial cells with recombinant plasmids] *all the proteins that have similar molecular weight with that of our protein and add them in separate cultures of human cancer cells'* (claim); *'Whatever protein kills the cancer cells, will be the one that our gene codes for*' (justification). S3 suggested that they planned a search for the protein through a search for its (possible) anti-cancer action. In other words, she suggested an 'indirect' search once more.
- S2: *'We do not know for sure that the protein of our gene can kill cancer cells. This is something that needs more research in order to be proved. That's why we are supposed to work with cloning the gene and then studying the protein'*. S2 challenged the previous argument (its justification in particular) by appealing to uncertainty.
- The peer-group arrived at a dead-end and so they started exploring the idea of searching directly for the gene. This new idea led to new claims, justified by drawing on background knowledge about the heating's effect on DNA, the cDNA notion (i.e. DNA synthesized from an mRNA template with the aid of reverse transcriptase) or the hybridization of partially complementary DNA sequences. More specifically, peers took a three-step approach to articulate a hybridization-based experimental procedure. The first step concerned the gene itself, as shown below.
- S2: *'It would be difficult to find the gene by using a probe* (claim), *because the gene is double-stranded'* (justification). S3: *'It wouldn't be difficult, because the gene can get single-stranded if we heat it'*. S2 argued for difficulties in finding the gene, whereas S3 challenged S2's argument by appealing to background knowledge. The second step concerned the gene's probe. More specifically:

- S3: *'The probe should be a single-stranded DNA* (justification), *and so we can make it by using the mRNA of the protein* (claim)'. S1: *'So, we'll have a cDNA probe'*. Peers tried to think about how to produce a molecule that could serve as a probe for their gene. Finally, the third step concerned the 'gene-cDNA' hybridization. More specifically:
- S3 underlined that *'the gene and its cDNA probe won't be fully complementary* (claim), *but this won't be a prob*lem (claim) *since partially complementary molecules can actually stick to each other* (justification)'. Moreover, she suggested that they *'should also label the probe'* (claim), so that *'the 'gene-cDNA' hybrid will be easier to spot'* (justification).
- S2: *'A partially complementary probe could stick not just to the target gene but probably somewhere else, too'* (justification); *'This is not good for us'* (claim). S2 appealed to uncertainty: she suggested that using a probe which is only partially complementary to the gene, would not necessarily lead them to it.
- Facing a dead-end for a second time, peers decided to re-think the suggested procedure in more detail. Their new attempt did not actually help them to go any further. However, it did reveal their weakness to clearly conceptualize the 'gene-locating' problem as a problem that needed to be solved in vitro: for instance, they suggested *'heating bacterial cells in order to denature DNA'* or *'introducing cDNA in the bacterial cells the same way as plasmids'*. Finally, the discussion came to an end without consensus about a complete experimental procedure for locating the target gene.

In sum, students were engaged in argumentative discourse once more. They provided justifications to support their claims either about overall solutions to the 'gene-locating' problem or about solutions' partial steps. Moreover, they came up with oppositions, concessions and of course challenges, which points to the social aspect of argument's use; in other words, to its use as a shared tool for coping with the task at hand. Finally, students drew upon several epistemic operations such as evaluating consequences, recognizing assumptions, appealing to background knowledge, uncertainty, data or goals, in order to build or critically reflect upon the arguments that shaped their four-part thinking strategy which ended with both 'indirect' and 'direct' gene searches.

8.4 Discussion

The dialogues analyzed in this chapter were carried out by a group of three female second-year biology students and concern different steps of constructing or using a DNA library. The first one was performed in the context of a decision-making task; this provided peers with two alternative versions of a molecular tool for the construction of a DNA library and required from them to decide which version they should use. The second discussion was performed in the context of a task about articulating an experimental procedure; this provided peers with specific information and required from them to articulate an experimental procedure for achieving the goal of locating a target gene in the DNA library. Students' thinking strategies for solving these tasks that lied in a specific science domain, are actually domain-general.

More specifically, the decision-making strategy of the peer-group consisted of an evaluation of each option ('bottom') in terms of whether its use could lead or not to the achievement of a specific goal ('up'). In fact, it included (a) a bottom-up acceptance of one option, and also (b) a bottom-up rejection of its alternative. This strategy has emerged in order to address the need for deciding about the cloning vector in a genetic engineering context, but it is actually a strategy of 'evaluating options' appropriateness' that may very well apply to decision-making tasks arising in everyday life. Peers could also use a top-down strategy by re-inventing the most appropriate option through a 'goal reframing' process; i.e. through a process of 'breaking' the experimental goal ('top') to a series of prerequisites the last of which would point to the specific option ('down'). This strategy is also relevant to everyday life, especially when there are not any predefined options (Ergazaki et al., 2007). Finally, it is worth noticing that bottom-up strategy employed in the discussion presented here, took into account each of the options. Peers did not accept one option by just rejecting its alternative, although they could within the specific task. Instead, they developed a separate reasoning strand for each option and came up with a decision based on a complete rather than a partial evaluation. This may be considered as an indication of a critical approach to the required decision-making. Critical thinkers tend to decide what to believe or to do by reasonably reflecting on different options in a thorough, evidence-based evaluation process underlied by rational criteria (Ennis, 1987, 2018; Siegel, 1989).

Students' thinking strategy towards an experimental procedure for locating a target gene in a DNA library cannot be considered as strictly domain-specific either. Although the first two parts of the strategy (familiarizing with the search scope and narrowing it down) were encouraged by the scaffolding questions of the task, the other two (performing an indirect or a direct target search) were suggested by the students themselves. All apply to everyday life, too. In fact, 'looking-for-a-needle-in-a-haystack' situations are quite common in everyday life and may be handled by trying to locate either the 'needle' one really wants to find or the 'thread' that will possibly lead them to the 'needle'; and it may be helpful if they get familiar with the search scope and narrow it down as much as possible before starting their search.

The part of the strategy that concerns the indirect search is probably more worth discussing, since it presupposes to establish a connection between what really needs to be found (i.e. the real target) and what is going to be searched just for leading to the real target. Establishing this connection requires knowledge and is crucial for the effectiveness of the strategy. A not well-grounded connection results in a not effective strategy. In fact, this happened when the peer-group decided to search for the protein in order to locate the gene that codes for it, without examining the gene-protein connection in the light of knowledge about the differences in the molecular mechanisms of eukaryotic and prokaryotic cells as they should. This might also happen in everyday life situations. If the presence of the target (i.e. the presence of the 'needle', if we go on with the 'looking-for-a-needle-in-a-haystack' example), was correlated with the presence of something else that is probably easier to spot but cannot really function as a reliable probe for the target (e.g. it was connected with the presence of a 'thimble' rather than the presence of the needle's 'thread'), then the strategy would not be effective enough.

In sum, the thinking strategies that peers employed in the context of both tasks seem to apply to or even originate in everyday life. Their usefulness in a scientific context underlines how relevant everyday thinking may be to thinking within science, and vice versa (Kuhn, 1996; Kuhn & Pearsall, 2000). Students' ability to cope with domain-specific problems such as those presented here may be enhanced by drawing on everyday thinking strategies, whereas the latter strategies may be eventually elaborated and refined due to their parallel use in a scientific context. Highlighting a relationship like this in explicit ways might contribute towards bridging the 'science-everyday life' gap in students' minds, decreasing their stress and increasing their motivation for learning science.

Students' thinking strategies emerged within argumentative discourse. The members of the peer-group did provide justifications for or against their claims about the options they evaluated for addressing the decision-making in the first task; and likewise, they did provide justifications for or against the experimental handlings they suggested for addressing the 'gene-locating' problem in the second task. On the other hand, students often mobilized non-typical argumentative operations like challenges, oppositions and concessions, which underlines that they didn't use argument just as an individual thinking tool but actually as a shared tool that engaged them in true dialogue. According to Kuhn (2019), students' dialogic practice itself can be considered as an instance of critical thinking. Moreover, students' arguments were based on several epistemic operations like evaluating consequences, recognizing assumptions, appealing to background knowledge, data, goals, contingency, analogy, or uncertainty. By activating these, students actually established rational criteria in their discourse. This pinpoints to critical thinking as well, since rationality and reflection are core elements not just in argumentation but also in critical thinking (Ennis, 1987, 2018; Siegel, 1989).

So, the arguments and their epistemic content did set the stage for critical examination of the issues that arose while solving the tasks. For instance recognizing the false assumption that a gene and its protein go always together, indicates that peers were thinking critically enough to reflect on their overall strategy, realize its ineffectiveness and finally shift to a different one. It is also worth noticing however, that the lack of background knowledge seemed to be an obstacle in students' critical reflections. For instance, lacking knowledge about the loop formation that makes the hybridization of partially complementary molecules possible, proved to be key in peers' failure to complete the experimental proposal they have been developing through a thinking strategy that otherwise could possibly be effective. It is well-known that building new knowledge does require previous knowledge no matter how good one's reasoning devices may be, but giving students the opportunity to reflect on this idea through concrete examples might be beneficial for them.

References

Boyd, N. L. (2019). Using argumentation to develop critical thinking about social issues in the classroom: A dialogic model of critical thinking education. In S. P. A. Robinson & V. Knight (Eds.), *Handbook of research on critical thinking and teacher education pedagogy* (pp. 135–149). IGI Global.

Christenson, N., & S., Rundgren, N. C., & Zeidler, D. L. (2014). The relationship of discipline background to upper secondary students' argumentation on socioscientific issues. *Research in Science Education, 44*(4), 581–601.

Driver, R., Newton, P., & Osborne, J. (2000). Establishing the norms of scientific argumentation in classrooms. *Science Education, 84*(3), 287–312.

Ennis, R. H. (1987). A taxonomy of critical thinking abilities and dispositions. In J. B. Baron & R. J. Sternberg (Eds.), *Teaching thinking skills: Theory and practice* (pp. 9–26). W. H. Freeman.

Ennis, R. H. (2018). Critical thinking across the curriculum. *Topoi, 37*(1), 165–184.

Erduran, S. (2008). Methodological foundations in the study of science classroom argumentation. In S. Erduran & M. P. Jimenez-Aleixandre (Eds.), *Argumentation in science education: Perspectives from classroom-based research* (pp. 47–69). Springer.

Erduran, S., Simon, S., & Osborne, J. (2004). TAPping into argumentation: Developments in the application of Toulmin's argument pattern for studying science discourse. *Science Education, 88*, 915–933.

Ergazaki, M., & Zogza, V. (2005a). Reaching consensus through adversarial discourse: Peers' argumentative activity on predicting experimental outcomes in the context of genetic engineering. In M. Ergazaki, J. Lewis, & V. Zogza (Eds.), *Trends in biology education research in the new biology era* (pp. 119–129). Patras University Press.

Ergazaki, M., & Zogza, V. (2005b). Fishing for the target gene: Peers' discursive activity on screening bacterial clones in a gene-cloning experiment. In R. Pinto & D. Couso (Eds.), *Proceedings of the fifth international ESERA conference on contributions of research to enhancing students' interest in learning science* (pp. 1350–1353). European Science Education Research Association.

Ergazaki, M., & Zogza, V. (2005c). From a causal question to stating and testing hypotheses: Exploring the discursive activity of biology students. In K. Boersma, M. Goedhart, O. De Jong, & H. Eijkelhof (Eds.), *Research and the quality of science education* (pp. 407–417). Springer.

Ergazaki, M., Dimitriadis, G., & Zogza, V. (2007). Constructing the profile of effective cloning vectors in decision-making tasks: Conceptual tools, epistemic operations and emerging reasoning patterns. *Journal of Science Education, 8*(1), 24–27.

Jimenez-Aleixandre, M. P., & Crujeiras, B. (2017). Epistemic practices and scientific practices in science education. In K. S. Taber & B. Akban (Eds.), *Science education* (pp. 69–80). Brill Sense.

Jimenez-Aleixandre, M. P., & Erduran, S. (2008). Argumentation in science education: An overview. In S. Erduran & M. P. Jimenez-Aleixandre (Eds.), *Argumentation in science education: Perspectives from classroom-based research* (pp. 3–27). Springer.

Jiménez-Aleixandre, M. P., & Puig, B. (2012). Argumentation, evidence evaluation and critical thinking. In B. Fraser, K. Tobin, & C. McRobbie (Eds.), *Second international handbook of science education* (pp. 1001–1015). Springer.

Jiménez Aleixandre, M. P., Bugallo, A., & Duschl, R. A. (2000). "Doing the lesson or doing science": Arguments in high school genetics. *Science Education, 84*, 757–792.

Kelly, G. J., & Licona, P. (2018). Epistemic practices and science education. In M. Matthews (Ed.), *History, philosophy and science teaching*. Springer.

Knorr-Cetina, K. (1999). *Epistemic cultures: The cultures of knowledge societies*. Harvard University Press.

Kuhn, D. (1991). *The skills of argument*. Cambridge University Press.

Kuhn, D. (1993). Science as argument: Implications for teaching and learning scientific thinking. *Science Education, 77*(3), 319–337.

Kuhn, D. (1996). Is good thinking scientific thinking? In D. Olson & N. Torrance (Eds.), *Modes of thought: Explorations in culture and cognition* (pp. 261–281). Cambridge University Press.
Kuhn, D. (2019). Critical thinking as discourse. *Human Development, 62*, 146–164.
Kuhn, D., & Pearsall, S. (2000). Development origins of scientific thinking. *Journal of Cognition and Development, 1*, 113–127.
Kuhn, D., Hemberger, L., & Khait, V. (2016). *Argue with me. Argument as a path to developing students' thinking and writing* (2nd ed.). Routledge.
Makhene, A. (2017). Argumentation: A methodology to facilitate critical thinking. *International Journal of Nursing Education Scholarship, 14*(1).
Mason, L. (1996). An analysis of children construction of new knowledge through their use of reasoning and arguing in classrooms discussions. *Qualitative studies in Education, 9*(4), 411–433.
Pontecorvo, C., & Girardet, H. (1993). Arguing and reasoning in understanding historical topics. *Cognition and Instruction, 11*(3&4), 365–395.
Resnick, L., Salmon, M., Zeitz, C., Wathen, S. H., & Holowchak, M. (1993). Reasoning in conversation. *Cognition and Instruction, 11*(3&4), 347–364.
Sandoval, W. A., Enyedy, N., Redman, E. H., & Xiao, S. (2019). Organising a culture of argumentation in elementary science. *International Journal of Science Education, 41*(13), 1848–1869.
Siegel, H. (1989). The rationality of science, critical thinking and science education. *Synthese, 80*, 9–41.
Siegel, H. (1995). Why should educators care about argumentation? *Informal Logic, 17*, 159–176.
Toulmin, S. (1958). *The uses of argument*. Cambridge University Press.
Vygotsky, L. S. (1978). *Mind in society: The development of higher psychological processes*. Harvard University Press.
Zohar, A., & Nemet, F. (2002). Fostering students' knowledge and argumentation skills through dilemmas in human genetics. *Journal of Research in Science Teaching, 39*, 35–62.

Chapter 9
Fostering Critical Thinking About Health Issues: Facts of Success and Failure in the Case of Homeopathy

Araitz Uskola

Science education aims to train citizens to take part in the decisions that affect them. Feinstein (2011) proposed that when making decisions about socioscientific issues, people act, in the best of the cases, as *competent outsiders*. One way to train students to become such *competent outsiders* is by evaluating claims based on the available information (Feinstein et al., 2013), that is, by introducing argumentation practices that foster critical thinking in science lessons (Jiménez-Aleixandre & Puig, 2012, Chap. 1, this book). Critical thinkers are those justify their decisions appealing to scientific evidence, show breadth of thinking, willingness to change their opinion (Ennis, 1996; Paul & Elder, 2006) and take into account what experts say (Norris, 1995).

Pseudosciences constitute an appropriate everyday context for promoting argumentation and critincal thinking in science classrooms. This chapter discusses the results of a research program that analyzed the opinion and justifications posed by students about the effectiveness of homeopathic products and about a directive that considers them to be medicine.

9.1 Health-Related Pseudosciences in Science and in Science Education

One of the main objectives of science education is to prepare citizens, rather than to be experts in science, to take part in decisions affecting their health, their diet, the appropriate use of new materials and technologies or the use of energy, among others

---

A. Uskola (✉)
Universidad del País Vasco/Euskal Herriko Unibersitatea, Facultad de Educación, Leioa, Spain
e-mail: araitz.uskola@ehu.eus

B. Puig, M. P. Jiménez-Aleixandre (eds.), *Critical Thinking in Biology and Environmental Education*, Contributions from Biology Education Research,
https://doi.org/10.1007/978-3-030-92006-7_9

(OECD, 2016). Health-related controversies are present in the media and were found to be the more frequent scientific controversies in the Spanish press (Díaz & Jiménez-Liso, 2012). Among these, one of the most controversial fields is that of the set of alternative therapies to conventional medicine, many of which can be considered pseudosciences (Lack & Rousseau, 2016), that is, they make "claims that fail to conform to accepted standards in science regarding openness to peer review, replicability, transparent methodology, and the potential for falsifiability" (p. 39).

Promoters of alternative therapies base on issues different from those on which conventional medicine is based (Lake, 2005). Thus, they allude to the origin, to the history of the treatments, while scientists allude to the composition of the substances, to how their compounds can act in the body and to the empirical evidence of their effect. On the other hand, alternative therapies try to relate their compounds to what is considered "natural", which as Lake (2005) showed, belongs together with "pure" to a linguistic metaphor to which a positive value is attributed, that is, in many cases it is identified "natural" with "good". In addition, pseudoscientists use technical buzzwords such as "quantum" to give an image of scientificity and they make reference to people's testimonies (anecdotes) but not to scientific evidence (Lack & Rousseau, 2016).

According to Yates and Chandler (2000), pseudoscience advocates seek to reject analyses based on critical thinking, encouraging belief in "all opinions are valid" regardless of the evidence behind them. Thus they attempt to give an image impression of scientific controversy where there is none, for example by publishing their work in journals of questionable quality (Lack & Rousseau, 2016).

This image of open and equidistant debate can be reinforced by the attention paid to the defenders of non-scientific positions in the media. In Spain, even in mainstream media, pseudosciences have increased their presence in recent years (Cortiñas-Rovira et al., 2015) but they are not critically evaluated (Fernández-Muerza, 2004).

In a study carried out with 49 Spanish scientific journalists, Cortiñas-Rovira et al. (2015) found that 35% did not see any danger in the presence of pseudoscientific information in the media. On the other hand, 45%, most of them with over 10 years' experience, considered the pseudosciences to represent a threat and saw the need to warn of the deception and to address them in formal education. Metin et al. (2020) and Preece and Baxter (2000) suggested that in science classrooms students should have the opportunity to engage with these pseudosciences. Yates and Chandler (2000) found that teachers-in-training had little skepticism about pseudosciences, highlighting that it is sometimes considered anecdotal or not very harmful to have that kind of beliefs. These authors point out that today, when the Internet offers so much information and so much disinformation, it is more necessary than ever to work in the classroom on critical thinking and the evaluation of information.

Several studies have examined gullibility and lack of scepticism among students. Preece and Baxter (2000) analysed the skepticism shown by 2159 secondary students regarding pseudo-scientific issues, such as reflexology, homeopathy or astrology. They found that the students were gullible, as their mean skepticism score was

2.8 in a 1–4 scale. This concern for students' lack of skepticism about pseudo-scientific issues was the focus of a study with primary school teachers in training, with the revealing title *Where have all the skeptics gone?* (Yates & Chandler, 2000). They provided eight statements considered incredible by experts and philosophers. A total of 232 teachers in-training responded, with an average of 3.5 ideas rejected; only four participants rejected all of them. Gullibility was also found by Metin et al. (2020) in eighth graders that believed that crystals could be used for medical purposes.

It could be expected that as people's scientific knowledge increases they will use it more proficiently in their decisions. However, research on the use of scientific knowledge in public affairs shows that there are other factors, besides knowledge. Feinstein (2011) reviewed studies about how decisions about socio-scientific issues are made, concluding that people selectively integrate scientific ideas with those from other sources to make decisions that are personally and socially meaningful. At best, they act as *competent outsiders,* people who have learned to recognize when science is relevant for their needs and interests and to interact with expert scientific sources to help them achieve their goals (Feinstein, 2011). Feinstein et al. (2013) discussed how to contribute to the formation of *competent outsiders* from within science education. From the scientific practices that science education aims to promote (Osborne, 2014), they pointed out to evaluating statements based on available information as one of the most useful to work on for non-scientists, therefore proposing working on argumentation and debates on socio-scientific issues in the classroom.

9.1.1 Socio-Scientific Issues in the Classroom

One way to foster critical thinking and scientific competence is to work on socio-scientific issues or controversies in science education (Erduran & Jiménez-Aleixandre, 2008). These are considered suitable for learning scientific concepts, since the student places the content in a broader context that gives it meaning (Sadler et al., 2007).

Kuhn (1993), concerned with fostering the relationship between scientific and everyday thinking, proposed using socio-scientific issues in science classrooms. She recognized that there may be a paradox in that scientific thinking may be developed even better by socio-scientific issues than by scientific ones, since in the latter, students may feel inhibited by a strong belief in their ignorance. That is, in scientific subjects students would show less confidence in themselves, and they would have a worse disposition for critical thinking (Barak et al., 2007). Decision making on socio-scientific issues, thus, constitutes a valuable context to develop use of data and critical thinking skills in science education (e. g. Albe, 2008; Kortland, 1996; Patronis et al., 1999).

9.1.2 Use of Data and Critical Thinking

The role of evidence-based evaluation of information is central in decision making (Feinstein et al., 2013). In a case in which inaccurate information may be involved, the capacity to evaluate information based on evidence will allow to discern whether the piece of information meets the conditions to be considered scientific or whether it can be deemed pseudoscience (Bell & Lederman, 2003). Kuhn (2010) stressed the importance of developing argumentation skills in science classrooms, even if no conceptual knowledge is developed. In trying to characterize scientific thinking, Kuhn (1993) observed that although conceptual knowledge played an important role, its acquisition did not fully characterize the development of scientific thinking. Kuhn accorded importance to being able to distance oneself from one's beliefs in order to evaluate them, so that she pointed out that the progress people made had to do with how they came to the conclusions. Thus, she characterized scientific thinking as argument. Arguing, evaluating statements based on available information, assessing the credibility of information, and differentiating well-founded opinions from those that are not are skills closely related to critical thinking (Jiménez-Aleixandre & Puig, 2012). Various definitions of critical thinking are coincident on this point.

According to Ennis (1996), critical thinking is logical and reflective thinking focused on deciding what to believe and what to do. He considers that the competence for critical thinking includes both skills, such as ability to analyse, evaluate and make inferences, and disposition towards critical thinking. Becoming competent requires lifelong learning to leave behind natural self-centered thinking (Paul & Elder, 2006). Critical thinking scholars point out that one of its constituents is the reason given when stating something (Ennis, 1996), the information on which a statement is based (Paul & Elder, 2006).

But there are other aspects that scholars deem important, namely taking into account others´ points of view and the consideration of experts. Johnson (2009) studied the assessment of critical thinking and pointed out that part of the problem lies in the diversity of definitions of critical thinking. For him the argument plays a central role. In an argument, critical thinking becomes apparent both in how the options are justified, and in the dialectic skills. He conceived that critical thinkers were the opposite of dogmatic ones. Therefore critical thinkers should be willing to receive criticism, consider the weaknesses of their position, and contemplate under what conditions they would change it or even do so. Along the same lines, Lack and Rousseau (2016) defined how a sceptic acts. Similarly, Paul and Elder (2006) stated that breadth of thinking, in other words, looking at the problem from different points of view is a standard for critical thinking. The adoption of multiple perspectives was also proposed as a criterion for evaluating arguments, for instance by Sadler and Donnelly (2006) to evaluate student argumentation in interviews. Likewise, Felton et al. (2009) and Kortland (1996) evaluated if students dealing with socio-scientific controversies where aware of the limitations of the option they were defending or of the advantages of opposite options.

Although listening to the arguments of experts may seem inconsistent with critical thinking, Norris (1995) pointed out that in science education people are not intellectually independent in a strict sense, they have to trust experts. Acting with a reflective skepticism allows people to judge who the experts are in a given context and to evaluate the evidence on which the claims are based.

9.2 Methods, Educational and Regional Context

9.2.1 Controversies About Homeopathy in Spain and Europe

Numerous studies have shown that homeopathy has no greater effectiveness than the placebo (Lack & Rousseau, 2016; Ministerio de Sanidad, Política Social e Igualdad, 2011). Despite this, in Europe there is a ruling, *Directive 2001/83/EC of the European Parliament and of the Council* (European Parliament, 2001) that considers them to be medicinal products, requiring their regulation as such. The regulation differentiates between medicinal products with therapeutic indications and without it. This regulation began more than a decade later in Spain and triggered the rise of voices from Medicine and Science to criticize it. An European manifesto against pseudo-therapies was drafted. In 2018 2008 products were presented to be considered as medicines, 12 of them with therapeutic indication. There is no product evaluation data at the time of writing this.

Some recent changes may suggest that the positioning towards homeopathy is changing in the Spanish government and in the Spanish media. It seems that its treatment in the media is more negative that it used to be (Martí-Sánchez & Roger-Monzó, 2018). The Spanish government submitted a proposal to the European Parliament in 2018 to amend Directive 2001/83/EC, but it was rejected, on the grounds that no other country had seen problems with it (European Commission, 2018). In 2019 the Royal Academy of the Spanish Language (Real Academia Española) which rules in cooperation with academies across the 21 Spanish speaking countries, after 167 years of defining homeopathy as a "healing system", decided to change it, now defining it merely as a practice. With regard to citizenship, the barometer of the Center for Sociological Research (Centro de Investigaciones Sociológicas, 2018) first included questions related to pseudosciences and homeopathy in 2018. The results showed that people had misconceptions about pseudosciences and did not clearly reject them. A Survey about the social perception of science and technology (Quintanilla et al., 2019) found that a positive attitude towards science and a high level of scientific knowledge was compatible with confidence in certain non-scientific practices in the area of health, namely homeopathy and acupuncture. According to the National Institute of Statistics (Instituto Nacional de Estadística, 2013), the higher the level of studies, the greater the use of homeopathic products. The preceding data show that while there is a change in how institutions and media treat pseudosciences, it seems that this change has not been transferred to how people think about them.

9.2.2 Research Questions and Design

This chapter analyses the skepticism and critical thinking dimensions in the products and in the process developed by students while making a decision about a socio-scientific issue. The research questions are:

RQ1. At what extent do the initial positions about homeopathy mirror the recent changes in the attitude of Spanish institutions and differ from those obtained 5 years before?
RQ2. How do the positions about homeopathy of students change after participating in teaching activities that include critical reading, making dilutions and group discussions?
RQ3. How do students justify their opinion about the effectiveness of homeopathic products and about naming homeopathic products medicine?
RQ4. At what extent do students show breadth of thinking and willingness to change in their decisions?
RQ5. In which way do students refer to experts´ knowledge in their arguments and decisions?

Skepticism and critical thinking of students can be analysed in the decisions made by the students and in how they justify them. The analysis of the decisions themselves was taken into account in research questions RQ1 and RQ2 (Feinstein et al., 2017; Lack & Rousseau, 2016; Metin et al., 2020; Preece & Baxter, 2000; Yates & Chandler, 2000). Research questions RQ3, RQ4, RQ5 capture the elements that characterize a critical thinker discussed above, namely the emergence of the justification and the use of evidence (Ennis, 1996; Jiménez-Aleixandre & Puig, 2012; Paul & Elder, 2006), the breadth and willingness to change (Johnson 2009; Lack & Rousseau, 2016; Paul & Elder, 2006), and the consideration of experts (Feinstein et al., 2017; Kolstø et al., 2006; Norris, 1995).

Three studies were carried out with students in the degree in Primary Education at a Spanish public university (S1, S2 and S3) and one (S4) with high school students. The information about the studies is summarized in Table 9.1. Part of the results of S1 and S2 were published (Uskola, 2016, 2017).

Pseudonyms were used for students. In S1 and S2, students were numbered, using A for girls and O for boys. In S3 and S4 each student was assigned a pseudonym beginning with the letter of their group.

Data sources for answering the research questions consisted of written answers to open questions on the effectiveness of homeopathy (Q1 *Do you think homeopathic products are effective in curing diseases?*) and its consideration as a medicine (Q2 *There is controversy today about whether homeopathic products should be considered as medicines in the legislation. Would you be for or against considering them as medicines?*), collected before and after the teaching activities. Besides, data from audio and video recordings of 12 group discussions in S3 and S4 also were used. Diverse data analyses were made to answer each research question. They are described along with the corresponding results.

Table 9.1 Participants and activities in the four studies

	S1	S2	S3	S4
Academic year	2014/15	2014/15, 2015/16	2019/20	2019/20
Number of participants	42 (71% girls)	134 (71% girls)	34 (50% girls) 9 groups (A-I)	10 (80% girls) 3 groups (J-L)
Academic context	4th year of degree in Primary Education	1st year of degree in Primary Education	4th year of degree in Primary Education	High school
Age	21–22	18.8	22.4	16
Activities	– Critical reading of a press release (Oliveras et al., 2013)	– Close-ended questionnaire – Two open-ended questions	– Close-ended questionnaire – Two open-ended questions	– Close-ended and open-ended questions
	– 20D dilutions of dye, sugar and salt – Search for information		– 20D dilutions – Reading press releases – Identification and evaluation of data for and against homeopathy – Group discussions	– 20D dilutions – Reading a brief – Identification and evaluation of data for and against homeopathy – Group discussions
	– Final individual decisions		– Final individual decisions	– Final individual decisions

9.3 Results: Skepticism About Homeopathy

In this section findings related to RQ1 and RQ2 are examined. First we compare, on the one hand shifts in the initial position of students in 2014 and in 2019 (RQ1); and on the other hand differences between positions before and after the teaching sequence (RQ2).

The results of the analysis of the answers to the open question Q1 are summarized in Fig. 9.1.

It can be seen that, at the beginning, in S1 and S2, more than 80% of the students believed in the effectiveness of homeopathy. Five years later, in S3 and S4, the results were different. Although the sample is limited and certainly cannot represent society, this difference may be in part related to the change in the attitudes of Spanish media and institutions. These results are somehow unexpected taking into account the findings from recent surveys described in the previous section.

After participating in the activities, the majority did not believe that homeopathic products are effective, especially in S3 and S4, in which they reached over 80%.

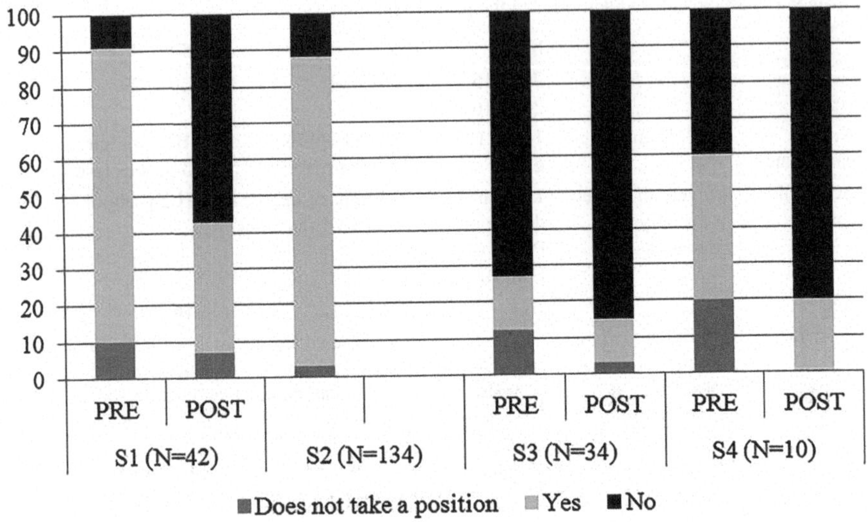

Fig. 9.1 Percentage of students who said that homeopathic products are effective or not, for each study, and in the pre and post activities situations

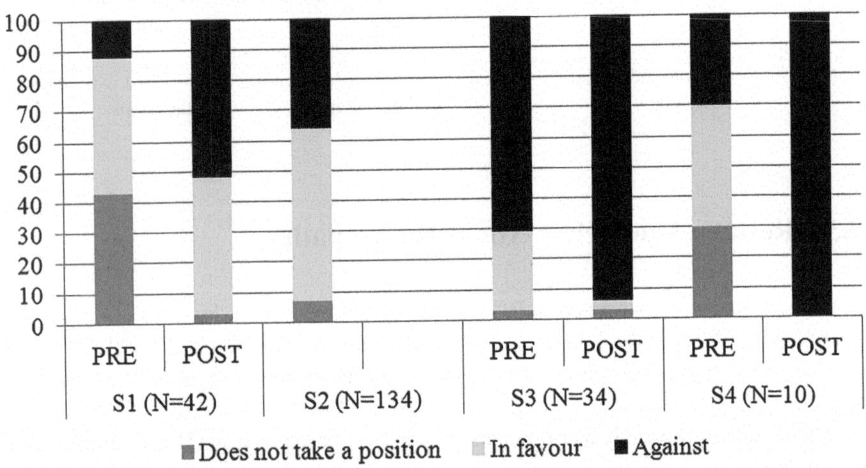

Fig. 9.2 Percentage of students who were for and against homeopathic products being considered medicines, for each study, and in the pre- and post-sequence situations

The results of the analysis of the answers to the open question Q2 are shown in Fig. 9.2.

At the beginning, as seen in the Figure, in S1, S2 and S4 more students were in favor of considering homeopathic products as medicines. At the end, in S1 the students against considering them as such constituted just over half, a position much more pronounced in S3 and S4, with a percentage of more than 90%.

Taking into account the responses to Q1 and Q2, the results shows that, on the one hand, in response for RQ2, the positions about homeopathy did change after participating in the activities, so that they showed a higher level of skepticism. On the other hand, the results also show that the changes in the attitude of institutions maybe are transferring to society in a higher level that survey data suggest, and that changes may help so that this healthy skepticism increases.

9.4 Results About Critical Thinking: Justifications and Use of Evidence

In this section findings related to RQ3 are examined, based on the comparison of the justifications put forward by students before and after the activities. First, we compare justifications for Q1 and, secondly, for Q2.

9.4.1 Effectiveness of Homeopathic Products

For the analysis of justifications for Q1, a scale of 4 levels, shown in Table 9.2 was established.

Conceptual errors in level 1 include confusing homeopathic products with natural and traditional remedies or thinking that the placebo effect is exclusive to these products. Some examples to illustrate the coding are:

S1O11(Pre): If the illness is not serious, I think so [that homeopathic products are effective]. In fact, natural medicine has been used since ancient times and has had positive results many times. (Level 1).

Level 3 includes explicit references to scientific evidence, justifications referring to the composition of products or to expert opinion/action (Norris, 1995).

S1O10(Post): No, after seeing the process that is followed to manufacture them, I have realized that homeopathic products are only lactose and sucrose. They can have a placebo effect, but any product can produce this effect. (Level 3).

Table 9.2 Levels for the categorisation of justifications in Q1

Level	Description
3	Justifications including scientific evidence
2	Justifications alluding to anecdotal testimonies and/or personal experience
1	Justifications with conceptual errors
0	No justification

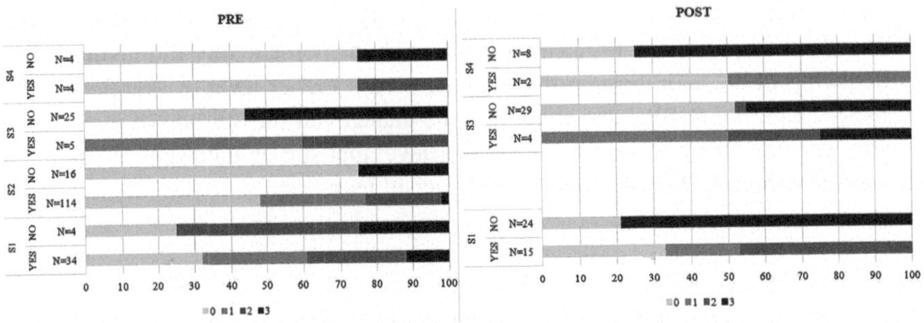

Fig. 9.3 Percentage of students, in relation to the total in each positioning, at each level of justification, when justifying their opinion about the effectiveness of homeopathic products (Q1)

Figure 9.3 shows the percentages of students who formulated justifications at the different levels, separating those believing in the effectiveness of homeopathic products from those who didn't. For this analysis, we considered the level of the best justification given by each student. It should be noted that in order to study differences between the justifications of students who believed and those who did not believe in the effectiveness of homeopathy, only those positioning themselves in one way or another were considered.

As seen in the Figure, prior to the sequence, when the students stated that they believed in homeopathy, on many occasions (29% in S1 and S2 and 60% in S3), they did so by resorting to errors such as considering that they are "natural" products. In fact, the results of the close-ended questionnaire conducted in S2 and S3 showed that 76% of the students believed that using medicinal herbs belongs to the practice of homeopathy, and that 60% believed that homeopathic products are composed solely of natural products. Therefore, as in other studies (Lake, 2005), many students considered homeopathy to be natural and identified "natural" with something good.

Moreover, in the answers to the open-ended questions Q1 and Q2 in S2, 26% of students spontaneously expressed that natural is "healthier, more effective and better than chemical". The difficulties they had with this issue were in some cases extended to the group discussions, for example in group H and J.

June: Yeah, then they put sugar in it and all you take is sugar with water.
Joana: I'm in favor of natural things.
Janire: Homeopathy is not natural.
Joana: It is more natural than medicine.

The other large group of justifications alluded to by students who believed in the effectiveness of homeopathic products were personal first-hand experience and the testimonies of people, i.e. everyday knowledge.

S2A25: Yes, I believe that homeopathic products are effective in curing diseases because I take them and they cure me. It is true that they are slower than conventional medicine but their effect is longer lasting and they have no side effects. (Level 2).

As in other studies with high school students (Metin et al., 2020; Patronis et al., 1999) and professional trainees (Albe, 2008), participants, especially those who defended homeopathy, seemed to assign great validity to personal experience and close testimonies.

Students who did not believe in the effectiveness of homeopathy provided justifications mostly at the highest level (3), based on scientific evidence, for instance the high dilution of the products.

June (S4, Post):	No because they only contain water and sugar and they are not scientifically proved. (Level 3).

In S1 and S4, at the end of the activities, students who did not believe in the effectiveness of homeopathy used more and better justifications. However, in S3 52% of students did not use any justification.

In the group discussions of S3 and S4, students alluded to the need for scientific evidence and used high level justifications:

Fernando:	The result [of dilution] is water.
Flora:	All of them [effectiveness tests for homeopathic products] are going to be negative.
Fernando:	We don't know. The conclusion is that only 12 want to do it [effectiveness test]. For me it's quite revealing.
June:	That's placebo.
Joana:	I don't think so, I think it does have an effect.
June:	If it does, why hasn't it been proven scientifically?
Janire:	It's like what we did with the water yesterday. After 20 times of filling and emptying, do you think that had salt in it? Well, it's the same with homeopathy.

9.4.2 Consideration of Homeopathic Products as Medicines

For the analysis of justifications for Q2 a scale of five levels, shown in Table 9.3, was established.

Figure 9.4 shows the percentage of participants who formulated justifications at each level, depending on whether they were for or against the consideration of homeopathic products as medicines. For this analysis, the highest level reached by each student was taken into account.

After participating in the activities, as seen in the Figure, the percentage of students who didn't justify their opinion (level 0) decreased. Besides, levels 3 and 4, which include reference to the characteristics that define medicine, were reached by most students who positioned themselves in one direction or the other. This shows that they had internalized what a medicine should accomplish. Thus, effectiveness was taken into account by 55% of the total number of students who had

Table 9.3 Levels for the categorisation of justifications in Q2

Level	Description
4	Justifications including medicine defining criteria and scientific evidence
3	Justifications alluding to the criteria that define medicine in Spanish legislation (quality, effectiveness, safety)
2	Other justifications (e.g. economy)
1	Justifications referring to origin of the products
0	No justification

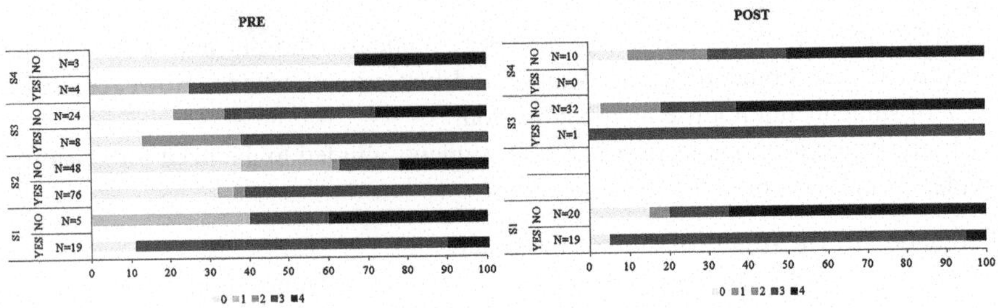

Fig. 9.4 Percentage of students, in relation to the total in each positioning, at each level of justification, when justifying their opinion about considering homeopathic products as medicine (Q2)

indicated their position. In fact, correlations were found between the belief in the effectiveness of homeopathy and positioning against its consideration as a medicine.

However, interesting exceptions were found: for example, students who did believe that homeopathy has some effect but considered that it must be scientifically proven to be considered a medicine; or students who did not believe it is effective but approved of it being classed as a medicine as, in this case, it would have to pass the required safety checks.

The difference between the results in levels 3 and 4 is that, in the case of level 4, the increase after the sequence was considerable among those students who were against considering homeopathic products as medicines, of which more than 50% referred to scientific evidence, to the need for the characteristics to be scientifically proven. It is interesting that this allusion to scientific evidence also appeared in the discussions of two of the three groups of 16 year olds in S4:

June: I believe that if it is not scientifically proven, it should be not in pharmacies.

Kirmen: I think it is bad that 2008 [homeopathic products] have requested testing to qualify as medicine, 12 of these also for recognition as medicine with approved therapeutic indication, and we are still waiting for the results. And they are on sale. For me, an untested medicine should not be on sale.

9.5 Results Concerning Critical Thinking: Breadth of Thinking and Willingness to Change

In this section we examine results related to RQ4, based on how students took into account the counter-position or even change their opinion.

Figure 9.5 shows the percentage of students that although taking a position attended to opposing arguments.

As seen in the Figure more students took into account the opposing justifications after the activities. Still, in most cases the percentage of students that did it so was less than 40%. The exception were the arguments put forward in S3 to justify the belief in the effectiveness of homeopathy. In this case students demonstrated that they were paying attention to opponents at a high degree, reaching 80%. It is striking that these were precisely the students that scored lowest in justications as seen in Fig. 9.3.

Table 9.4 shows the changes in positioning with respect to the situation at the beginning of the sequence for S1, S3 and S4.

The percentage of students who changed their position in these studies was relatively high, especially among those who initially defended the effectiveness of homeopathy and its consideration as a medicine.

Even students who acknowledged taking homeopathic products themselves questioned their effectiveness after the experiment. This can be seen in the

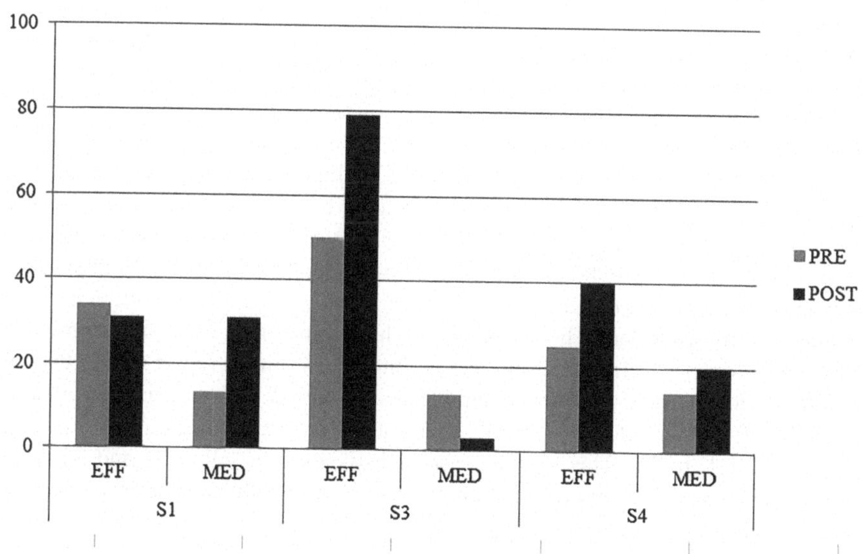

Fig. 9.5 Percentage of students, in relation to those who positioned themselves, who referred to the opposing evidence

Table 9.4 Percentage of students, with respect to the total sample, who changed position. All changes were taken into account both for students who initially took the opposite position and for those who had not positioned themselves initially

	S1 (N = 42)		S3 (N = 34)		S4 (N = 10)	
	Effectiveness	Medicine	Effectiveness	Medicine	Effectiveness	Medicine
To NO	50	48	21	24	40	70
To YES	2	21	9	3		

discussion of group K, when Karmele said “I have allergies and I take homeopathy but I didn’t know what it was until now (. . .) After the experiment I question it a little bit.”

9.6 Critical Thinking: Expert Consideration

In this section findings related to RQ5 are examined, based on how students appealed to experts´ knowledge in their arguments.

The experts´ viewpoints were referred to in a total of 18 initial and final responses by students from the four studies as justifications for adopting a particular opinion. Although their percentage was low (7%), these responses are worthy of discussion.

On the one hand, in some cases, considering what experts say caused confusion in students. We must take into account that some of those who are expected to act as trustworthy and intellectually dependent experts (Norris, 1995), don’t in fact act that way. Nor do they necessarily base their actions on evidence in the case of alternative therapies. For example, the Spanish Ministry of Health itself, from its position of authority, may have made it difficult to judge the information by calling the set of non-conventional therapies “natural therapies”, while recognizing that “using this terminology may mistakenly encourage the public to think that these therapies use more natural means than conventional medicine, when this is not necessarily the case” (Ministerio de Sanidad, Política Social e Igualdad, 2011, p. 5). As noted, this may lead one to believe that they are also “good” (Lake, 2005). Moreover, in the case of homeopathy, considering homeopathic products as medicines may also imply that they have proven effectiveness, and this despite a government report stating that “it is difficult to interpret that the favourable results found in some trials are distinguishable from the placebo effect” (Ministerio de Sanidad, Política Social e Igualdad, 2011, p. 72). This was the case with six students, all in S1 and S2, who showed how the actions of certain institutions or doctors influence the acceptance of pseudosciences such as homeopathy.

S1A1(Pre): Yes, if the Health Department considers them to be medicines, it will be because they have been proven to be valid for curing diseases.

S2A22: I believe that homeopathic products can be effective in curing diseases. In fact, some experts who know a lot about medicine recommend homeopathy.

The other 12 students used the reference to experts to rule out homeopathy.

The role of experts was highly considered in S3 and S4. Indeed, 50% of the groups (A, C, F, G, J and K) talked frequently about their confidence in experts. More precisely, A, C, F, and K addressed the fact that calling these products medicine and selling them in pharmacies as such can generate confusion among citizens:

Karmele: We know this because we have done the experiment; but the person who goes to the pharmacy does not know about it.

Kirmen: And that's what they use, they use the fact that people don't know what they are getting into.

Karmele: Well, that's why not [they shouldn't be considered medicine].

Karla: What has made me doubt is this, that there are more than 10,000 products in Spain that are prescribed. If there are so many products and so many are sold, will they really cure something? I doubt it. But then you think about it, and I can't believe that this will cure anything. It's nonsense. But to accept this? Does it do anything? It makes me think. Either they're laughing at us and they're laughing at the sick who want to be cured or it really does do something.

Koro: I believe in what the doctors say.

Karla: There are many lives at stake, but it is allowed, so...

In fact, in S3, 22% of the justifications against considering homeopathic products as medicines at the end of the sequence refer to this point:

Aitor: Clearly not. If you call them medicine, you can create confusion. You have to make it clear to people what they are taking and what the consequences are.

Carmen: No. Because people may think they are as effective as real medicine. That can be dangerous.

Fernando: I think that in the sense that they are not effective, they cannot be medicines, because in society the word "medicine" is very much linked to its effectiveness.

9.7 Implications for Education

The students that participated in the various studies proved to be quite gullible and initially not sceptical, as was found in other studies (Metin et al., 2020; Preece & Baxter, 2000; Yates & Chandler, 2000). It must be highlighted that there was a difference in the initial positions (RQ1) taken by S1 and S2 on the one hand (more than 80% of the students believed in the effectiveness of homeopathy) and those initially taken by S3 and S4 on the other (15–40% believed in its effectiveness). As discussed before, the difference between their positions/opinions, which were recorded 5 years apart, may be attributable in part to may be attributable in part to the change in attitude of the institutions in Spain. Indeed, in S1 and S2, some students consistently referred to the decisions of the Spanish institutions as justification for their own view that homeopathic products are effective and/or medicines. In contrast, in S3 and S4, such decisions were used to justify exactly the opposite.

As far as justifications are concerned, the students who were less sceptical about homeopathy used two main types of justifications: either they considered these products to be natural or they based their opinions on their own testimonies or those of people close to them. The activities carried out with the students in S1, S3 and S4 were aimed, firstly, at creating a cognitive conflict so that students would be faced with their own misconceptions; secondly, at considering other types of justifications, such as scientific evidence and expert opinion; and thirdly, at evaluating the consistency of the different types of justifications. It was expected that, after participating in the activities, the students would be more sceptical (RQ2), and that they would develop more critical thinking skills (RQ3, RQ4, RQ5).

One of the achievements of the activities carried out with the students was to clarify what homeopathy consists of, since one of the factors that lead them to be rather credulous was their lack of knowledge. After the activities students were less gullible (RQ2). One of the main sources of error was that they confused homeopathy with "natural" (60% of the 168 students). In addition to this confusion, it was observed, similarly to Lake (2005), that students consider that being natural is equivalent to being good, and that not being natural has negative connotations. Reading the leaflet of homeopathic products and making dilutions helped to show exactly what they are and also to demonstrate that they are not necessarily natural products. However, it may still be necessary to discuss with students the meaning of certain terms, such as "natural" and "pure", and the different use that is made of them in the scientific context and in everyday life (Lake, 2005). This would help students to be more aware of the meaning of "natural" and the attributes that are given to it. The same would be applicable to the term "chemical", usually negatively charged. Indeed, it became an issue of discussion in some groups. In H, Humberto stated that "But that creates a chemical reaction in your body so it's a chemical".

After participating in the activities, the students used higher-level justifications (RQ3), making reference to scientific evidence. Differences were found depending on the student's position, and those that were sceptical of homeopathy also showed a higher level of critical thinking, using justifications based on scientific evidence. This is shown, in the case of these students, by the increase in level 3 of the belief in its effectiveness, and in level 4 for its consideration as medicine,. The use of more scientific justifications was facilitated by the information given to students about homeopathy and by the dilutions made in the laboratory, but, in the case of S3 and S4, some activities were added to improve the results of S1. In S1, many students continued to believe in the effectiveness of homeopathy, in spite of the dilutions they made, and still based their positioning on personal experience.

The context of the activities can be relevant for the weight that personal experience is given in decision-making. Kolstø (2001) raised eight aspects, unrelated to content, to be considered when dealing with socio-scientific issues in science education, some of them closely related to the consideration of evidence.

He highlighted the difference between the consideration of what is evidence in the scientific context and in the everyday context, and stressed that, in the everyday context, anecdotal evidence is also valued. Educators, then, must try to get students to value the role of evidence not only in the process of building scientific knowledge, but also in the performance of critical thinking in their everyday contexts.

In S3 and S4, students had to consider the validity of both types of evidence (anecdotal and scientific) in line with Kolstø's (2001) proposal. Furthermore, after considering the data individually, they had the opportunity to discuss them in a group. In these discussions, as seen in the previous sections, high-level justifications were given for rejecting both the belief in the effectiveness of homeopathy and its consideration as a medicine. In the discussions, students had the opportunity to evaluate the information they were given (in both senses), as can be seen, for example, in Fernando's intervention in Sect. 13.5.1.

The results of S4 are comparable to those of S3 despite the age difference. Although in S4 they started from a greater lack of knowledge, in the end many changed their minds and rejected homeopathy. Their discussions were rich and the justifications were based on scientific evidence; they were also very aware of the consequences of calling homeopathic products medicines. The results for S4 can be considered acceptable, whereas those for S3 provoke much thought. S3 students are about to become practicing teachers, and a greater degree of development of the critical thinking skills could be expected of them. Indeed, many of them (52%) didn't even think it necessary to justify their scepticism towards homeopathy after the activities. Maybe some reflection is needed about how to encourage to a greater degree this type of thinking all throughout education. However, it is true that the S3 students did stand out from the rest in one component of critical thinking: the consideration of other points of view (RQ4). In fact, 83% of those who did not believe in the effectiveness of homeopathy discussed issues that could be used to believe in it (e.g. placebo effect, testimonies). This reflection on the justifications of the opposing view is considered to characterize both a sound argument (Felton et al., 2009; Kortland, 1996; Sadler & Donnelly, 2006) and critical thinking (Johnson, 2009; Paul & Elder, 2006).

If the results of the level of the justifications for Q1 (Fig. 9.3) and Q2 (Fig. 9.4) are compared with those of the consideration of the opposing justifications (Fig. 9.5), it seems that in many cases, when students performed well in justifying, they did not do so well in considering the other point of view and vice versa. For example, in S3 the students used less justifications for their arguments about effectiveness, but they attended to opposing justifications more than in other cases. Overall, students had difficulties in developing arguments that meet all the quality criteria.

Discussions in S3 and S4 were found to be very productive. Students referred to the high dilution of homeopathic products, the importance of scientific evidence, the role of experts, and, linked to the latter, the consequences of calling homeopathy a medicine The use of such arguments are characteristic of high-level critical thinking (Ennis, 1996; Paul & Elder, 2006) and of skepticism (Lack & Rousseau, 2016).

Indeed, some students put forward justifications that showed signs of critical thinking, and even modified their beliefs (Table 9.2), becoming more skeptical. This is considered unusual (Albe, 2008) and would correspond to the maximum degree of open-mindedness or breadth of thinking (Ennis, 1996; Paul & Elder, 2006). At the end of the activities, the majority did not believe that homeopathic products are effective, especially in S3 and S4 with over 80% adhering to this view.

A further important issue in the case of primary school teachers is that they need to develop not only their own critical thinking skills, but also those of their future students. To this end, they obviously first have to develop the skills in themselves (Windschitl, 2003) but it is also essential that they genuinely value them. In S3 the participants were asked about dealing with pseudo-scientific issues in their classrooms. 93% found this an interesting topic. Of these, 44% justified this interest based on the fact that they are current issues, 59% linked it to the educational context and curriculum, and 63% explicitly referred to the development of critical thinking:

Fernando: In addition to encouraging students´ critical thinking, they will develop knowledge that will be valuable to them for their future. And what is education, if not the teaching of valuable concepts? Teachers should not give their opinion and should use data on the subject. In this way, students will develop healthy habits and their own opinion on this very contemporary subject.

Participants also put forward ideas as to how they would go about dealing with pseudo-scientific issues in class. Here it can clearly be seen that they have internalized some keys to critical thinking, such as justifying statements, contrasting and evaluating different sources of information, even though throughout the sequence of activities, the concept of critical thinking was not explicitly worked on.

Flora: Educational centres must guarantee integral development, which includes critical thinking. I think they should not be taught what [products] to take or not, but to act critically in the face of new information and to base their arguments on significant and reliable sources of information, and to contrast them with other sources.

It can be said that the activities designed have shown to be valid for developing students´ critical thinking and for enabling them to act in their daily lives as *competent outsiders* (Feinstein, 2011), making decisions about their health, and so fulfilling one of the objectives of science education (OECD, 2016). In the case of these future primary teachers, moreover, one can say that they were aware of the importance of educating critical thinkers and *competent outsiders* themselves in the future.

Acknowledgments This research was funded by the University of the Basque Country UPV/EHU *Research Grants: Mod. II Groups* (codes GIU19/008 and PPGA20/14).

References

Albe, V. (2008). Students' positions and considerations of scientific evidence about a controversial socioscientific issue. *Science and Education, 17*, 805–827.

Barak, M., Ben-Chaim, D., & Zoller, U. (2007). Purposely teaching for the promotion of higher-order thinking skills: A case of critical thinking. *Research in Science Education, 37*, 353–369.

Bell, R. L., & Lederman, N. G. (2003). Understandings of the nature of science and decision making on science and technology based issues. *Science Education, 87*(3), 352–377.

Centro de Investigaciones Sociológicas. (2018). *Barómetro de Febrero 2018* [Barometer of February 2018]. http://www.cis.es/cis/export/sites/default/-Archivos/Marginales/3200_3219/3205/es3205mar.pdf. Accessed 20 Feb 2020.

Cortiñas-Rovira, S., Alonso-Marcos, F., Pont-Sorribes, C., & Escribà-Sales, E. (2015). Science journalists´ perceptions and attitudes to pseudoscience in Spain. *Public Understanding of Science, 24*(4), 450–465.

Díaz, N., & Jiménez-Liso, M. R. (2012). Las controversias sociocientíficas: temáticas e importancia para la educacion científica [Socio-scientific controversies: Themes and relevance for science education]. *Revista Eureka sobre Enseñanza y Divulgación de las Ciencias, 9*(1), 54–70.

Ennis, R. H. (1996). *Critical thinking*. Prentice Hall.

Erduran, S., & Jiménez-Aleixandre, M. P. (Eds.). (2008). *Argumentation in science education: Perspectives from classroom-based research*. Springer.

European Commission. (2018). *Answer to a parliamentary question*. http://www.europarl.europa.eu/doceo/document/E-8-2018-004948-ASW_EN.html. Accessed 20 Feb 2020.

European Parliament. (2001). *Directive 2001/83/EC of the European Parliament and of the Council of 6 November 2001 on the Community code relating to medicinal products for human use.* https://eur-lex.europa.eu/legal-content/EN/TXT/HTML/?uri=CELEX:32001L0083andfrom=ES. Accessed 20 Feb 2020.

Feinstein, N. (2011). Salvaging science literacy. *Science Education, 95*, 168–185.

Feinstein, N. W., Allen, S., & Jenkins, E. (2013). Outside the pipeline: Reimagining science education for nonscientists. *Science, 340*(6130), 314–317.

Feinstein, N. W., Bromme, R., Barzilai, S., & Baram-Tsabari, A. (2017). What counts as success in public engagement with science? (Blog Post). *Public Understanding of Science*. https://sagepus.blogspot.com/2017/08/what-counts-as-success-in-public.html. Accessed 20 Feb 2020.

Felton, M., García-Milá, M., & Gilabert, S. (2009). Deliberation versus dispute: The impact of argumentative discourse goals on learning and reasoning in the science classroom. *Informal Logic, 29*, 417–446.

Fernández-Muerza, A. (2004). *Estudio del periodismo de información científica en la prensa de referencia: el caso español a partir de un análisis comparativo* [Study of scientific information journalism in the reference press: The Spanish case from a comparative analysis]. Doctoral dissertation, Universidad del País Vasco.

Instituto Nacional de Estadística. (2013). *Encuesta Nacional de Salud [National Health Survey] 2011–2012*. http://www.ine.es. Accessed 20 Feb 2020.

Jiménez-Aleixandre, M. P., & Puig, B. (2012). Argumentation, evidence evaluation and critical thinking. In B. J. Frasser, K. G. Tobin, & C. J. McRobbie (Eds.), *Second international handbook for science education* (pp. 1001–1016). Springer.

Johnson, R. H. (2009). The implications of the dialectical tier for critical thinking. In J. Sobocan & L. Groarke (Eds.), *Critical thinking education and assesment: Can higher order thinking be tested?* (pp. 55–74). The Althouse Press.

Kolstø, S. D. (2001). Scientific literacy for citizenship: Tools for dealing with the science dimension of controversial socioscientific issues. *Science Education, 85*(3), 291–310.

Kolstø, S. D., et al. (2006). Science students' critical examination of scientific information related to socioscientific issues. *Science Education, 90*, 632–655.

Kortland, K. (1996). An STS case study about students' decision making on the waste issue. *Science Education, 80*, 673–689.

Kuhn, D. (1993). Science as argument: Implications for teaching and learning scientific thinking. *Science Education, 77*, 391–337.

Kuhn, D. (2010). Teaching and learning science as argument. *Science Education, 94*, 810–824.

Lack, C. W., & Rousseau, J. (2016). *Critical thinking, science, amd pseudoscience*. Springer.

Lake, D. (2005). About being pure and natural: Understandings of pre-service primary teachers. *International Journal of Science Education, 27*(4), 487–506.

Martí-Sánchez, M., & Roger-Monzó, V. (2018). La percepción social de la homeopatía en la prensa digital española: Un análisis semántico [The social perception of homeopathy in the Spanish digital press: A semantic analysis]. *Panace@, XIX*(47), 115–123.

Metin, D., Cakiroglu, J., & Leblebicioglu, G. (2020). Perceptions of eighth graders concerning the aim, effectiveness, and scientific basis of pseudoscience: The case of crystal healing. *Research in Science Education, 50*, 175–202.

Ministerio de Sanidad, Política Social e Igualdad. (2011). *Análisis de situación de las terapias naturales* [Situation analysis of natural therapies]. https://www.mscbs.gob.es/novedades/docs/analisisSituacionTNatu.pdf. Accessed 20 Feb 2020.

Norris, S. P. (1995). Learning to live with scientific expertise: Toward a theory of intellectual communalism for guiding science teaching. *Science Education, 79*(2), 201–217.

OECD Organisation for Economic Cooperation and Development. (2016). *PISA 2015 assessment and analytical framework: Science, reading, mathematic and financial literacy, PISA*. OECD Publishing. https://doi.org/10.1787/9789264255425-en. Accessed 20 Feb 2020.

Oliveras, B., Márquez, C., & Sanmartí, N. (2013). The use of newspaper articles as a tool to develop critical thinking in science classes. *International Journal of Science Education, 35*, 885–905.

Osborne, J. (2014). Teaching scientific practices: Meeting the challenge of change. *Journal of Science Teacher Education, 25*, 177–196.

Patronis, T., Potari, D., & Spiliotopoulou, V. (1999). Students´ argumentation in decision-making on a socio-scientific issue: Implications for teaching. *International Journal of Science Education, 21*(7), 745–754.

Paul, R., & Elder, L. (2006). *Critical thinking* (2nd ed.). Pearson Prentice Hall.

Preece, P. F. W., & Baxter, J. H. (2000). Scepticism and gullibility: The superstitious and pseudoscientific beliefs of secondary school students. *International Journal of Science Education, 22*(11), 1147–1156.

Quintanilla, M. A., Escobar, M., & Santos-Requejo, L. (2019). La actitud global hacia la ciencia en las comunidades autónomas [The global attitude towards science in the Autonomous Communities]. In FECYT (Ed.), *Percepción Social de la Ciencia y la Tecnología en España 2018* [Social Perception of Science and Technology in Spain 2018] (pp. 86–105). FECYT.

Sadler, T., Barab, S., & Scott, B. (2007). What do students gain by engaging in Socioscientific Inquiry? *Research in Science Education, 37*, 371–391.

Sadler, T. D., & Donnelly, L. A. (2006). Socioscientific argumentation: The effects of content knowledge and morality. *International Journal of Science Education, 28*(12), 1463–1488.

Uskola, A. (2016). ¿Los productos homeopáticos pueden ser considerados medicamentos?: Creencias de maestras/os en formación [Homeopathic products can be considered medicine? Beliefs of elementary preservice teachers]. *Revista Eureka sobre Enseñanza y Divulgación de las Ciencias, 13*(3), 574–587.

Uskola, A. (2017). Escepticismo del profesorado de Primaria en formación hacia las pseudociencias: Influencia de las concepciones erróneas en el caso de la homeopatía [Skepticism of Elementary preservice teachers about pseudosciences: Influence of erroneous conceptions in the case of homeopathy]. *Profesorado Revista de currículum y formación del profesorado, 21*, 391–408.

Yates, G. C. R., & Chandler, M. (2000). Where have all the skeptics gone?: Patterns of New Age beliefs and anti-scientific attitudes in preservice primary teachers. *Research in Science Education, 30*(4), 377–387.

Windschitl, M. (2003). Inquiry projects in science teacher education: What can investigative experiences reveal about teacher thinking and eventual classroom practice? *Science Education, 87*, 112–143.

Part III
Research About Critical Thinking in Environmental and Sustainability Education

Chapter 10
Teaching Science in Chilean Environmentally Degraded Areas: An Analysis from a Critical and Ecofeminist Perspective

Corina González-Weil, Valeria León, Delia Cisternas, Gabriel Caro, and Roberto Morales

10.1 Introduction: Socio-Environmental Conflicts and Science Education

The current *planetary emergency* that we are in (IPCC, 2018) has its origin in the socioeconomic system, specifically in the accumulation of capital and the search for short term particular benefits, carried out under the premise of a continued growth that is non-sustainable (Foster & Clark, 2012). In addition, demographic explosion, migration towards urban areas and *hyperconsumerism* of the more developed societies, continues to grow as if the capacities of the Earth were infinite (Vilches & Gil Pérez, 2007). This has generated unsustainable imbalances, which translate into pollution, destruction of resources, loss of biodiversity and cultural diversity, hyperurbanization and desertification. This, in addition to making our planet more inhabitable, increases inequality, extreme poverty, conflict and violence (Bencze & Carter, 2011).

For South America, the neoliberal system has perpetuated the role of the *producer of raw materials,* such as mineral deposits, gas, oil, agricultural, forestry and fishing resources, initiated in our continent with the arrival of Christopher Columbus (Lander, 2014). South America has mainly become a supplier of primary resources to attend to the demands of the capital, through the logic of neo-extractivism (Machado, 2012). This consolidates a development model based on the over-exploitation of natural resources, most of them non-renewable, as well as on the expansion of production towards new areas, generating a breakdown of the regional economies, and the displacement of rural, farmer and Indigenous communities (Merchand, 2016).

C. González-Weil (✉) · V. León · D. Cisternas · G. Caro · R. Morales
Centro de Investigación en Didáctica de las Ciencias y Educación STEM, Pontificia Universidad Católica de Valparaíso, Valparaíso, Chile
e-mail: corina.gonzalez@pucv.cl

B. Puig, M. P. Jiménez-Aleixandre (eds.), *Critical Thinking in Biology and Environmental Education*, Contributions from Biology Education Research,
https://doi.org/10.1007/978-3-030-92006-7_10

With the excuse of economic growth, the accelerated process of appropriation of natural resources has caused environmental deterioration in South America, which translates in forest loss, deterioration of natural habitats, endangered species and exaggerated use of fertilizers (Acosta & Machado, 2012). These environmental and ecological degradations not only affect nature but the whole social structure, modifying the lifestyle and everyday practices of the inhabitants of degraded areas, the communities in rural areas being the most affected. This has resulted in an explosion of socio-environmental conflicts related to access and control of natural resources that involve Indigenous and farmer movements, as well as mobilization and citizen participation, which is strongly repressed and criminalized (Svampa, 2012).

Many of these movements have women as protagonists. From this perspective, the practices of resistance towards neo-extractivism agree with certain principles that – not being exclusive of this view – can be grouped under the term *ecofeminist* (Svampa, 2015). Among them we can highlight acknowledgement of interdependence and ecodependence (Escribano, 2017), overcoming of anthropocentric visions by placing the sustainability of life at the center (Pérez Orozco, 2014), creation of an ecological equality culture and the value of relationships and collaborative work for survival (Bernardos et al., 2020). In this way, feminism has earned prominence during recent decades. In Chile, the movement called *Women of Sacrifice Zones in Resistance of Quintero – Puchuncaví*, born in 2016, is a clear example of resistance, by which women that inhabit highly degraded areas –about which the government claimed that they need to "sacrifice"– seek to protect the life, health and environment of their communities through organizational, educational and judicializing practices (Bolados & Sánchez, 2017).

Within the context of environmentally degraded territories, science education has a particularly relevant role in the education of youth, especially when addressed from a placed-based perspective. It would contribute to develop territorial awareness and critical thinking (CT), as well as competences for citizen participation. In addition, some ecofeminist principles can be a contribution to science education. Feminist scholars warned about the risk of teaching in isolation from real world, with knowledge that cannot be located (Haraway, 1991). Therefore, a perspective of science education that is territorial, critical and that is addressed on the basis of ecofeminist principles becomes relevant, especially in degraded socio-environmental contexts.

However, just as socio-environmental conflicts have divided society, so has the school-business dependence. Whether it be because of the schools' desire to obtain job placements for their most socially vulnerable students, or because businesses involved are important job markets for parents, a subordination scenario is created. In it, not all school members dare to take a critical stance, which hinders the action of teachers. In this context, a teacher seeking to educate from a socio-critical perspective will be forced to challenge current political values and educational objectives – oriented towards nature exploitation, economy and consumerism – shifting to objectives aligned with sustainable development (Edwards, 2016).

In this chapter, we analyze, from a critical and ecofeminist standpoint, through the analysis of interviews and written records, the experience of two teachers and of a group of science teachers taking part in an in-service teacher education program. In all three cases, teachers work in the Aconcagua Valley – central area of Chile – in environmental degraded territories.

10.2 Socio-Environmental Context and Theoretical Perspectives

10.2.1 Socio-Environmental Degradations in the Aconcagua Valley

According to the Chile Map of Socio-environmental Conflict (INDH, 2020), the Aconcagua Valley is affected by nine socio-environmental conflicts, many of them also mentioned in the Global Atlas of Environmental Justice (EJatlas, 2020). Located in the Region of Valparaiso (Chile), the valley is crossed by the Aconcagua River, which rises in the Andes Mountains and runs into the Pacific Ocean. At one end of the Valley, towards the mountains, we find the provinces of San Felipe and Los Andes. In this area, the great mining industry directly affects the watercourses, either due to contamination by spilling, or by glacier destruction (INDH, 2020). There is a great amount of water –2.000 liters per second– used in the mineral extraction process (OLCA, 2015).

On the same valley, towards the coast, are situated the cities of Quintero and Puchuncaví. This territory has been systematically impacted since the 1960s due to the setting of an industrial park that gathers various companies related to the smelting and refining of copper, five thermoelectric centrals, a port and gas maritime terminal, fuel, asphalt, cement, among other chemicals (Liberona & Ramírez, 2019). This resulted in strong atmospheric pollution, which has caused severe episodes of mass intoxication, leading to the temporary closing of La Greda School in Puchuncaví (INDH, 2020). It also caused water pollution, both of runoff and groundwater, affecting consumption of well water in rural areas, and seawater pollution, due to spilling and beaching of oil and charcoal (Saravia et al., 2016). Due to the loss of marine biodiversity caused by pollution, small-scale fishermen in the Puchuncaví-Quintero bay could not continue their jobs. All these reasons led to name this territory as a *Sacrifice Zone*.

In addition, the General Direction of Waters, has decreed all continental Provinces of the region as zones of water scarcity. Rainfall has decreased throughout the Aconcagua Valley (OLCA, 2015). Therefore, the water problem, its scarcity and pollution, represents an environmental degradation across this territory. MODATIMA (2014) frames this issue within the denominated "Water conflict" originated in the neoliberal economic model, through the Constitution and the Water Code of Chile (Ministerio de Justicia, 1981), implemented under Pinochet's dictatorship, that grants the administration and use of this common resource to private companies. Since then, mining, industrial and agro-industrial activities have aggravated the situation, generating socioenvironmental degradations in and around the territories where they take place.

From a social dimension, this has had direct impact on small-scale activities related to farming, agriculture and fishing for family subsistence. Because of the deterioration of both nature and quality of life, socio-environmental movements and ecological organizations have emerged, with the purpose of defending territory and waters. These groups have taken on the task of disputing power spaces through the generation of collaboration networks within the valley.

10.2.2 Critical Thinking and Critical Scientific Literacy in the Context of Environmentally Degraded Areas

In a context of environmental degradation, it may be expected a politically engaged science education, emphasizing transdisciplinarity, promotion of values and an orientation towards action. These actions aims are caring for ourselves and our surroundings, in a way that makes it possible for us and the planet to survive (Sadler, 2011; Bencze, 2017). In this way, we propose that science education should develop a critical perspective in which an activist disposition is promoted, with the purpose of increasing social and environmental justice, a vision termed *critical scientific literacy* (Sjöström & Eilks, 2018).

Critical scientific literacy involves the promotion of critical thinking, understood as the competence of developing independent opinions and the ability to reflect about the world that surrounds us and our participation in it (Jiménez-Aleixandre & Puig, this book). This vision of CT combines components of argumentation – such as evaluating knowledge from evidence and having the disposition to look for reasons and question authority– with components of social emancipation and citizenship – such as being able to develop an independent opinion, and critically analyzing the discourses that justify the inequalities and power relationships – (Jiménez-Aleixandre & Puig, this book). This perspective of critical scientific literacy and promotion of CT requires science teachers to be critical and transformative. This means viewing science as a product of social, cultural and political contexts, and understanding the school as a space for the development of ideas and generation of contextualized school knowledge, as well as their own role as agents of transformation (Moura & Silva, 2018).

In the context of degraded territorial contexts, being a critical and transformative teacher involves a high commitment to the territory, for instance the identification of local socioenvironmental problems as scenarios for contextualized science education that transforms the community. In rural contexts, students can take advantage of their local knowledge to make sense of school knowledge, but also what students learn in school influences, improves and increases the available knowledge in the community (Roth, 2010). A curriculum that is alive, situated, addressing local socio-environmental problems, helps students to get involved with their own development and visualizes the utility of what they are learning for their territory. It also allows them to evidence the impact that science can have in our lives, and generate the need for ecojustice (Roth, 2010).

10.2.3 Ecofeminist Principles and Transformation from the School

From the dialogues between deep ecology and feminism, *Ecofeminism* allows us to reflect on domination, exploitation and appropriation of nature, interpreting human relationships and the subjection of women with this perspective. This is part of the foundation of the capitalist and patriarchal system (Federici, 2018) which invisibilizes the tasks associated to reproduction and care, such as children's care, supply for basic needs, health promotion, emotional support, or social participation. Therefore, we deem adequate to adopt the perspectives of critical Ecofeminism (Puleo, 2016) and Ecofeminism of the South (Svampa, 2015) to question the current development model, which affects environmentally degraded territories. It results appropriate then, to consider these ecofeminist principles, as a contribution to building processes of socio-environmental transformation from the school:

Acknowledgement of interdependence between people, and with nature (ecodependence) (Escribano, 2017): it opens the possibility to question social relationships and its forms of reproduction in school. It invites us to understand human interdependence and the generation of relationships for survival and mutual support.

Overcoming the anthropocentric vision and placing life sustainability at the center: The principles of feminist economy (Pérez Orozco, 2014) incorporate an integration of the production-reproduction tasks from the ethics of care, along to judging valuable all kinds of knowledge (Korol, 2016).

Creation of a new world vision anchored in the ecologic culture of equity, in the face of current social injustice scenarios that result in the violation of human rights and the destruction of nature (Escribano, 2017).

Highlight the importance of relationships for survival and wellbeing, to address objectives in a collaborative manner and overcome individualism (Bernardos et al., 2020).

10.3 Experiences of Chilean Science Teachers in Environmental Degraded Areas: From In-Service Teacher Education to Citizen Education

The three experiences that we analyze take place in the Aconcagua Valley, in the region of Valparaiso. We have been working for several years with schoolteachers, specifically from public schools that serve the more vulnerable areas of the population. Many of them live in environmental degraded areas, where many people do not have running water at home, and if they do, it is contaminated. Case 1 refers to teachers from Los Andes province, the mountain area of the valley, that were part of a continued education program that promotes inquiry oriented towards the use of territory as scenario for critical scientific literacy. Case 2 focuses on a teacher from the same area, suffering a mega-draught, aggravated by the settling of monocultures and mining activity. Case 3 refers to a teacher from Puchuncaví, a coastal area of the valley, severely impacted by industrial pollution. From a territorial, critical and ecofeminist perspective, we address these three experiences, and ask these questions:

1. How do teachers relate to their territory and how is this relationship identified in the accounts of their educational practices?
2. What sense do they give to their practices and how do their accounts substantiate the promotion of critical thinking aspects in their students?
3. Which principles of ecofeminism can be identified in teachers and their narratives of their educational practices?
4. What are the difficulties of teaching science in an environmentally degraded territory?

In order to answer them, we analyze individual and group written records and teachers' interviews. The names are pseudonyms.

10.3.1 Case 1: Understanding Together the Sense of Educating in Sciences Through the Relationship Between Science Education and Territory

Since 2015, the Chilean Ministry of Education promotes ICEC (Scientific Inquiry for Science Education), an In-Service Teacher Education Program for the public system. It seeks the professional development of science teachers through collaborative work promoting the generation of learning communities. In our region, the Pontificia Universidad Católica de Valparaíso is one of two in charge of the program. Throughout these 5 years, we have sought to integrate the territorial component and the use of socio-scientific issues, encouraging participants to identify problems of their territory and use them for their teaching. Within this context we focus on three sessions (4 h each), in which 29 teachers of the Aconcagua Valley were invited to reflect on problems of their community and think together about possible ways to address them in teaching. For the purposes of this analysis, we examine individual and group written records. We highlight the use of narratives about their professional practice, the collaborative reflection regarding teaching practices that generates learning and the use of socio-scientific issues and planning tools, for the design of teaching sequences with a territorial focus.

10.3.1.1 How Is Science Teaching Understood in a Professional Education Community?

Activity 1: In 7 groups of 4–5 participants, teachers discussed about the sense of science education, around the questions: (a) What is for me the purpose of science education? (b) How can knowing science help my students in their everyday lives? (c) How can I link science teaching with the context of my students? and (d) How can I link science with current social events[1]?

About the purpose of science education and the role of science knowledge in students' lives, teachers consider that they include: (a) Generating in students an understanding of themselves and their environment, helping them acknowledge that science is present in their everyday lives. (2) Training students to develop the abilities of discovery, research and reflection, promoting curiosity and bringing them closer to the world of science (3) Promoting CT and the ability to apply what has been learned in order to solve problems of everyday life and (4) Promoting environment care. Teacher 4 in Group 2 offers an example:

> [knowing science] *Helps them to be citizens that understand their environment, the processes and events that affect their everyday life, it helps them to make better decisions, to develop abilities they can apply in other fields, and also to understand how their actions have an impact on the planet.*

Regarding the link between science education and the students' environment, considering the social crisis in our country, teachers mention that: (1) Teaching must be based on students' knowledge, considering their interests and experiences and use their contexts as a setting, carrying out activities that relate what they learned with their life. (2) Teaching must be oriented towards students learning to take care of themselves (e.g., nutrition), and of the environment, promoting awareness in the good use of natural resources (as water). (3) Reflection must be promoted among students, as well as social awareness, CT and better knowledge of national issues, so they can become agents of change in their communities. When talking about the national situation, teachers refer to specific topics, such as the water crisis, the quality of the health system and the pollution of the environment.

Group 6 concludes: "[as teachers we must] *Achieve significant learning, considering the context of each student so we can educate agents of change that are responsible for their environment*".

In addition, Teacher 2 of Group 3 proposes: *"That students achieve their own critical thinking about issues and base it on information they manage in order to defend their ideals"*.

---

[1]On October 18, 2019, a strong social outbreak began in Chile, an expression of great citizen discontent triggered by a high level of inequity and abuse. This outbreak was characterized by massive demonstrations and riots, as well as great police repression that led to the loss of human lives and massive eye losses. This situation led to the suspension of on-site classes in Schools and a reformulation of the teaching role.

10.3.1.2 What Is the Focus of the Groups' Lesson Plans? Views for a Teaching Sequence on "Water"

During the second session, teachers analyzed, in groups, the problem "What do I make for dinner?" The activity allowed teachers to relate an everyday decision to aspects of science learning, such as pyramids of energy and biomass, the salmon culture in Chile and socio-environmental significant crises such as the red tide and its relation to tons of salmon thrown to sea in the Island of Chiloé. Then, they constructed collective definitions for critical scientific literacy and used an analysis tool in order to generate classroom proposals, considering their reflections. After working in groups they presented their results in a plenary session.

Activity 2: *Use of an analysis tool of the content to be taught for the creation of learning proposals with a focus on critical scientific literacy.*

In order to describe teacher proposals, we examine the answers of three groups that proposed similar topics related to water. The analysis tool of the content to be taught was a schema that includes 6 elements related to critical scientific literacy (CSL) and pedagogical content knowledge (PCK). It required teachers to discuss around: (1) Selected concept, (2) Key idea, (3) Relation with social, political and economic aspects, (4) Learning difficulties and student's interests, (5) Teaching strategies and resources and (6) Possible actions and strategies for evaluation. A summary of the results of the three groups is presented in Table 10.1:

10.3.1.3 Teachers Report Their Own Learning

At the end of the third session 26 (of 29) teachers answered a "ticket out" that include the following questions: *What did I learn today?* and *How does my learning modify my classroom practice?*. Figure 10.1 summarizes the analysis of the categories in their answers: Curriculum, Teacher, Context and Student, with the references to each. How teachers view: (a) themselves: they stress the importance of fostering a *critical scientific literacy focus* and they state the importance of analysis tools such as the one used for the analysis of content, to intentionally use activities promoting students' critical and scientific literacy; (b) the pedagogical context: they refer to a vision of science teaching centered in the relations of their students to the *political, cultural and social world*; in their narrative they refer to socio-scientific issues as vehicles that allow their students to develop scientific thinking to be able to respond in the aforementioned scenarios; (c) their students: considered as *decision-making* subjects; teachers should consider *students' interests* and *daily life situations* to support a student-centered science learning; (d) the curriculum: they account for two dimensions, the concept of *"key idea"*, however, when highlighting their own learning, they only mention the concepts or *"keywords"* learned during the session.

Table 10.1 Teacher analysis of the content "Water"

Selected concept	Teachers propose that the concept to be taught be simply *water*, or the problem associated to it (*draught*)
Key idea	Proposed in three ways as a: concept: *"Care of the Environment"*; brief scientific idea: *"Water is a fundamental element for life"*; socio-environmental problem: *"We are running out of water, therefore, it is necessary to take care of its use and do so in a responsible manner, recycle and preserve the available water sources"*.
Relation with social, political and economic aspects	Learning about water should involve an understanding of the context: "*It is necessary to become aware of inequalities regarding access to water*", as well as to understand that "*there are no policies that regulate the use of water in our territories*". Teachers point out that it is important to consider real situations, such as "*the use of large amounts of water for mining, to the detriment of other essential uses*" or "*Animal feed shortage*". It is also important to develop attitudes, such as awareness and collaborative work that lead students to take a position on environmental problems and promote water care actions for the local community.
Learning difficulties and student's interests	References to three dimensions: (a) Students' previous ideas that hinder learning: *"Water is an inexhaustible resource"* or *"students do not place freshwater or seawater within the water cycle"*. (b) Attitudes: lack of awareness or interest from the students and their families (c) Pedagogical context or local territory: little availability of safe water in their territories, social reality and levels of water care.
Teaching strategies and resources	Mentions of experience – based learning, the use of concrete material (models), and inquiry projects, like the design and application of a survey to know about water use habits in the community or the comparison of images of the river at different years
Possible actions and strategies for evaluation	Proposal of a call to action to students, through communicating the results of their inquiring: *"Presenting conclusions at the School" "Radio program and use of social networks"*. Propose actions oriented towards citizen participation: *"Creation of a water care brigade"*, *"Publicly denounce products that use more water for their production"*, *"Design posters about the good use of water at home"*.

Teachers also mention in their responses that they would require more courses or tools to relate the curriculum with the formative goal of critical scientific literacy. We interpret this as declarative statements that would reach greater development when they elaborate proposals to address real territorial issues.

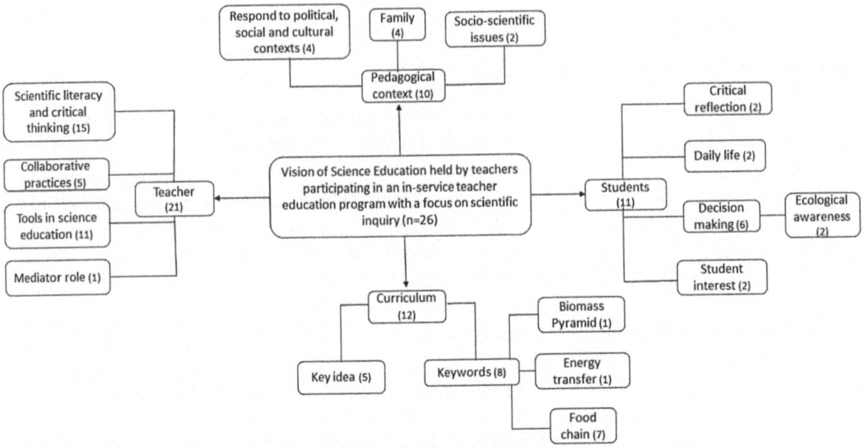

Fig. 10.1 Vision of Science Education held by teachers participating in an in-service teacher education program with a focus on scientific inquiry. (n) indicates the number of participants (N 26). The categories of the teachers' responses in their "ticket out" are displayed, indicating the presence of the subcategory in their writings

10.3.2 Case 2: To Include Students, the Experience of a Teacher in the City of Los Andes

> *The truth is that I could not find a more beautiful profession, where what you give comes back to you, as it happens in teaching [. . .] To do what I do, makes me very happy [. . .]. I feel that I don't work alone. If I have questions I can ask around, now I have a network of colleagues to go to. . .* (Paula, teacher from Los Andes).

Paula is a high school biology teacher currently working in a school in the city of Los Andes; their students are between 13 and 18 years old. She lives with her 8-year old daughter and her 81-year old mother. She describes herself as *busquilla,*[2] since she finds different job opportunities such as selling clothes, making jewelry or nail art, whatever brings extra income to the family. Among her academic achievements, she has participated twice in the ICEC Program, in addition to an in-depth course on inquiry. Paula has carried out scientific inquiry projects with her students, achieving recognition at city and national levels. She is currently leading and coordinating the ICEC teacher science network in her city. She is a mother, daughter, sister, aunt, friend, teacher, homemaker, guide and partner. Throughout time she took on the role of supporting and guiding the teacher science community, showing a disposition to collaborative work focused on service and peer accompanying. She keeps a trusting relationship with her students, centered on dialogue and their interests. She describes herself as a caring and understanding teacher when it comes to the different interests of the kids, especially teenagers.

[2]In Chile, a person that is ingenious and persistent regarding the search for sources of income.

10.3.2.1 Highly Degraded Territory and Citizens' Response

Paula's description of her territory is based on comparison: she explicitly mentions marked differences between the space she now inhabits and the one of her childhood. The accounts of her grandmother, mother and her childhood memories describe an environmental deterioration of her territory *"[...] the river does not catch your attention as it did when I was a child, when you travel to San Esteban or San Felipe and go through the bridges, you can see only drops of water coming [...] before, the water flow was such that you could hear it from blocks away [...] there is a completely different vegetation, it doesn't rain like before, [...] the water has a strange smell, like that of a disinfectant [...]"*. She refers to urban growth, the decrease of the water flow of the Aconcagua river, which crosses her city, and pollution, "*[...] the problem we have with the mining company,* [is that] *they have their "good neighbor" slogan but in fact they are not so good neighbors because they produce waste and always discharge their tailing to the river"*. The exploitation of natural resources, such as water, directly affected by pollution and extraction, appears as the greatest impact in the Aconcagua river basin, where the draught takes its toll on the most defenseless ones, such as the small-scale farmers that cannot compete with big producers of the area, as avocado exportation. Water became an economic good when Chile turned into the only country in the world to privatize this public resource, under the concept of *Right to Water*. Paula tells the story of a relative that can pay for their right to water only once a week, for a couple of hours. "[...] *Suddenly we arrived at a canal and my cousin says: This is where these guys keep the water! The canal was almost overflowing, and my cousin receives a very small flow of water once a week. She and her husband work, and if the water arrives and they are not there, they lose it."* In this context, the teacher has promoted, along with her friends and neighbors, different actions at home and neighborhood levels, such as the installation of filters to purify greywater. She also mentions how the members of the group have organized irrigation "[...] *we give each other information among neighbors [...] it is hard for me because I don't have a filter, but I still go and water the crops using cans, it is good exercise"*. The community is organized around different activities, such as recycling and the maintenance of plastic collection containers, something she values and encourages.

Among the initiatives around the water problem, Paula mentions collaborative work carried out with teachers from schools in San Felipe and Los Andes, part of the ICEC Program. Together they designed projects specifically related to the Aconcagua river. "[...] *in fact, all teachers from this area had water-related research.* [My colleague] *wanted to investigate heavy metals found in the water in an area of the river and I was researching "how irrigation with river water affects plants". The river was important to us"*. As a group, they started to review their research documentation, which allowed them to gain deeper knowledge into the water issue. This research led them to become a part of a city initiative in which they expressed their rejection to a mining project, which through the expansion of excavation works, endangered 18 rock glaciers and five white glaciers, which, because of their closeness to the mining company sites, became a threat to them. This caused a generalized rejection from these local groups, and the mining company decided not to carry out the project. "*[...] we made a lot of noise with that; we moved a lot.* [...] *they closed the group [...] in fact the woman that led the group had to disappear for a while [...] we even held secret meetings*". Paula's participation, along with other teachers in her High School ended when the school principal forbade them from being a part of this investigation "[...] *at that time a colleague and I were very much committed to continue with the investigations, but our school principal at the time told me: be careful, do not mess with them* [the mining company]". Considering this, Paula put her research project aside.

10.3.2.2 Classroom Practices in the Local Context and Development of Critical Thinking

In her teaching, Paula considers the narrative itself to be a process that creates meaning. She tells stories to her students, which somehow allows them to connect with the territory they inhabit, as a way to develop CT. "[...] *I try to integrate critical thinking as much as possible* [...] *I like to tell stories in my classes, Did you know that?... when I was a kid...my friend told me, etc....what I do is transport them to a comparison of what I am telling them and reality.* [...] *when the student feels close to something, not just something that is being told to them but that they live daily, they start to get involved in their learning*". Paula states that she encourages scientific inquiry as a methodological approach in her teaching. "[...] *I love when they ask questions! It is difficult for them at the beginning*". To get students involved with territorial environmental problems is a way to direct their interest towards inquiry, "[...] *I feel that when you teach you have to not only deliver content but also the tools so they can continue to develop* [...] *we have to be very realistic as teachers and bring the content to their reality and current situation* [...] *sometimes there is something that catches their attention and they discuss it in wonderful debates.* [...] *I don't mind if I feel I am straying away from the content, I let them continue since I feel there is much gain in this*". Inquiry can be developed from different approaches, however if it is linked to the territory learning would be more significant. Paula carries out research with her students, such as to establish relationships between the contamination of the Aconcagua river water that is used for irrigation and the growth of vegetables. The research lines developed in the classroom invite students to participate in different aspects of scientific research and the understanding of how scientific knowledge is generated, which is a crucial aspect for the development of critical scientific literacy. "[...] *Students in general are used to receive everything already done, and to be told what to do or what not to do. I like to bring about this in class and have them think and research. It is important that kids experience it*".

10.3.3 Case 3: Critical Scientific Literacy Outside School

We have an industrial ring that started with a few thermoelectrical units and this has grown exponentially. Suddenly we have an industrial system with many companies and most of them polute. This has impacted the whole ecosystem: air, earth and sea.

Collaborative work is fundamental in order to generate something that will impact the communities, something that is relevant. For this, each actor has to get involved in the action (Marisol, teacher from Puchuncaví).

Marisol is an elementary and special education teacher in the School Integration Program (PIE, for its name in Spanish) in a school located in the Las Ventanas community, in the city of Puchuncaví. She is also a singer in a traditional folk group from La Greda, enjoys Latin American music and playing the guitar. There, she says, she has connected with students: "*Being a teacher and a woman is a great challenge, especially in areas of high vulnerability. To connect with them is very important, to their lives. What this job entails is much deeper than providing knowledge and content, and implies values that go beyond this*". The vulnerability of this territory, she tells us, is marked by environmental deterioration, which has directly affected the health of boys and girls, resulting in a high rate of students with learning difficulties associated to their neurodevelopment, along with respiratory problems, asthma and cancer. After these consequences were made evident, in addition to mass intoxication episodes which affected the La Greda school, among others, the population has achieved greater ecological awareness. This is directly related to the role of some social movements, comprised by teachers and other inhabitants, who have promoted CT through raising awareness about the socio-environmental conflict and the deterioration caused by industries located in the area. At the same time, she says that youth has had a great participation, joining this struggle in a significant way.

10.3.3.1 Building Collaborative Work in the Community

As a consequence of socio-environmental damage caused by the development of the industrial ring, in which most companies are highly contaminating, Marisol has made relevant social and political decisions, joining a group of residents of the Quintero – Puchuncaví bay, formed 15 years ago, called *"Communities for the Right to Life"*. According to her account, this group responds to a form of collective self-organization in the face of the contamination that during the 1990s *"had no filter"*. They began to work for the environmental literacy of the population, raising awareness about these problems: *"we would go to the communities, the small towns, neighborhood councils and we showed them videos so that people could be aware of what was happening to them"*. This task continued and encouraged the generation of socio-environmental movements; among them the "*Women of Sacrifice Zones in Resistance*" network has emerged strongly. This group of women has had a center stage role in the visibilization of the conflict and the violation of rights in this territory. Among their characteristics we can highlight the generation of support networks through different organizations, Human Rights, International Courts, the support of scholars and experts in health issues, ecosystems, biology and chemistry, NGO's that, together with the population's environmental education and literacy, have led them to present the socio-environmental degradation happening in the territory to different media and at different scales.

10.3.3.2 Critical Scientific Literacy Outside of School in Highly Deteriorated Territories

For this Puchuncaví teacher, the integration of the pedagogical practice in the fight against pollution has been a difficult path to travel. First, because her school has a direct relationship with the polluting companies, through a system that facilitates the future entrance of students as workers of these industries: *"here we have our differences, it is difficult, so we have to leave it here, we cannot go further. But we can move forward through the different organizations outside the school"*. This is the first obstacle hindering implementation of critical scientific literacy initiatives within the School. Especially in territories that have been highly degraded and intervened by the industrial businesses, where their dynamics of introducing themselves within the communities, and compensating them, are characterized by material and financial incentives: *"That's why is so hard* [...]*... here it is hard to fight against the powerful"*. From this perspective, education becomes a form of resistance of a political nature, and at the same time, action-oriented educational practices are faced with limitations. In spite of this, she states that outside school, students participate of rallies and protests around the city that end at the doors of the industries. These spaces enable to articulate and highlight the importance of different knowledge among the citizens, where they complement the artistic dimension through music, dance, theater interventions, with discussion groups – assemblies with fishermen and meetings with scientists, dialogues with communication media, among others.

10.3.3.3 Education of Critical Citizenship in Interdisciplinary Spaces

About critical citizenship, Marisol suggests that Citizenship Education can be one of the fields through which topics related to socio-environmental contamination and degradation could be addressed from a CT perspective in school, considering *"the importance of the role they can play in decision making when they elect their authorities and how these elections can affect the interests of the community"*.

She emphasizes the importance of integration of knowledge in collaborative and interdisciplinary spaces, through the participation of boys and girls in talks, *"Presentations by experts, doctors, environmentalists, human rights representatives"*, artistic events, rallies, demonstrations, workshops on environmental issues. These allow students to *"express their feelings on the environmental topic"*. On the other hand, she considers essential to strengthen student organization, and empower them on this topic. Marisol states that collaborative work can generate changes that impact communities *"and for this every actor must get involved in the action"*. In this sense, it is possible to observe the socio-environmental problem through a space where each collaborator can contribute with their own abilities, generating together a work of value.

10.4 Discussion and Implications

Teaching science in a territory marked by socio-environmental degradation faces teachers with certain dilemmas, both inside the school, and in the development of their role as a political subject of the territory the reside in. Often, the promotion of actions motivated by critical scientific literacy is limited in schools that receive financial incentives from polluting companies, or in students that have families whose only income comes from them. This hinders the development of CT, which also involves students' formation of independent opinions (Jiménez-Aleixandre & Puig, this book), given that censorship, sometimes explicit, exists in educational spaces. Overcoming these obstacles requires, firstly, to address socio-environmental problems from a socio-scientific perspective, that will allow its understanding as a contextualized problem, enhancing a territorial approach that is anchored in the local reality of each school. Secondly, the incorporation of ecofeminist perspectives to educational practices could support the promotion of attitudes, such as collaboration, equity, wellbeing and a harmonic relationship with nature, opening new paths for transformation (Rodríguez & Herrero, 2017).

10.4.1 Educational Practices Contextualized to the Territory

The teachers that collaborated with this study possess a high sense of territorial awareness, associated to a critical vision about the socio-environmental conflicts in their territories. This leads them to promote, in different ways, critical thinking and science teaching from a situated knowledge (Camacho-González, 2020). For the teachers participating in the in-service science teacher education program, we identify the need to know their students and the territory they inhabit in order to design contextualized learning opportunities, so they could apply what they have learned to everyday life. In the case of Paula, territory becomes a setting for the development of scientific practices, which goes hand in hand with educating citizens that are able to understand the world and make everyday decisions.

We can see that in the three cases strong ideas emerge related to critical scientific literacy and teaching for life within the local context. At the same time, participation in professional learning communities supports teacher practice on the use of new knowledge and the promotion of change in collaboration with others (Taotao Long et al., 2019). In this sense, the Inquiry program discussed in the first case creates dynamic spaces that invite to (1) question one's own vision about science education (2) build a collective vision of science education (3) make science education territory-related, based on collaboration. These spaces seem to have great relevance for promoting socially active professional trajectories. The exchange of experiences and the construction of common visions allows for the development of didactic proposals that expand the initial visions of teachers.

10.4.2 Promotion of Critical Thinking Inside and Outside of the Classroom

The development of CT can be identified in different levels in their narratives. In the case of the teachers in the program, the importance of promoting CT in students is declared in a generic way. This is associated to reflection about the problems of the territory, both locally and nationally, in order to become agents of change. This agrees with the aspects pertaining to critical thinking proposed by Jiménez-Aleixandre & Puig (this book), in relation to developing the ability to reflect about the world surrounding us and our participation in it. In the accounts offered by Paula, development of CT is rather associated to the development of scientific abilities and assessment of knowledge from evidence, which the teacher promotes through inquiry projects that use the territory as the scenario and propose to evaluate the impact of environmental issues *in situ*.

For Marisol the purpose of her teaching practice is directly related to her conception as a political subject within the area she resides in. She takes out of the classroom proposals that promote the development of their own opinions through CT, among other performances (Jiménez-Aleixandre & Puig, this book). This includes networking with other community actors, scientists, experts and social leaders that strengthens students' own opinion, through access to knowledge including different points of view. This would support the development of a sense of children and youth as agents of transformation of their local reality (Moura & Silva, 2018), encouraging some eco-social values related to the consolidation of an organized critical citizenship (Rodríguez & Herrero, 2017).

According to Jiménez-Aleixandre & Puig (this book), the selection of the different socio-scientific topics that teachers consider appropriate to include in their teaching, would support the development of scientific argumentation. In the case of Paula from Los Andes, she selects, on one hand, the issues that emerge from the extraction and contamination of natural resources such as water, and on the other, she uses a methodology with a focus on scientific inquiry, which allows to obtain evidence for the questions proposed by students. According to the authors, one of the fundamental aspects of argumentation and critical thinking develops through the appropriation of scientific practices. The teacher fosters debates in her classes where students can choose between one position or another.

10.4.3 Principles of Ecofeminism That Facilitate the Development of Critical Thinking

As a concluding thought, we propose that it is relevant to take into account those principles of ecofeminism that would help in the development of a critical scientific literacy based on the analyzed cases. Our analysis points to the importance of collaborative spaces, understood as the foundation of the weave of life itself, both inside and outside the classroom. From an ecofeminist perspective, the recognition of the relationships for survival, makes it possible to replace an anthropocentric vision for another one that places life sustainability at the center (Pérez Orozco, 2014), valuing, above all, local knowledge, support networks and interdependence within the territory. In turn, the incorporation of multiple knowledges that facilitate the development of critical thinking (Jiménez-Aleixandre and Puig, this book), would allow to problematize the ways of life imposed by the development model (Korol, 2016), especially in territories affected by neo-extractivism (Machado, 2012; Merchand, 2016). This opens new reflection spaces around the ecological crisis, encouraging thinking about environmental justice. Therefore, as these principles are incorporated to the educational practices, it is possible to build new ways towards an ecological culture of equality (Escribano, 2017), for the creation of a transforming educational practice.

Acknowledgments We thank all the teachers who shared their experiences with us, and who, despite the circumstances, do their best to improve the education of children and young people who inhabit environmentally degraded territories. Moreover, they see in them the hope of a generation that establishes fairer relationships with society and the environment.

References

Acosta, A., & Machado, D. (2012). Movimientos comprometidos con la vida. Ambientalismos y conflictos actuales en América Latina. *OSAL, 13*(32), 67–94.

Bencze, L., & Carter, L. (2011). Globalizing Students Acting for the Common Good. *Journal of Research in Science Teaching, 48* (6), 648–669.

Bencze, L. (2017). STEPWISE: A framework prioritizing altruistic actions to address Socioscientific issues. In L. Bencze (Ed.), *Science and technology education promoting wellbeing for individuals, societies and environments* (pp. 19–45). Springer.

Bernardos, A., Martínez, I., & Solbes, I. (2020). Vínculos y comunidad: Introducción de la ética del cuidado en educación a través de estructuras participativas. *Diversidade e Educação, 8*, 195–212.

Bolados García, P., & Sánchez Cuevas, A. (2017). Una ecología política feminista en construcción: El caso de las "Mujeres de zonas de sacrificio en resistencia", Región de Valparaíso, Chile. *Psicoperspectivas, 16*(2), 33–42.

Camacho-González, J. P. (2020). Educación Científica, reflexiones y propuestas desde los feminismos. *Revista Científica, 38*(2), 190–200.

Edwards, J. (2016). *Socially-critical environmental education in primary classrooms*. Springer.
EJatlas. (2020). *The Global Atlas of Environmental Justice*. www.ejatlas.org. Accessed 7 Mar 2020.
Escribano, R. (2017). Tejiendo lazos entre el ecofeminismo crítico y la educación para el desarrollo. *Géneros, 21*(2), 151–164.
Federici, S. (2018). *El Patriarcado del Salario. Críticas feministas al marxismo*. Traficantes de Sueños.
Foster, J., & Clark, B. (2012). The planetary emergency. *Monthly Review, 64*, 1–25.
Haraway, D. (1991). Conocimientos situados: la cuestión científica en el feminismo y el privilegio de la perspectiva parcial. In D. Haraway (Ed.), *Ciencia, cyborgs y mujeres: La reinvención de la naturaleza* (pp. 313–346). Ediciones Cátedra.
INDH. (2020). *Mapa de conflictos socioambientales en Chile*. https://mapaconflictos.indh.cl/#/. Accessed 10 Mar 2020.
IPCC. (2018). Global warming of 1.5 degrees. https://report.ipcc.ch/sr15/pdf/sr15_spm_final.pdf. Accessed 15 Mar 2020.
Korol, C. (2016). *Feminismos populares. Pedagogías y políticas*. El Colectivo.
Lander, E. (2014). El Neoextractivismo como modelo de desarrollo en América Latina y sus contradicciones. Heinrich Böll-Stiftung. https://mx.boell.org/sites/default/files/edgardolander.pdf. Accessed 20 Nov 2019.
Liberona, F., & Ramírez, H. (2019). Antecedentes y reflexiones sobre la zona de sacrificio de Quintero y Puchuncaví. *Cuadernos médico sociales, 59*(1), 21–31.
Machado, H. (2012). Los dolores de Nuestra América y la condición neocolonial. Extractivismo y biopolítica de la expropiación. *OSAL, 13*(32), 51–66.
Merchand, M. (2016). Neoextractivismo y conflictos ambientales en América Latina. *Espiral, 23*(66), 155–192.
Ministerio de Justicia de Chile. (1981). *DFL 1122 Fija texto Código de Aguas*. https://www.leychile.cl/Navegar?idNorma=5605. Accessed 8 Mar 2020.
MODATIMA (Movimiento de Defensa por el acceso al Agua, la Tierra y la Protección del Medioambiente). (2014). *Recuperación del agua y defensa de la vida*. http://modatima.cl/2014/10/19/recuperacion-del-agua-defensa-la-vida/. Accessed 15 Apr 2020.
Moura, B. A., & Silva, C. C. (2018). Critical and transformative teachers: A rationale for history and philosophy of science in teacher education. In M. E. B. Prestes & C. C. Silva (Eds.), *Teaching science with context, science: Philosophy, history and education* (pp. 3–13). Springer.
OLCA (Observatorio Latinoamericano de Conflictos Ambientales). (2015). *Expansión de Codelco Andina 244: Una amenaza a los Glaciares, el Agua y la Vida del Valle del Aconcagua*. https://www.ocmal.org/wp-content/uploads/2015/07/Expansion-de-Codelco-Andina-244.pdf. Accessed 13 Mar 2020.
Pérez Orozco, A. (2014). *Subversión feminista de la economía. Aportes para un debate sobre el conflicto capital-vida*. Traficantes de Sueños.
Puleo, A. (2016). *Ecofeminismo para otro mundo posible*. Ediciones Cátedra.
Rodríguez, V., & Herrero, Y. (2017). Claves para la construcción de un proyecto educativo transformador. In E. Assadourian (Ed.), *Educación ecosocial, cómo educar frente a la crisis ecológica* (pp. 331–347). Icaria Editorial.
Roth, W. (2010). Local matters, EcoJustice, and community: Opportunities of village life for teaching science. In D. J. Tippins et al. (Eds.), *Cultural studies and environmentalism* (pp. 51–82). Dordrecht.
Sadler, T. (2011). Situating socio-scientific issues in classrooms as a means of achieving goals of science education. In T. Sadler (Ed.), *Socio-scientific issues in the classroom* (pp. 1–10). Dordrecht.
Saravia Ramos, P., Armingol Jaime, K., & Garland Castro, B. (2016). El derrame de petróleo en Quintero, V región de Chile. Una mirada desde las organizaciones sociales. *Población & Sociedad, 23*(2), 179–206.

Sjöström, J., & Eilks, I. (2018). Reconsidering different visions of scientific literacy and science education based on the concept of Bildung. In Y. J. Dori, Z. Mevarech, & D. Baker (Eds.), *Cognition, metacognition, and culture in STEM education* (pp. 65–88). Springer.
Svampa, M. (2012). Consenso de los commodities, giro ecoterritorial y pensamiento crítico en América Latina. *OSAL, 13*(32), 15–38.
Svampa, M. (2015). Feminismos del sur y ecofeminismos. *Nueva Sociedad, 256*, 127–131.
Taotao, L., Guoqing, Z., Xuanyang, Y., Rongchi, Z., & Qianqian, C. (2019). Bridging the belief-action gap in a teachers' professional learning community on teaching of thinking. *Professional Development in Education*. https://doi.org/10.1080/19415257.2019.1647872
Vilches, A., & Gil Pérez, D. (2007). Emergencia planetaria: necesidad de un planteamiento global. *Educatio Siglo XXI, 25*, 19–49.

Chapter 11
Spatial and Temporal Dynamics in Climate Change Education Discourse: An Ecolinguistic Perspective

Asli Sezen-Barrie, Joseph A. Henderson, and Andrea L. Drewes

11.1 Re-imagining Climate Change Education for Next Generations

Modern, human-caused climate change is altering the physical world faster than prior, non-human caused climatic changes, and it has already shown disastrous impacts through floods, droughts, extinctions, and more intense storms (Mann, 2012). With a high degree of certainty, scientists state that human activity since the Industrial Revolution is the main reason behind *modern* climate change and its impacts (Dessler, 2011). Younger generations are increasingly aware of the global climate change problem as accumulated scientific evidence continues to show the human impact on the climate (Hamilton et al., 2019). Despite the scientific knowledge available, the climate change problem seems to be getting worse (Gardiner, 2006). The worry is that coming generations will suffer more from the impacts of modern climate change, and this has led to children leading calls to action. During her speech at the 24th Conference of the Parties to the United Nations Framework Convention on Climate Change, then 15 year-old climate activist Greta Thunberg said: "You say you love your children above all else, and yet you are stealing their future in front of their very eyes" (Thunberg, 2018).

A. Sezen-Barrie (✉)
School of Learning and Teaching & Research in STEM Education Center, University of Maine, Orono, ME, USA
e-mail: asli.sezenbarrie@maine.edu

J. A. Henderson
Department of Environment & Society, Paul Smith's College, Paul Smiths, NY, USA

A. L. Drewes
Department of Graduate Education, Leadership, and Counseling, Rider University, Lawrenceville, NJ, USA

B. Puig, M. P. Jiménez-Aleixandre (eds.), *Critical Thinking in Biology and Environmental Education*, Contributions from Biology Education Research,
https://doi.org/10.1007/978-3-030-92006-7_11

This call toward social action is important to us as both climate change education scholars and as parents of young children who are more likely to struggle with the impacts of climate change. The three of us had the opportunity to work on a climate change education project (National Science Foundation Grant #1239758) across schools in Maryland and Delaware in the mid-Atlantic region of the United States. In these settings, we have seen many educators who are committed to learn and teach climate change and who care about their students, the next generation. However, our work has shown how despite climate change targeted professional development, challenges remain and questions linger for educators enacting climate change teaching in the classroom (Drewes et al., 2020; Shea et al., 2016). But more importantly, our work has also shown that despite educators' intentions to teach evidence-based explanations on how human activity causes modern climatic changes, many teachers were also ready to accept the inaccurate, nonscientific debates distributed through the media (Sezen-Barrie et al., 2019). Therefore, we argue that to be informed citizens on the climate change problems that the next generations will face, it is not sufficient to read about climate change, but learners must also develop critical thinking skills to judge the scientific knowledge being read. This chapter focuses on how students develop critical thinking skills about climate change evidence, impacts, and solutions to make local and individual meaning of what is ultimately a global and unequally distributed collective action problem.

11.2 Critical Thinking and Scientific Learning

In his book, *How We Think,* John Dewey highlighted critical thinking as "active, persistent and careful consideration of any belief or supposed form of knowledge in the light of the grounds that support it, and the further conclusions to which it tends" (1910, p. 6; 1933, p. 9). Dewey's writings brought attention to the role of philosophy in developing a system to critique our belief systems and our suppositions about everyday phenomena. Dewey highlighted the scientists' habits as more than objectifying the eternal truth and formed the basis of focusing on critical thinking for educating citizens. Later, Glaser in 1941 mentioned that despite having a literate electorate in the United States, the public lacks the skill of evaluating "critically what they read" (pp. 4–5). In his definition of critical thinking, Glaser highlighted evidence based thinking as well as the use of logical inquiry and reasoning. Since the 1980s, critical thinking has been a popular topic in research (e.g., Glaser, 1985; National Council on Education Standards, & US Dept. of Education, 1992). Researchers saw critical thinking not only as a higher order skill to tackle academic tasks (e.g., Halpern, 1998), but also as a skill with political meaning that helps citizens critically question claims (e.g., Fenton & Smith, 2019; McLaren & Giarelli, 1996).

Critical thinking has a crucial place in scientific research. The ideas and theories in science become stronger as they are defended when they are critically questioned by peers during lab meetings, symposia, and other scientific meetings (Latour & Woolgar, 1986). Since our goal in science education is to show students the authentic work of scientists, critical thinking should be part of science learning activities (Osborne, 2014). Unfortunately though, science is taught in schools "more as a dogma – a set of unequivocal, uncontested, and unquestionable facts – more akin to the way people are indoctrinated into a faith than into a critical, questioning community" (Osborne, 2014, p. 54). In order to change this view of teaching science, recent research has a strong emphasis on the practice of scientific argumentation (e.g., Driver et al., 2000; Erduran & Jiménez-Aleixandre, 2008; McNeill & Krajcik, 2009). If we want instruction on scientific argumentation in science learning environments to provide an epistemic window into the world of scientists, teachers and students need to develop an understanding of the distinct epistemic practices of disciplinary cultures of science (Kelly et al., 2000; Knorr Cetina, 1999). These disciplinary practices can determine the nature of critical questions we raise to evaluate the arguments. For example, we can ask about repeated observations on how the length of a pendulum impacts its swing counts. On the other hand, to evaluate the strength of plate tectonics theory, we will need to ask what geologic activity (volcanic, topographic, seismic) we see common on the plate boundaries (Sawyer et al., 2005).

Drawn from Dewey (1910) and Glaser (1941), we see critical thinking as the ability to judge the beliefs, assumptions, and propositions while considering the role of evidence in the (re)construction of scientific knowledge. Therefore, we agree with the science education scholar's view that teaching based on evidence is at the center of scientific learning (Osborne, 2011). In order to evaluate the strength of evidence and ask critical questions, we should not only understand the culture of science, but the cultures of each subfield of science (Kelly et al., 2000; Knorr Cetina, 1999). In this chapter, we focus on how climate science heavily relies on the distinct practices of spatial and temporal reasoning to build stronger claims.

11.3 The Spatial and Temporal Nature of Climatic Change

Global climate changes are usually cyclical phenomena on our planet. There are times during the Earth's history where climates have warmed, and times where the planet's climates have cooled. Modern, anthropogenic climate change, however, is a break from this natural pattern. We know from the accumulated body of scientific research that rates of modern global warming are above normal background levels due to human activities on the planet, and we also know with near-certainty that the burning of fossil fuels via industrialization is the primary cause (IPCC, 2018). Much of the carbon dioxide and methane in the atmosphere is the result of past human activities, and much of the carbon released now into the atmosphere will remain there for decades to come (Stager, 2011). CO_2, for example, stays in the atmosphere approximately 200 years (Archer, 2005). This long lifetime of CO_2 makes it more challenging for us to reverse what we have inherited from the past for the future generations.

Further, not all carbon is consumed in an even fashion, as those humans in the so-called "developed" world tend to produce higher levels due to their wealth (Oxfam International, 2015). Even though the commonplace term "global climate change" is often used in scholarship on climate change education, the spatial and temporal reality is complicated by the uneven social and physical dimensions of the issue (Henderson & Drewes, 2020).

Moreover, it is often hard for the lay public to understand the impacts of climate change given their spatial and temporal distance from lived experience (McDonald et al., 2015). For example, sea level rise is a process that plays out over decades or even centuries. Temporal and spatial thinking is an important practice of geoscientists' work as they interpret their observations or data by considering the history of Earth. Understanding geological time – also referred to as "deep time" in the history of Earth – geoscientists pay attention to conditions like Earth with no life, Earth with no humans, a hothouse Earth, or a snowball Earth, for example. The historical interpretation of past time is critical in the nature of climate data. In order to claim that the climate is changing, scientists have to be able to talk about what it was in the past (Edwards, 2010).

While scientists – especially geoscientists – are often trained to think about the intergenerational and interspatial dimensions of climate change, most people, including science teachers, do not think this way (Devine-Wright, 2013; Sezen-Barrie, 2018). One result of this conceptual short-sightedness is that leaders with short term horizons have little incentive to plan for longer-term or more widespread impacts, thereby discounting the impact of climate change on future generations (Weisbach & Sunstein, 2008). Further, while climate scientists have achieved consensus on human-caused climate change (Cook et al., 2016), there remains the "scientific uncertainty about the precise magnitude and distribution of effects" (Gardiner, 2006, p. 401). In other words, while geoscience and geoscience education are quite good at explaining past and current events across space and time, exactly how these play out in the future remains uncertain. Dealing with climate change in the present means interpreting the past to inform possible futures, and this means moving from a descriptive "is" orientation to a normative "ought" orientation in terms of climate change education activities (Drewes et al., 2018). It also means expanding the educational sphere of concern outward toward distant others and places around the world in order to truly understand the uneven causes and effects of climate change across space (Henderson, 2015).

The role of spatial and temporal awareness while critically evaluating climate change data and claims is essential (Edwards, 2010). When we underemphasize these skills of geosciences in our education system, confusion about causes of climate change are highly possible in science classrooms. Plutzer and colleagues' large-scale study with 1500 public middle and high school teachers in the U.-S. showed that about 31% of the teachers are confused about the scientific consensus that the climatic changes we witness today are due to human activity such as industrialization and increased air travel. Many of these teachers think the other side of the controversy is equally credible which inaccurately explains that the current climatic changes are due to natural causes such as solar cycles (Plutzer-et al., 2016). Simply presenting two sides of the argument in the case of climate change eliminates opportunities for critical thinking where students can judge strong vs. weak claims and pieces of supporting evidence. On the other hand, if teachers want to create opportunities for critical thinking, they will need to scaffold students to look at climate data in temporal and spatial scales to gain an appreciation of the nuances beyond a two-sided debate.

For this chapter, we present one set of data from our project that looks at how students make meaning of climate change phenomena through their social and cultural experiences. Since temporal and spatial reasoning is crucial in understanding climate change and the science behind it, we explored how the temporal and spatial aspects of climate change appear in middle school students' explanations of climate change issues. In the following sections, we will explain our methods and findings on the following question:

> How do students locate themselves and others across space and time relative to climate change phenomena in sociocultural learning contexts?

11.4 Modes of Inquiry

This research is designed as a qualitative study that utilizes the methods of discourse analysis with an ecolinguistics perspective to evaluate the stories we live from an ecological lens (Stibbe, 2015). We accomplish our analysis by leveraging theories from sociolinguistics (Blommaert, 2010) and spatial literacy (Leander & Sheehy, 2004; Scollon, 2013) to examine how students make sense of the climate change phenomenon. We are interested in how particular phenomena (i.e., causal mechanisms, effects, solutions) manifest in sociocultural contexts of middle school students. We used Blommaert's (2010) concept of indexical ordering and Scollon's (2013) concept of conceptual mobility to examine regimes of linguistic interaction in climate change education. In sociolinguistic theory, indexicality refers to how signs point toward objects in sociocultural contexts. For example, a student saying that humans caused global climate change signals their membership in a broader social community, and at the global scale over time. The student points to a broader collective ("humans") that existed in the past ("caused"). Further, social actors actively mobilize ideas across space and time as they learn, and not all climate change concepts achieve cultural salience depending on how they are contextualized in educational spaces (Drewes et al., 2018). Analytically deploying these concepts allowed us to see how particular climate change concepts moved both across scalar space (e.g., local, regional, global) and through time (e.g., past, present, future) in student explanations of phenomena.

11.4.1 Context and Data Collection Process

This study is a component of MADE CLEAR (Maryland and Delaware Climate Education, Assessment, and Research) which was an NSF funded project that focused on the implementation of comprehensive climate change education across Maryland and Delaware. A prominent part of this project was a Climate Academy which brought together educators for a year-long professional development experience focused on improving climate content knowledge and pedagogy.

While the current study in this chapter does not focus on outcomes of teacher professional development, we have previously engaged with this classroom to explore those related learning outcomes (see Drewes et al., 2018 for further details). Prior to the start of one enactment of climate change teaching, we worked with the teacher, Emma (a pseudonym), to recruit and interview 13 middle school students (ages 12–14) from her classes to explore how sociocultural dimensions of middle school students' daily life are highlighted in their explanations of modern climatic changes. Emma was a past participant in a MADE CLEAR Climate Academy and demonstrated deep engagement and interest throughout the professional development sessions. The school study site was a public charter school. In the U.S., the charter schools are public schools that receive funding from the government, but are independent from their state or local school system for most of their rules and regulations (Bettinger, 2005). The charter school in our study, is located close to an urban area, was in close proximity to our research team, and serves a racially and socio-economically diverse student body. The school hosts about 15% African American students, approximately 25% White students, and more than 50% Hispanic students.

With these 13 middle school students, we conducted a "sociocultural interview" prior to Emma's formal instruction on climate change to emphasize students' initial perceptions of climate change and the sources of their information in regards to climate change. The interviews were designed as responsive, semi-structured interviews (Rubin & Rubin, 2011) and included questions about the sources that shape students' conceptions of climate change and how these sources influence not only students' scientific or novice explanations, but also their agency, and emotions. The reason for expanding our analysis to agency and emotions is due to recent call by scholars on the role of these constructs for developing deep learning of controversial science topics such as climate change (Calabrese Barton, 2008; Lombardi & Sinatra, 2013). Agency is simply defined as an individual's ability to shape the world around them (Shanahan, 2009). Drawn from this definition, science education scholars see agency as including learners' ability to use their understanding of science to develop or have a willingness to take action on societal problems (Basu et al., 2009; Calabrese Barton, 2008). These scholars highlight that the action can be at different levels, such as individual or community. Narrowly, in our data, we look at students' agency in terms of how students positioned individual-self, their communities, and others when they suggested climate change action. The students' decision to take action on climate change issues can be influenced by their emotions (e.g., such as worry, frustration, and hope) about impacts and solutions of climate change (Lombardi & Sinatra, 2013). Some scholars suggest that developing optimistic views about solutions can increase students' feeling of agency and lead to improved critical thinking and an action orientation toward climate change solutions (e.g., Ojala, 2016). The emotions can be expressed in variety of ways during learning such as semantically by using emotion words (This is surprising), raised intonation in oral discourse (REALLY!), and facial impressions (contraction of the brow muscles). In other cases, emotions might not be explicit in discourse unless it is intentionally prompted for the research study (Sezen-Barrie et al., 2020). Therefore, we will be limited to analyzing students' emotions that are visible through discursive interactions.

During the interview, students watched a short video entitled *Climate Change Basics* (https://www.youtube.com/watch?v=ScX29WBJI3w). This video was developed by the U.S. Environmental Protection Agency (EPA) to review societal problems that cause climate change and the impacts felt by many species on Earth. This video was followed by interview questions on how students relate the climate change mechanisms, impacts, and solutions to their life, to their communities, and to our planet Earth. Including watching the video, each interview lasted between 12 and 20 min.

11.4.2 Data Analysis Approach

Informed by the sociolinguistic approach, we organized our data into transcripts by sequence units, that are "cohesive and thematically tied interactions" (Kelly & Chen, 1999). We used Dedoose, a collaborative qualitative analysis software, to organize our transcripts and code each sequence unit. We took an abductive approach to our qualitative discourse analysis. According to Tavory and Timmermans (2014):

> Abductive analysis is a qualitative data analysis approach grounded in pragmatism and aimed at theory construction. Abduction refers to the process of producing theoretical hunches for unexpected research findings and then developing these speculative theories with a systematic analysis of variation across a study. This approach depends on iterative processes of working with empirical materials in a relationship with a broad and diverse social science theoretical literature. (p. 131)

The "pragmatic" step in our analysis was to frame our initial codes based on the current seminal documents and literature on climate science and climate change education. For example, by looking at the most recent standards and learning progression studies, we decided to examine four core ideas of climate change that is mechanism, human activity, climate change effects, and mitigation/adaptation (Breslyn et al., 2017). Table 11.1 shows the codes we determined through this pragmatic approach: Climate Change Core Ideas, Temporal Indexicality, Spatial Indexicality, Pronouns, and Emotions.

Table 11.1 Codes and subcodes, explanations and examples

Code	Explanation	Subcodes	Examples
Climate Change Core Ideas	These codes correspond to the 4 core concepts of climate change in the Next Generation Science Standards (NGSS, 2013) and represented in the developed learning progression for climate change (Breslyn et al., 2017)	Impacts of Human Activity Mechanisms Effects Mitigation & Adaptation	How people are putting *more carbon dioxide into the air* and giving it a lot of heat to the Earth Humans *should start using natural resources*
Temporal Indexicality	Climate scientists rely on historical data dating back to pre-industrial times (NASA Earth Observatory, 2016) to make claims that modern climate changes we are experiencing now are due to human activity (Karl & Trenberth, 2003). Then, they need to make future projects to better adapt and mitigate (IPCC, 2018)	Past Present Future	Well at home, I know my mom *is always* like, "Turn off the lights, we need to save energy." They could destroy ecosystems which would *then eventually affect* us
Spatial Indexicality	Socially and culturally structured "real" and "imaginary" geographies can shape our sensemaking of climatic changes (Leander & Sheehy, 2004). Climate scientists also look at both worldwide data (Edwards, 2010) and observe local impacts of climatic changes (Najjar et al., 2000) to make stronger claims about climate change	Personal Humans/Humanity Environment Family/Home/House Friends School Local/regional State Country Planetary/World	We are *close to the ocean* Letting more of *the sun's* rays in
Pronouns	The use of pronouns can be interpreted as identity construction (Goffman, 1974, 1981). Use of singular pronouns (I, You) can be seen as individualistic while plural pronouns can be interpreted as collective perspectives (Bramley, 2001)	She/he God I/me It (object) No one Everyone (exclusive) They (objects) They/them (people) Us/we/our (inclusive) You	Well if *you* use the bathroom and you wash your hands and maybe your friend comes in and the water is running So *people* can be playing outside
Emotions	Studies call for attention to students' emotions in science	Good (moral)	*It is better* to use solar energy

(continued)

Table 11.1 (continued)

Code	Explanation	Subcodes	Examples
	learning (Sinatra et al., 2015) and environmental action (Ojala, 2016). Recent studies show links between our emotions and judgements about climate change (Lombardi & Sinatra, 2013) and actions we consider to solve the climate change problem	Concerned Displeasure Fear/worry Bad (moral) Joy/Fun Irony/sarcasm Surprise Hope Love	My friend *like to be sarcastic* and say this happened because of global warming

Table 7.1 provides explanations to each of these codes to show the related source. Other than these documents and the literature, we used an ecolinguistics approach (i.e., looked at students' stories about climate change) to look for new subcodes such as school, a spatial index, and irony/sarcasm as an emotional response to climate change. We then utilized an iterative process with our empirical data from students' interviews to revise the subcodes. To ensure the internal reliability of our codes, each student interview was assigned to two researchers and when there were disagreements or confusions on any sequence unit, the questions were brought to the whole group where three researchers discussed the issues until 100% agreement was achieved (Saldaña, 2015). Drawn from an ecolinguistics approach, we wanted to look at aspects of climate change that were made salient or ignored across all 13 students by calculating frequencies for each subcode (Stibbe, 2015). In order to understand how temporal, spatial, and relational sensemaking of climate change occur on each core idea of climate change, we paid attention to frequencies at the co-occurrence of subcodes within each core idea of climate change (Table 11.1).

11.5 Epistemic, Relational, Emotional Aspects of Climate Change in Students' Explanations

During our analysis, we looked at the temporal and spatial indexicality of climate change core ideas from middle school students' life experiences that are socially and culturally constructed in their relationships with their friends, families, neighborhood and through their perceptions of themselves, others, and their world. Here we present our findings from the frequency analysis of co-occurrences to show what temporal, spatial, personal/relational, and emotional aspects of climate change were made salient (or erased) in regards to students' explanations of the core concepts of climate change. We then present examples from our discourse analysis of the students' excerpts to understand how these aspects were highlighted.

11.5.1 *Temporal Indexicality for Climate Change Core Ideas*

Temporal indexicality was coded 109 times out of 396 total codes. When we look at the frequency results for the temporal indexicality, we notice that "Effects of Climate Change" is the most highlighted aspect of climate change (61 out of 109) while "Mitigation and Adaptations Strategies" is the least apparent (7 out of 109). In general, we also see that students position the solution strategies for mitigation and adaptation (0 out of 109) and human activity impacts infrequently as a past event (1 out of 109) (Fig. 11.1).

11.5.1.1 Effects of Climate Change in the Future

When we look at the co-occurrences, a prevalent time frame in students' responses is the future when students talk about the forthcoming or expected effects of climate change that we or our Earth will experience (39 out of 109). Students' examples for effects of climate change varied among the issues of sea level rise, increase in dissemination of diseases, flood and droughts, extinction of living things including humans, decrease in outdoor play areas (e.g., soccer and baseball), and heat waves. For example, the excerpt below explains how the increasing effects of climate change will impact the lives of people and other living things who inhabit their state in the future:

> Ocean levels *will* rise so when you go to the beach it will be higher and it *might affect* some people that live around the water because it might form a flood and damage their house. Ocean levels *will* rise so. . .it *might affect* people because, like some people here in [name of nearby town and state], it is mostly water near us and if the water levels rise that is going to affect them, probably some animals, and the environment and their homes.

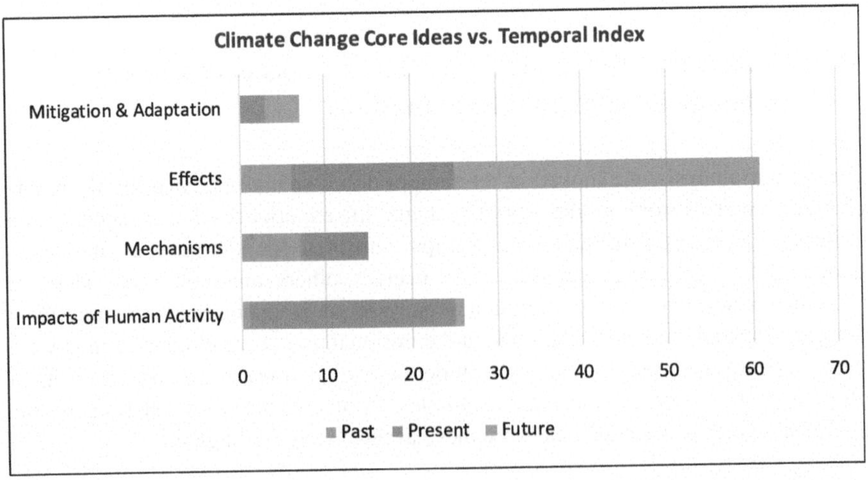

Fig. 11.1 The frequencies of core ideas of climate change vs. temporal indexicality

In another example, we see that the student is talking about climate change causing diseases and that these diseases will get worse and affect him and people around him:

> Because if people of my generation start getting these diseases and making the earth worse, then it *is going to* effect me by making my family and friends of mine, or anyone, at risk of dying, and organisms that help people like me get fed and alive.

11.5.1.2 Humans Are Altering the Climate Change NOW

Another common co-occurrence we noticed is that middle school students mostly explained the impacts of human activity on climate change situated in the "present" time (24 out of 109). In other words, students were considering "present" as being the timeframe to see how humans cause climatic changes. Within our data, we saw that students highlighted wasting energy at home and school, and poor transportation choices as the causes for modern climatic changes. In the example below, we see that the student talks about her parents' behavior that leads to "wasting energy" in the present time:

> There *is* a lot of light being used and sometimes *they just keep* their computers on or their TVs on. *They keep* their lights on and just forget about it, *they are* wasting energy and it *is* coming in and there *is* fossil fuel being burnt like coal fossil fuel and gas till it *is polluting* the air and causing climate change. *They are* wasting energy.

11.5.2 Spatial Indexicality for Climate Change Core Ideas

The spatial indexicality was coded 91 out of 396 total codes. The frequency analysis of the spatial indexicality and core ideas of climate change showed that students highlighted the global/planetary scale of climate change (32 out of 91). Interestingly, students only referred to the state or the country on the mechanisms of climate change, i.e. the causes of the modern climatic changes. Students highlighted the local aspects of climate change while they are talking mostly about their region and their home environment. We will look at example excerpts for both local and global sensemaking of climate change. The "environment" is also heavily mentioned (21 out of 91) in regards to other living organisms and their habitats, both in global and local references to climatic changes (Fig. 11.2).

11.5.2.1 Global Perceptions: Planet Is Impacting and Affected by the Climate Change

Students made many references to Earth, people in general, and other living organisms who are living on Earth as both impacting and being affected by modern climate changes (28 out of 53). In the excerpt below, the student takes a stance that people of the world are making Earth a worse place for other living things:

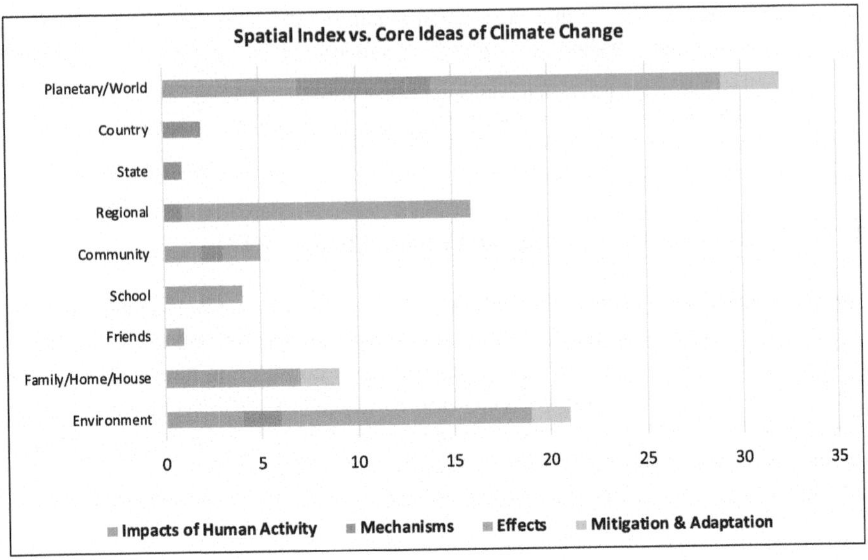

Fig. 11.2 The frequencies of spatial indexicality vs. core ideas of climate change

> I think it is horrible that *people [everywhere]* are changing *the earth* in a bad way and making living creatures and organisms extinct and harming themselves in the process. . . .

In another example, a student mentioned that people like to "travel for fun" and not only that travelling might get harder, but that jobs and businesses are affected globally by climatic changes:

> People might travel for fun and climate change can make it hard for them to travel because of flooding and heat waves. I think *jobs [everywhere]* can be affected because they can't get money because of the things that happen, they may not be able to make it to work, and *businesses [all around]* might be broken down or weathered away because of it and they might lose their jobs because of it.

11.5.2.2 Local Perceptions: Region & Family Is Impacting and Affected from the Climate Change

Although not as frequent as the references to global scale, local aspects of climate change often were expressed by middle school students. These references were mostly made about their region (the places they have often been) and their family (25 out of 35). For instance, one student talked about her parent's choice to drive her to school and therefore causing the exhaust fumes to contribute to climate change:

> Well *my mom or dad* drives me to school every day because I live 30 minutes down the road and my bike has a flat tire. Then the buses also have gas in it because it has a lot of stops and then they have to go back and go to the gas station and get more gas and there is exhaust coming out of the pipe.

Another student talked about how his soccer field would be affected if the climatic changes cause the lake nearby to flood:

> Well like natural things, I like to play soccer, so if I try to play in fields. . .in one *place I play there* is this little lake and if it *affects the lake* then it will soon flood too and then I won't have anywhere to play.

11.5.3 Positional and Emotional Sensemaking of Change Core Ideas

In our attempt at understanding sociocultural dimensions of climate change, we also explored how students relate climate change events to themselves or to others or to no one (Harré, 2015). When looking at the pronouns students used in their explanations of climate change (171 references out of 396 total codes), we noticed that students varied in their ownership such as I/Me, We/Us/Our (79 out of 171) or dissociations such as You, They/Them (92 out of 171). In either case, we also notice that students tend to see climate change causes, impacts, and solutions as more of a collective action (we, they) than individualistic (I, you) (Fig. 11.3).

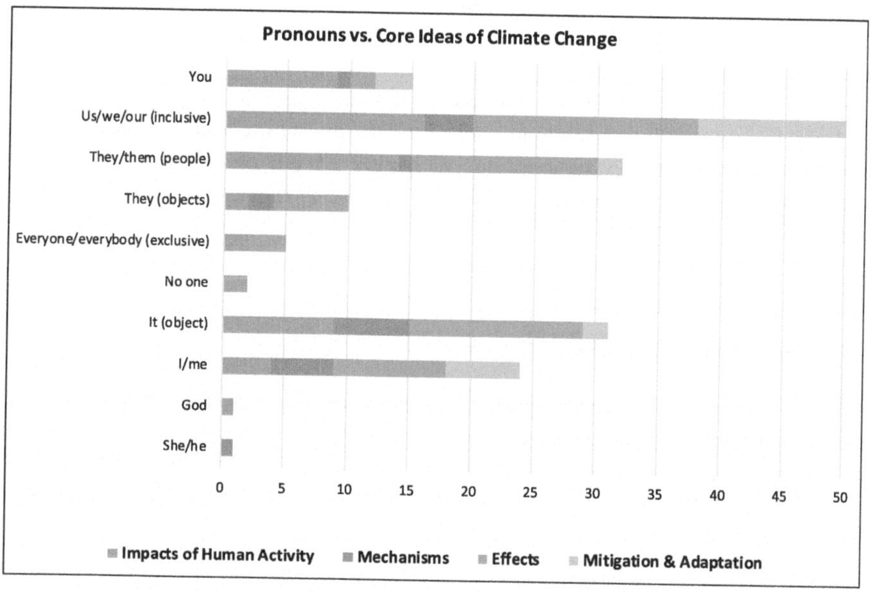

Fig. 11.3 The frequencies of pronouns vs. core ideas of climate change

11.5.3.1 Collective Problem and Action

The excerpt below shows an example from a student who mostly refers to We/Us/Our in his explanations of climate change. The student here talks about alternative energy sources as a solution to climate change and "we" need to act on these solutions because the use of electric and hydro cars will cost less for "a lot of us":

> If *we* use something like electricity or hydro cars, as my fellow student did his essay on, this could help the environment more with the climate change and even though the electric cars are higher in terms of cost of the car, the fuel economy is better and I know *a lot of us* around the world are very worried about gas prices rising but with electric vehicles and hydro cars, it costs less, much less.

Although not as common, some students owned the climate change problems but talked about climate change as more as an individualistic problem: "*I* could probably make the world less at stake for global warming *I* guess."

Another student mostly used "they" and "you" when she was talking about what is causing the impacts of climate change we observe today:

> Sometimes *they* just don't want to blame it on *themselves* so *they* just say global warming did it and you see it a lot on TV. *They* say global warming did this but *they* did it, don't blame everything on global warming. And a lot of people say that this happens because of global warming but I don't think everything happens because of global warming... sometimes it is just *you*.

This excerpt also shows that the student was emphasizing the human aspect of climatic event, by opposing the idea that "they" blame it on "global warming", however, keeps herself outside of the problem.

We looked at the emotional sensemaking semantically with emotion words such as love, fear and with interjections, such as wow, ahh. Our coding of emotions did not show high frequency of emotions with only 25 out of 396 total codes. We noticed that the most common emotion was surprise in regards to mechanisms and causes of climate change (Fig. 11.4).

11.5.3.2 Unexpected Mechanisms and Effects of Climate Change

After watching the video on Climate Change Basics, students found that the mechanisms leading to climate change or effects people will experience are unexpected and therefore expressed being surprised. In the example below, the student talked about how even the excessive use of electronics at home can be the contributor to the climatic changes:

> I was very *surprised* at what I was seeing. There were obviously some things we all know about cars and fossil fuels but I mean, I wasn't really thinking about... I mean I know that electronics get hot when you use them a lot and they feel a little warm but I never thought of that [using electronics a lot] as being a fossil fuel kind of and heating up the planet.

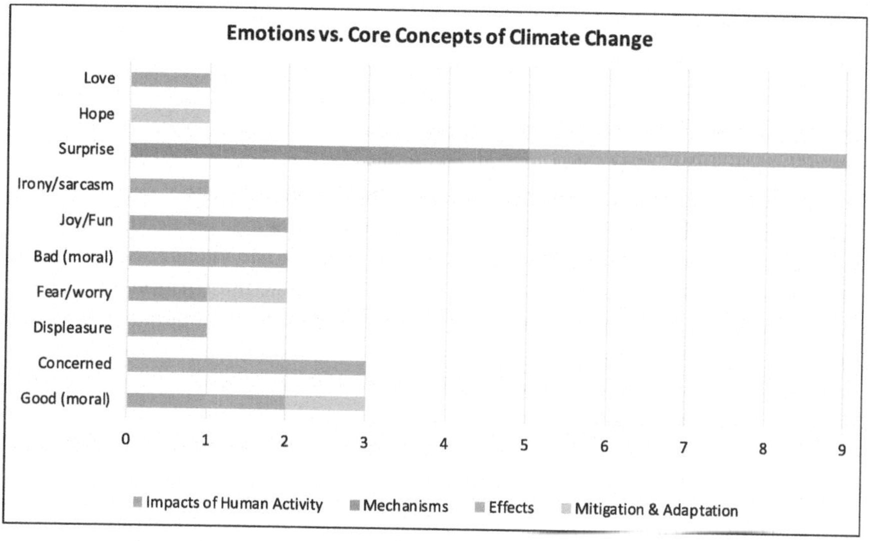

Fig. 11.4 The frequencies of emotions vs. core ideas of climate change

In another example, we interpreted a student's use of "wow" that she was surprised about how much ice has been melting and will have damaging impacts on climate change. We found it interesting that this emotion led to an analogy between climate change impacts and the scene from the "Hunger Games" movie where children fight for resources.

> It's destroying ice sheets so maybe in a couple hundred years or less we might get the Arctic... the Antarctic might be completely gone, it will be underwater and that will rise us up and [state name]in general is by a lot of water so I think a lot of this... what they call coastal states might be flooded and we will have to move in more. *Wow!* I just thought of the Hunger Games, they flooded up and now they have to live somewhere in the Rockies and they call it... because the water... it is because of us.

11.5.3.3 Empathy and Concern for the Environment

Students also mentioned that their moral dilemmas, worries or fears are due to the love or joy they feel about the world and the environment. For example, the student below highlights that because he loves animals, he feels bad about their extinction due to human caused climate change:

> Well I saw that animals are getting extinct which is *really bad* because I *love* animals, especially the ones that are most gone. If global warming keeps happening and the animals that live in the icy areas that are really cold, they won't have anywhere to live and they will soon die.

11.6 Conclusions

Critical thinking is at the core of scientific practices (NRC, 2000; Schwab & Brandwein, 1962). If we want students to experience how science is done, we need to consider ways to support students in developing critical thinking skills. As climate change education researchers, we see critically thinking about socioscientific issues as especially important due to two main reasons. First, we need students to critically question the information they obtain about climate change. Students can easily get exposed to unreliable and scientifically inaccurate information through the media, or elsewhere. Second, we need all students, independent of pursuing a STEM career, to critically evaluate their options to act on this massive problem. We paid particular attention to middle school students (ages 12–14) in our work because other studies showed that we can have a better influence on adolescents whose worldviews are still developing (Stevenson et al., 2014). Due to such flexibility in worldviews, we see this as an opportunity where middle school students can improve a habit to critically think about climate change claims. Plutzer and Hannah (2018) recently found that many secondary school science teachers in the United States engage students in inquiry-based debate environments only for claims that are empirically settled among vast majority of scientists and scientific organizations. On the other hand, Berbeco and colleagues (Berbeco et al., 2014) suggested inquiry-based debates around climate change questions that still has competing claims in scientific communities and those questions that are within the scope of students' abilities. For instance, instead of asking if humans cause climatic changes, it will be more authentic to ask how will climatic changes impact lobster migration. Debating around claims with clear scientific consensus might limit critical thinking as students can find answers through respected scientific organizations. If students tackle authentic scientific questions with multiple reasonable claims, they will then work with foundational knowledge and build on those by engaging in epistemic practices to evaluate the claims. Recent work showed that these real and relevant learning experiences have the potential to engage youth to have a socially active role in solving the climate change problem (Stapleton, 2019).

We expect that middle school students will use critical thinking while learning about climate change if they understand climate change causes, impacts, and solutions across varying times scales (such as pre-industrial revolution, 50 years later) and various spaces (such as their homes, states, country, and globe). The focus on spatial and temporal practices and human-nonhuman relationships on the impacts and solutions of climate change is due to epistemic culture of the climate scientists' work. Climate scientists are formed and encultured into these epistemic cultures with distinct interpretation of scientific practices to decide on norms for warranting knowledge (Kelly et al., 2000; Knorr Cetina, 2007). Such distinct interpretations in the climate change considers a balance between "matters of fact" that are risk-free body scientific knowledge and "matters of concern" that are knowledge entangled with risks, values, and concerns (Latour, 2004, pp. 23–25). Different than the traditional laboratory sciences that privileges matters of fact, climate science works at the merge of matters of fact and matters of concern. Therefore, warranting climate science knowledge needs to consider the past human activity which led to current impacts of climate change and how human activities of today will impact the next generations. Moreover, critical evaluation of evidence for what caused climatic changes and what actions will be effective needs to be evaluated in local and global contexts (Vallabh et al., 2016).

In our study, we found that most students dissociated themselves from past climate causality and most of them focused on the present. Some students also talked about the future mostly when they talked about the effects on climate change on people or other living things. While doing so, students see the future belongs to their generation and show care for the climate change problem. However due to lack of understanding of historic interpretation of past climate data, they were not able to ask critical questions on the mechanisms of climate change or the impacts of human activity. We, therefore, see that it is crucial for students to learn about how the historic data about climate is collected and interpreted.

The middle school students in our study referred to both local and global aspects of climate change phenomena. However, we found it interesting that the students only mentioned their state or country when they talked about mechanisms of climatic changes. Studies show that mitigation and adaptation strategies can be more effective if they are regulated by statewide and nationwide policies (Engel, 2006). Moreover, although we noticed that students are using plural pronouns which can be seen as their view of collectivism in climate change problem and action, the examples they mentioned for solutions to climate change were more individualistic in nature such as recycling and consuming greener products. For students to critically think about real collective action, they need to drive from social science and humanities (McKeown & Hopkins, 2010). For example, students can understand and develop opinions on taxation to build an efficient public transportation system if they study economics (Lundholm, 2011). In addition, students can evaluate how climate change is impacting different geographies on Earth unequally if they study geography and sociology (Bohle et al., 1994). We therefore recommend that climate change should be thought of as more than "just science" (Drewes et al., 2018; Fahey et al., 2014) for students to be able to develop critical thinking skills.

Finally, middle school students expressed emotional stress at the distant and overwhelming global nature of climate phenomena, and coped via the scalar downshifting of solutions. This current positioning of students in making sense of climate change limits their ability to improve the necessary critical thinking skills to be active agents for taking part in solving the climate change problem.

Acknowledgements We like to acknowledge our colleagues Drs. Emily Hestness and Wayne Breslyn for their help with the data collection process and the late Prof. Dr. J. Randy McGinnis for his guidance and curiosity for exploring the sociocultural aspects of climate change education.

This material is based upon work supported by the National Science Foundation under Grant #1239758. Any opinions, findings, and conclusions or recommendations expressed in this material are those of the authors and do not necessarily reflect the views of the National Science Foundation.

References

Archer, D. (2005). Fate of fossil fuel CO2 in geologic time. *Journal of Geophysical Research: Oceans, 110*(C9), 1–6.

Basu, S. J., Calabrese Barton, A., Clairmont, N., & Locke, D. (2009). Developing a framework for critical science agency through case study in a conceptual physics context. *Cultural Studies of Science Education, 4*, 345–371.

Berbeco, M., McCaffrey, M., Meikle, E., & Branch, G. (2014). Choose controversies wisely. *The Science Teacher, 81*(4), 8.

Blommaert, J. (2010). *The sociolinguistics of globalization*. Cambridge University Press.

Bohle, H. G., Downing, T. E., & Watts, M. J. (1994). Climate change and social vulnerability: Toward a sociology and geography of food insecurity. *Global Environmental Change, 4*(1), 37–48.

Bramley, N. R. (2001). *Pronouns of politics: The use of pronouns in the construction of self and other in political interviews*. [Doctoral dissertation, Australian National University]. Open Research Library.

Breslyn, W., Drewes, A., McGinnis, J. R., Hestness, E., & Mouza, C. (2017). Development of an empirically-based conditional learning progression for climate change. *Science Education International, 28*(3), 214–223.

Calabrese Barton, A. (2008). Feminisms and a world not yet: Science with and for social justice. In W.-M. Roth & K. Tobin (Eds.), *The world of science education: Handbook of research in North America*. Sense Publishers.

Cook, J., Oreskes, N., Doran, P. T., Anderegg, W. R., Verheggen, B., Maibach, E. W., … Nuccitelli, D. (2016). Consensus on consensus: A synthesis of consensus estimates on human-caused global warming. *Environmental Research Letters, 11*(4), 048002.

Dessler, A. (2011). *Introduction to modern climate change*. Cambridge University Press.

Devine-Wright, P. (2013). Think global, act local? The relevance of place attachments and place identities in a climate changed world. *Global Environmental Change, 23*(1), 61–69.

Dewey, J. (1910). *How we think*. DC Heath.

Dewey, J. (1933). *How we think: A restatement of the relation of reflective thinking to the educative process*. DC Heath.

Drewes, A., Henderson, J., & Mouza, C. (2018). Professional development design considerations in climate change education: Teacher enactment and student learning. *International Journal of Science Education, 40*(1), 67–89.

Drewes, A., Rogers, M., & Petrone, C. (2020). Climate change professional development approaches 'MADE CLEAR': Looking back on one project and looking forward to the future. In J. Henderson & A. Drewes (Eds.), *Teaching climate change in the United States*. Routledge.

Driver, R., Newton, P., & Osborne, J. (2000). Establishing the norms of scientific argumentation in classrooms. *Science Education, 84*(3), 287–312.

Edwards, P. (2010). *A vast machine: Computer models, climate data, and the politics of global warming*. MIT Press.

Engel, K. (2006). State and local climate change initiatives: What is motivating state and local governments to address a global problem and what does this say about federalism and environmental law. *Urban Law, 38*, 1015.

Erduran, S., & Jiménez-Aleixandre, M. P. (2008). Argumentation in science education. In *Perspectives from classroom-based research*. Springer.

Fahey, S. J., Labadie, J. R., & Meyers, N. (2014). Turning the Titanic: Inertia and the drivers of climate change education. *Journal of Applied Research in Higher Education, 6*(1), 44–62.

Fenton, J., & Smith, M. (2019). 'You can't say that!': Critical thinking, identity politics, and the social work academy. *Societies, 9*(4), 71.

Gardiner, S. (2006). A perfect moral storm: Climate change, intergenerational ethics and the problem of moral corruption. *Environmental Values, 15*, 397–413.

Glaser, E. M. (1941). *An experiment in the development of critical thinking* (Vol. No. 843). AMS.
Glaser, E. M. (1985). Educating for responsible citizenship in a democracy. In *National Forum*. (Vol. 65, No. 1, p. 24). Honor Society of Phi Kappa Phi.
Goffman, E. (1974). *Frame analysis: An essay on the organisation of experience*. Free Press.
Goffman, E. (1981). *Forms of talk*. University of Pennsylvania Press.
Halpern, D. F. (1998). Teaching critical thinking for transfer across domains: Disposition, skills, structure training, and metacognitive monitoring. *American Psychologist, 53*(4), 449.
Hamilton, L. C., Hartter, J., & Bell, E. (2019). Generation gaps in US public opinion on renewable energy and climate change. *PLoS One, 14*(7), e0217608.
Harré, R. (2015). Positioning theory. In *The international encyclopedia of language and social interaction* (pp. 1–9). Wiley-Blackwell.
Henderson, J. A. (2015). Out of sight, out of mind: Global connection, environmental discourse and the emerging field of sustainability education. *Cultural Studies of Science Education, 10*(3), 593–601.
Henderson, J., & Drewes, A. (Eds.). (2020). *Teaching climate change in the United States*. Routledge.
Intergovernmental Panel on Climate Change [IPCC]. (2018). *Global warming of 1.5° C: An IPCC special report on the impacts of global warming of 1.5° C above pre-industrial levels and related global greenhouse gas emission pathways, in the context of strengthening the global response to the threat of climate change, sustainable development, and efforts to eradicate poverty*. Intergovernmental Panel on Climate Change.
Karl, T., & Trenberth, K. (2003). Modern global climate change. *Science, 302*(5651), 1719–1723.
Kelly, G. J., & Chen, C. (1999). The sound of music: Constructing science as sociocultural practices through oral and written discourse. *Journal of Research in Science Teaching, 36*(8), 883–915.
Kelly, G. J., Chen, C., & Prothero, W. (2000). The epistemological framing of a discipline: Writing science in university oceanography. *Journal of Research in Science Teaching, 37*(7), 691–718.
Knorr Cetina, K. (1999). *Epistemic cultures: How the sciences make knowledge*. Harvard University Press.
Knorr Cetina, K. (2007). Culture in global knowledge societies: Knowledge cultures and epistemic cultures. *Interdisciplinary Science Reviews, 32*(4), 361–375.
Latour, B. (2004). *Politics of nature. How to bring the sciences into democracy*. Harvard University Press.
Latour, B., & Woolgar, S. (1986). *Laboratory life: The [social] construction of scientific facts* (2nd ed.). Princeton University Press.
Leander, K. M., & Sheehy, M. (2004). *Spatializing literacy research and practice*. Peter Lang.
Lombardi, D., & Sinatra, G. M. (2013). Emotions about teaching about human-induced climate change. *International Journal of Science Education, 35*(1), 167–191.
Lundholm, C. (2011). Society's response to environmental challenges: Citizenship and the role of knowledge. *In Factis Pax, 5*, 80–96.
Mann, M. E. (2012). *The hockey stick and the climate wars: Dispatches from the front lines*. Columbia University Press.
McDonald, R. I., Chai, H. Y., & Newell, B. R. (2015). Personal experience and the 'psychological distance' of climate change: An integrative review. *Journal of Environmental Psychology, 44*, 109–118.
McKeown, R., & Hopkins, C. (2010). Rethinking climate change education: Everyone wants it, but what is it? *Green Teacher, 89*(Summer), 17–21.
McLaren, P., & Giarelli, J. (Eds.). (1996). *Critical theory and educational research*. SUNY Press.
McNeill, K. L., & Krajcik, J. (2009). Synergy between teacher practices and curricular scaffolds to support students in using domain-specific and domain-general knowledge in writing arguments to explain phenomena. *Journal of the Learning Sciences, 18*(3), 416–460.
Najjar, R. G., Walker, H. A., Anderson, P. J., Barron, E. J., Bord, R. J., Gibson, J. R., … Polsky, C. D. (2000). The potential impacts of climate change on the mid-Atlantic coastal region. *Climate Research, 14*(3), 219–233.

National Aeronautics and Space Administration [NASA] Earth Observatory. (2016). *Paleoclimatology: Ice cores*. Retrieved from http://earthobservatory.nasa.gov/Features/Paleoclimatology_IceCores/
National Council on Education Standards, & United States Dept. of Education. (1992). *Raising standards for American education: A report to congress, the secretary of education, the National Education Goals Panel, and the American people*. US Government Printing Office.
National Research Council [NRC]. (2000). *Inquiry and the national science education standards: A guide for teaching and learning*. National Academy Press.
NGSS Lead States. (2013). *Next generation science standards: For states, by states*. National Academies Press.
Ojala, M. (2016). Young people and global climate change: Emotions, coping, and engagement in everyday life. In N. Ansell, N. Klocker, & T. Skelton (Eds.), *Geographies of global issues: Change and threat. Geographies of children and young people* (Vol. 8). Springer.
Osborne, J. (2011). Science teaching methods: A rationale for practices. *School Science Review, 98*(343), 93–103.
Osborne, J. (2014). Teaching scientific reasoning: The ineluctable in search of the ineffable? In *Learning and becoming in practice: The International Conference of the Learning Sciences (ICLS) 2014* (Vol. 3). International Society of the Learning Sciences.
Oxfam International. (2015). *Extreme carbon inequality*. Retrieved from: https://www.oxfam.org/en/research/extreme-carbon-inequality
Plutzer, E., & Hannah, A. L. (2018). Teaching climate change in middle schools and high schools: Investigating STEM education's deficit model. *Climatic Change, 149*(3–4), 305–317.
Plutzer, E., McCaffrey, M., Hannah, L., Rosenau, J., Berbeco, M., & Reid, A. (2016). Climate confusion among US teachers. *Science, 351*(6274), 664–665.
Rubin, H. J., & Rubin, I. S. (2011). *Qualitative interviewing: The art of hearing data*. Sage.
Saldaña, J. (2015). *The coding manual for qualitative researchers*. Sage.
Sawyer, D. S., Henning, A. T., Shipp, S., & Dunbar, R. W. (2005). A data rich exercise for discovering plate boundary processes. *Journal of Geoscience Education, 53*(1), 65–74.
Schwab, J. J., & Brandwein, P. F. (Eds.). (1962). *The teaching of science as enquiry*. Harvard University Press.
Scollon, R. (2013). Geographies of discourse. In I. de Saint-Georges & J.-J. Weber (Eds.), *Multilingualism and multimodality: Current challenges for educational studies* (pp. 183–198). Sense.
Sezen-Barrie, A. (2018). Utilizing professional vision in supporting preservice teachers' learning about contextualized scientific practices. *Science & Education, 27*(1–2), 159–182.
Sezen-Barrie, A., Shea, N., & Borman, J. H. (2019). Probing into the sources of ignorance: Science teachers' practices of constructing arguments or rebuttals to denialism of climate change. *Environmental Education Research, 25*(6), 846–866.
Sezen-Barrie, A., Miller-Rushing, A., & Hufnagel, E. (2020). 'It's a gassy world': Starting with students' wondering questions to inform climate change education. *Environmental Education Research, 26*(4), 555–576.
Shanahan, M. (2009). Identity in science learning: Exploring the attention given to agency and structure in studies of identity. *Studies in Science Education, 45*(1), 43–64.
Shea, N., Mouza, C., & Drewes, A. (2016). Climate change professional development: Design, implementation, and initial outcomes on teacher learning, practice, and student beliefs. *Journal of Science Teacher Education, 27*(3), 235–258.
Sinatra, G. M., Heddy, B. C., & Lombardi, D. (2015). The challenges of defining and measuring student engagement in science. *Educational Psychologist, 50*(1), 1–13.
Stager, C. (2011). *Deep future: The next 100,000 years of life on earth*. Thomas Dunn Books.
Stapleton, S. R. (2019). A case for climate justice education: American youth connecting to intragenerational climate injustice in Bangladesh. *Environmental Education Research, 25*(5), 732–750.

Stevenson, K. T., Peterson, M. N., Bondell, H. D., Moore, S. E., & Carrier, S. J. (2014). Overcoming skepticism with education: Interacting influences of worldview and climate change knowledge on perceived climate change risk among adolescents. *Climatic Change, 126*(3–4), 293–304.

Stibbe, A. (2015). *Ecolinguistics: Language, ecology and the stories we live by*. Routledge.

Tavory, I., & Timmermans, S. (2014). *Abductive analysis: Theorizing qualitative research*. University of Chicago Press.

Thunberg, G. (2018). Speech at UN Climate Change COP24 Conference, Katowice, Poland, December 15. Retrieved from: https://www.gbnews.ch/greta-thunbergs-speech-to-the-world/

Vallabh, P., Lotz-Sisitka, H., O'Donoghue, R., & Schudel, I. (2016). Mapping epistemic cultures and learning potential of participants in citizen science projects. *Conservation Biology, 30*(3), 540–549.

Weisbach, D., & Sunstein, C. R. (2008). Climate change and discounting the future: A guide for the perplexed. *Yale Law & Policy Review, 27*, 433.

Chapter 12
Social Responsibility and Critical Disposition for Considering and Acting upon Conflicting Evidence in Argumentation About Sustainable Diets

Pablo Brocos and María Pilar Jiménez-Aleixandre

12.1 Introduction and Objectives: Critical Thinking and Conflicts

In our times, the effects of uncompromising industrial and economic growth, large-scale implementation of technological advances and demographic growth have led to energy, health, environmental and climate crises across the globe. The current situation urgently calls for the creation of new ways of thinking, producing, consuming, and acting. There is widespread consensus about the fact that, in this context, Critical Thinking (CT) is necessary for addressing the complex issues we are facing, and for promoting the articulation of innovative solutions to both persistent and emerging challenges.

Critical thinking is a complex, multifaceted practice, related to purposeful judgment and citizenship practice. The revised characterization of CT by Jiménez-Aleixandre and Puig (Chap. 1, this volume) serves as the theoretical foundation for this chapter. It is developed in four components: cognitive-epistemic skills, critical character, independent opinion, and critical action. Empirical research on CT has focused mainly on analysing the cognitive and epistemic skills, which is coherent with the traditional conceptualization of CT itself as a skill. Studying CT in terms of character dispositions, intellectual independence and agency entails both operationalizing and analytical difficulties (Abrami et al., 2008, 2015). Arguably, as a result of those methodological challenges and the later conceptualization of those three aspects as an integral part of CT, they are currently understudied. In this chapter we aim to provide an approach for exploring these three components in classroom discourse.

P. Brocos (✉) · M. P. Jiménez-Aleixandre
Universidade de Santiago de Compostela, Santiago de Compostela, Spain
e-mail: pablo.brocos@usc.es

B. Puig, M. P. Jiménez-Aleixandre (eds.), *Critical Thinking in Biology and Environmental Education*, Contributions from Biology Education Research,
https://doi.org/10.1007/978-3-030-92006-7_12

An issue of particular concern is how to equip teachers to foster the development of critical thinking in their students. CT is rarely explicitly taught; this may be related to an implicit assumption about its spontaneous development alongside disciplinary knowledge. Research shows that a mixed approach, integrating explicit instruction and modelling combined with practice yields better results (Heijltjes Van Gog et al., 2014). We argue about the importance of appropriate contexts for promoting CT: our position is that when conflict is embedded in task design it creates fruitful learning opportunities to engage in CT practice. Argumentation tasks about socio-scientific issues (SSI), such as the one presented in this study, provide privileged learning environments for fostering CT, since SSI involve conflicts which prevent reaching a solution that meets all interests. This acknowledgement is part of the development of CT about complex, real life issues. Arguing about SSI is considered relevant for the development of responsible citizens capable of taking part in future collective decision-making processes (Acar et al., 2010; Kolstø, 2001) and for sustainability (Tytler, 2012). In Education for Sustainable Development (ESD), the promotion of environmentally responsible behaviours is a priority. However, Oliveira and Ackerson (2015) emphasize that the construct of agency is an understudied topic. Stevenson and Stirling (2010) argue about the need of developing the capacity of appropriate and effective action, pointing to the need of reflecting about one's own experiences, assumptions, beliefs and values, as well as about the contextual factors that shape them.

Accordingly, this chapter examines the aforementioned three components of CT in pre-service teachers' discourse in the context of an environmental socio-scientific argumentation task, in which participants had to choose a diet and construct arguments about food choices. The task was framed as a dilemma between omnivorous and vegetarian diets, and it was designed drawing from recent environmental science literature, which provides evidence for the environmental benefits of diets with higher proportions of vegetables, and for the contribution of meat-based diets to global warming, resources depletion, and deterioration of public health (IPBES, 2020; Stehfest et al., 2009; Tilman & Clark, 2014).

The research objectives are:

1. To examine the participants' acknowledgement of conflicts between the evidence considered in their arguments and the options chosen, and how they account for and justify them.
2. To examine the dimensions related to taking actions towards sustainability in participants' discourse.

Specifically, we explore features related to critical character by examining, in their discourse, participants' disposition to consider evidence which would support non-mainstream options –e.g. vegetarian or vegan diets– and potentially challenge their previous beliefs and worldviews. We address the dispositions to independent thinking and social responsibility by examining the existence and acknowledgement of conflicts between the options chosen and the dissonant evidence considered in their arguments, and the ways in which these conflicts are accounted for and justified. We examine the component of critical action by analysing dimensions related to agency in participants' discourse, particularly the existence of spontaneous references to the need for modifying social or individual behaviours, and the identification of obstacles preventing change.

12.2 Theoretical Framework

The rationale draws from three bodies of knowledge, education for sustainable development (ESD); Critical Thinking framed in Criticality and action; and argumentation about Socio Scientific Issues (SSI) with a focus on the role of values and conflicts.

12.2.1 Education for Sustainable Development and Environmental Agency: Sustainable Food Choices

Education for sustainable development (ESD) is receiving increasing attention in science education, with a focus on the emergence of values (Garrison et al., 2015). Industrial civilization has caused environmental problems, such as ecosystem degradation, extinction of species and climate change. As David W. Orr puts it: "time is running out on the experiment of civilization" (2017, p. viii). Current worldwide efforts to advance ESD gravitate around the 2030 Agenda for Sustainable Development (United Nations, 2015), which includes 17 specific Sustainable Development Goals (SDG). Within this framework, ESD is understood in terms of competencies, which UNESCO (2017) defines as an "interplay of knowledge, capacities and skills, motives and affective dispositions" (p. 10). According to de Haan (2010), Rieckmann (2012) and Wiek et al. (2011), key competencies for sustainable development include: systems thinking, anticipatory, normative, strategic, collaboration, critical thinking, self-awareness, and integrated problem-solving competencies.

UNESCO (2017) claims that the most effective pedagogical approaches to achieve these objectives are learner-centred, action-oriented, and transformative; in other words, those which promote the capacity to put into question the ways in which students think about the world (Mezirow, 2000; Slavich & Zimbardo, 2012), in order to change the statu quo, which aligns with the objectives of transgressive learning (Lotz-Sisitka et al., 2015). This approach involves a participatory learning, which stimulate critical thinking (Tilbury, 2011; UNESCO, 2012) through the engagement in tasks that encourage discussion –in other words, argumentation–, analysis and application of values (UNESCO, 2012) and decision-making (Breiting & Mogensen, 1999). Addressing sustainability in education requires an approach emphasizing multiple perspectives, and the ability to establish relationships between processes, scales and contexts (Colucci-Gray et al., 2006). For this purpose, the analysis of local issues and contexts through an interdisciplinary perspective, as the one addressed in this study, has been suggested as a fruitful educational strategy (Laurie et al., 2016; UNESCO, 2012). These considerations about ESD are coherent with our CT approach, which emphasizes engagement with place-based and social issues (Brocos & Jiménez-Aleixandre, 2020a). Critical approaches to ESD reveal a discourse emphasizing objectivity and faith in technology, while "blackboxing" ideologies underlying decisions and policies (Ideland & Malmberg, 2015). We contend that the question of which diet should be promoted is also framed in particular ideologies and economic interests, such as intensive breeding or aquaculture companies, which should be taken out of this "black box", in order to allow assumptions, as for instance the image of a diet for "developed countries" linked to high meat consumption, being questioned.

The most crucial and pressing objective of ESD is the promotion of behaviour patterns that are environmentally responsible. As a researcher interviewed by Laurie et al. (2016) puts it: "Society does not need people that know how to save water. It needs people that actually do save water" (p. 237). In this regard, Oliveira and Ackerson (2015) point out that research on environmental education has made limited analytical use of the construct of agency. They define environmental agency as the capacity to act upon and resolve environmental problems. The efforts of environmental education for several decades have been focusing on education *for* the environment. However, there are not many studies focusing on action, on the development of commitment to act in ways that care for the environment, protect it, and defend it from degradation. Stevenson and Stirling (2010) discuss the need for developing the capacity for appropriate and effective action, characterizing learners' agency as involving three distinct forms of reflective, relational and transformative agency. They suggest the need for deep reflection on one's own experiences, assumptions, beliefs and values, as well as on the contextual factors that shape them. The task analysed in this study was designed with the intention of promoting this type of reflection.

We approach education for sustainability by addressing the development of sustainable food choices. The FAO (2012) defines a sustainable diet as "protective and respectful of biodiversity and ecosystems, culturally acceptable, accessible, economically fair and affordable; nutritionally adequate, safe and healthy; while optimizing natural and human resources" (p. 294). Most studies and policies about sustainability and climate change mitigation focus on the production and use of energy (Stehfest et al., 2009), and little attention has been paid to the environmental impact of diets until recent years, both in science education and in climate science. However, as Stehfest et al. show, the livestock sector accounts for 18% of the greenhouse emissions, as well as for 80% of land use. They suggest that a global transition to a low-meat diet would reduce the climate change mitigation costs by about 50%. Thompson et al. (2013), taking into account cultural traditions, propose changes in the amount and types of meat to be consumed. Other recent studies reach the same conclusions (Springmann et al., 2016; Tilman & Clark, 2014), also highlighting the public health benefits of plant-based diets. The scope of the potential health benefits of low-meat diets involve not only lower incidences of chronic disorders, but also infectious diseases, since research indicates that biodiversity loss and habitat disruption caused by human activities such as intensive breeding enable interspecies transmission of microbes, which increases the risk of emerging pandemics such as COVID-19 (IPBES, 2020). So, as a summary, environmental science research has recently pointed out the convenience of promoting nutrition patterns with a higher proportion of plant products, which would be beneficial for the environment, public health, and in the fight against hunger, which is aligned with the Sustainable Development Goals (UNESCO, 2017).

12.2.2 Critical Thinking Framed in Criticality and Action

The study is framed in the revised characterization of CT by Jiménez-Aleixandre and Puig (Chap. 1, this book), which endorses a shift from CT as solely or mainly argumentation and judgment formation, focused on skills and disposition, towards *criticality*, which sees CT as practice, focusing on *action* (Davies & Barnett, 2015). A detailed discussion of this shift is to be found on that chapter, and in this section we only summarize some of its main tenets. *Criticality* comprises critical thinking, critical reflection and critical action, thus combining the traditional view of CT as the evaluation of evidence and the disposition to consider a range of views, with perspectives from critical pedagogy and critical theory, emphasizing critical citizenship. This new approach would mean both engaging in critique of discourses concealing particular interests and power games, and participating in action concerned with environmental and social justice. Therefore, criticality is connected with agency, developing the capacity for appropriate and effective action. Criticality is also taking into account affective dispositions, emotional sense-making and identities for, as Hufnagel (Chap. 3, this book) points out, biology and environmental education issues are emotionally laden, and in CT identities are the reference points for emotions.

12.2.3 Argumentation About Socio Scientific Issues: The Role of Values and Conflicts

Conflicts make part of life, of decisions, of scientific endeavor; in a number of matters, as socio-scientific issues or environmental questions, it is almost impossible to avoid them. We argue that it is necessary to educate teachers –and students– by raising awareness about these challenges. For instance, taking actions in order to address environmental problems involves conflicts, both social –within structural dimensions of the issue, as illustrated in the case of diets– and personal –as with lifestyles. Reducing waste involves more effort than dealing with waste in a careless way; using public transportation instead of private cars entails personal sacrifices in terms of practicality. Thus, with food choices, eating less meat and more vegetables may involve facing some difficulties or renouncing to some pleasures. However, it may be necessary to make these decisions, for the sake of the environment, or for respecting animals' welfare.

Argumentation about SSI may reveal these conflicts; it addresses complex issues involving consideration of a wide range of dimensions, pieces of information and viewpoints (Aikenhead, 1985). SSI contexts provide opportunities to face open-ended questions from a multidisciplinary perspective. According to Morin et al. (2014), dealing with these issues involves considering complexities and uncertainties associated to problems without a clear-cut solution, and which reflect a variety of value systems and social representations. As Kolstø (2005) argues, in SSI argumentation decisions are not solely based on knowledge, being rather a result of the interaction between knowledge and values, to which we would add emotions. Values, then, are necessary in order to assess the desirability of the different potential consequences of alternative decisions. Drawing from Kolstø (2006), by values we mean those ideas a person appeals to as criteria or justifications used as the basis to evaluate the desirability of a given action, consequence or conclusion.

It also needs to be noted that there are different argumentation contexts, so, for instance, there are differences in argumentation in the context of evaluation of a causal explanation, and in the context of making a decision (Jiménez-Aleixandre & Brocos, 2018). Thus, we suggest that, in SSI contexts, shifts in acceptability of the options are related to dynamic interactions among emotions, appeals to scientific evidence, and other dimensions such as cultural identities or ethical concerns. We also suggest that in the case of the evaluation of explanations, the subject is often individual, but in decision making the acceptability is generally considered in a broader social context. The study that we present focuses on groups' discourse, rather than on individual discourse.

12.3 Methods and Instructional Context

The methodological approach is qualitative, appropriate to study processes (Creswell, 2013) and knowledge evaluation practices. It seeks to identify patterns through systematic analysis in order to understand the meanings of participants in their context (Merriam & Tisdell, 2016). Eighty-five pre-service primary teachers participated in the case study, in the context of an argumentation teaching sequence as part of a science methods course, constructing arguments in small groups ($N = 20$) about sustainable and healthy diets. Data collection included multiple sources, student teachers' written products (pre-test, portfolio reports, and final essays) and video recording of oral debates. The final written reports are the main focus of this study. Participants' discourse was analysed with rubrics constructed in interaction with data, according to constant comparative analysis (Glaser & Strauss, 1967). Groups are identified with a code formed by the seminar number and the small group number, from 1.1 to 4.5.

Table 12.1 Dimensions of teaching Critical Thinking in Higher Education

Intervention/strategy/practice	Characterization
Teaching CT as a content	CT is explicitly taught (there are objectives and lessons about CT) and is modelled in the classroom
Practice: Tasks intended for University students to develop CT	Tasks which promote CT skills and disposition development in students are designed and students engage in them
Transfer of practice: Tasks intended for University Students to transfer CT to their professional target (school pupils)	Tasks which promote transfer of CT skills and disposition development in future professional target

The context for the study is a 15 weeks science education course taught by the second author, which included six tasks about argumentation, implemented in a Faculty of Education in Galicia, a predominantly rural region in Northwest Spain, in which omnivorous diets are the norm. The design of the teaching sequence is discussed in greater detail elsewhere (Brocos & Jiménez-Aleixandre, 2020a). In the sequence, content about CT is framed in a mixed approach. Our preliminary proposal for what is involved in teaching CT in Higher Education, summarized in Table 12.1, distinguishes three dimensions, which can be conceived as types of practices. Research shows that explicit instruction combined with practice yields better results (Heijltjes et al., 2014).

- Teaching CT as content: It corresponds to the explicit facet; we characterize it in terms of objectives and lessons about CT, as well as modelling it during instruction.
- Practice: University students' engagement in tasks intended for them to develop CT's skills and dispositions. The task discussed here would correspond to this dimension.
- Transfer of practice: University students' engagement in tasks intended to promote transfer to their professional target, in our study the school pupils.

In task 6 of the teaching sequence about argumentation, participants were asked to construct, working in small groups, a written argument choosing a diet and justifying their choice in the available evidence and based in values, taking into account five criteria (cultural-personal, environmental, ethic, nutritional, economic). The task design provided students with five dossiers of information, each one of them focused on one of the five criteria considered. These dossiers included conflicting pieces of information that could be used to support opposing options. Thus, most information related to the environmental and ethic criteria would arguably point towards the adoption of vegetarianism over an omnivorous diet: for instance the significantly lesser water and carbon footprint derived from the production of plant-based food compared to those of animal origin; or the concerns about animal rights and wellbeing. On the other hand, the data related to Galician economy –which is highly dependent on livestock and fishing industry– and Galician food culture –which highlights the social and gastronomic prestige of animal products– would support opting for an omnivorous option.

12.4 Findings

12.4.1 Critical Character: Open-Mindedness and Consideration of Inconvenient Evidence

As a way to explore dimensions related to critical character, which involves open-mindedness and disposition to consider evidence which would contradict previous beliefs and threaten established worldviews, we pay attention to how participants collected and used evidence pertaining to the environmental and ethic criteria in their argumentative discourse. As stated earlier, most information included in the dossiers related to these criteria would point towards the benefits of vegetarian diets, which are, in the Galician context, marginalized options.

Environmental data were considered in 18 of the 20 written reports. Data about pollution, and efficiency in the use of land and energy were used in more than half of the written reports. From the 63 instances of environmental data identified, the overwhelming majority (58) were used to indicate that plant-based diets produce a lower environmental impact. The remaining 5 pieces of information were used by 4 groups to point out perceived negative outcomes of the mass adoption of vegetarianism, mainly related to potential pollution derived from agricultural practices. In another publication (Brocos & Jiménez-Aleixandre, 2020b) we present a detailed analysis of the use of environmental data in these reports.

The ethic criterion was addressed in 15 of the arguments. Most of them, 13 groups, used information related to animal wellbeing to highlight the ethical desirability of reducing or completely removing the proportion of animal products in the diet: "eating meat or animal products involves an underlying idea: animal suffering. Breeding animals in precarious conditions just to be killed and eaten afterwards implies inflicting pain in order to eat these animal products" (group 4.4). However, two of these 13 groups, besides expressing the ethical advantages of plant-based diets, also pointed out the fact that, in nature, "some animals eat each other" (group 2.4), which might be considered as a justification to accept human consumption of animals, although it is presented as an *argumentum ad naturam*, a kind of informal fallacy. The other two groups presented an equidistant position regarding the ethic criterion, considering that both vegetarian and omnivorous diets could be carried out in a more or less ethical way, and that one option should not necessarily be considered intrinsically better than the other one from an ethical point of view: "Analysing the fourth aspect, ethics, rather than choosing one [diet] or another in accordance to their advantages, we observe than in the case of the vegetarian diet instances of non-ethic production can also occur"(group 2.3).

These results indicate that most of the groups considered environmental and ethic information, and that they used this evidence to criticise the consumption of animal products and indicate advantages associated to plant-based diets, which implies putting their lifestyle and worldviews into question. Accordingly, we interpret these results as an indicator of open-mindedness and good disposition to consider evidence, even if it is disruptive.

12.4.2 Thinking and Social Responsibility: Making Decisions and Overcoming Conflicting Evidence

Assessing dispositional traits of CT such as independent thinking entails analytical difficulties, as stated earlier. We address this component of CT at group level, by analysing the decisions made by the participants and how they are justified. Particularly, we examine the existence and acknowledgment of conflicts between the evidence included in the written arguments and the conclusions defended, analysing how the conflict was justified and overcome. As conflict was embedded in task design, no option could meet all desirable criteria. Thus, the capacity of recognizing and making visible the shortcomings of the available options –and especially of one's own choice– is, we argue, a sign of intellectual honesty, and a good starting point for making a decision as an independent thinker who is capable of perceive that the issue is ill-structured. The way in which conflict is argumentatively justified and overcome indicates which aspects take pre-eminence over others, and as such, sheds light on how participants undertake and handle their decision-making process in terms of social responsibility.

First, we analyse the options chosen by each group (see Table 12.2). From the 20 written reports, 10 chose an option described by them as omnivorous diet with reduced intake of animal products, so, from a normative point of view, involving moderate changes; 7 selected an omnivorous diet with regular amounts of animal-based products, in other words, they followed the prevailing norm; 2 opted for an ovo-lacto vegetarian diet, an unconventional option; and one group chose a vegan diet, exclusively based on plant products, a diet which radically differs from social norms.

The examination of the reports revealed that in ten of them there was an acknowledgement of explicit conflict between the choice of a diet and the arguments displayed. In these reports the participants acknowledged and made transparent the trade-offs associated with each option. For instance, group 4.3, choosing an omnivorous diet, recognized the ethic and environmental advantages of vegetarian diets, and the difficulties involved in making a decision:

> Choosing a mixed [omnivorous] diet was a difficult task taking into account that, from the ethic and environmental point of view, the vegetarian diet would be the ideal one. [...]

Table 12.2 Conflict between decision and arguments in the written reports (N = 20)

Diet option	Omnivorous reducing animal products (O-)	Omnivorous (O)	Vegetarian (V)	Vegan (Vg)	Total
Number of reports	10	7	2	1	**20**
Conflicts between arguments and decision	6 EC	2 EC	2 EC		**10 EC**
	4 IC	4 IC			**8 IC**
		1 NC		1 NC	**2 NC**

Legend: *EC* Explicit Conflict, *IC* Implicit Conflict, *NC* No Conflict

> Despite this, the advantages of an omnivorous diet are still numerous, since the other dimensions (nutritional, economic, and cultural-personal) bring benefits.

Group 2.5, choosing a vegetarian diet, acknowledged the shortcomings of their choice and the negative economic consequences for Galicia of its full-scale adoption:

> As the last opposing factor that we find regarding this [vegetarian] diet, we can address the cultural and economic domain, focusing in our region, which is Galicia. Most of this cattle-breeding region's economy is based on the livestock, cannery, milk and fishing industries. This implies that a frequent usage of the vegetarian diet would reduce the economic level of our region.

In eight reports the conflict was implicit in the way the arguments were deployed. In them, conflict is inferred from the lines of reasoning developed in the argument, but contradictions and difficulties were not made visible. For instance, in this example from group 2.4, each criterion was evaluated independently, with no coherent crosslinks to justify the opposing claims and how they relate to their conclusion, an omnivorous diet:

> Cultural-personal dimension: [. . .] we choose an omnivorous diet because it is the one we have followed since we were little. [. . .] In this dimension [environmental] we find pros and cons for both the omnivorous and vegetarian diet. [. . .] Ethic dimension: we find mainly arguments in favour of a vegetarian diet, because [. . .] it is cruel to sacrifice animals for human consumption. [. . .] Conclusion: Seeing the pros and cons of every dimension this group chooses the omnivorous diet as the best option.

The choices were largely coherent with the arguments deployed in two reports, which do not acknowledge conflict. However, their arguments might be considered biased ("*cherrypicking*") in terms of the evidence presented, they failed to recognize the conflict regarding the issue, and they did not inform of the shortcomings of the elected options. Interestingly, these two groups elected extreme but opposite options, situated at both ends of the spectrum: omnivorous and vegan diets. In contrast, the results presented in Table 12.2 suggest that conflict is more likely to be recognized by groups choosing "intermediate" or compromising options: vegetarian (2/2 EC) and omnivorous diet with meat reduction (6/10 EC, 4/10 IC).

It should be noted that we do not evaluate the acceptability of the choices. We are not implying judgments about the approvability of the decisions –we do not imply that being a critical thinker means choosing a particular option, the one that better aligns with our position. Our objective is to examine how these pre-service teachers justify making a decision which conflicts with the evidence they present. A thematic analysis revealed some common patterns in the way in which the 16 groups, choosing either a standard omnivorous diet or with a reduction of meat, justified their conclusion, despite acknowledging the benefits of vegetarianism. These written arguments justify taking a conflicting choice using three types of justifications, which are not mutually exclusive:

1. *Anthropocentrism:* conflict between personal (we/our) and world rights; acknowledgement that some decisions could be better for the planet or for the wellbeing of other living beings, but they would come at a high economic or individual cost. They are present in 10 reports: 6 O, 4 O-. For instance, acknowledging that human preferences are placed above the other dimensions considered in this dilemma:

   > Most of the documents we were given, in relation with the different dimensions, support the vegetarian diet; however, from our point of view, we concluded that for human beings a mixed diet without excessive abuse of animal meat is preferable (group 4.3).

2. *Pragmatic*: practical difficulties for carrying out a vegetarian diet in the Spanish and Galician context, such as higher prices or scarce availability in restaurants, expressed in seven reports: 3 O, 4 O-. The following excerpts present pragmatic issues for eating vegetarian in restaurants, or for committing to a regular use of dietary supplements: "In current society is easier to be an omnivore [...] in catering, for instance, the supply of vegetarian meals is still inferior than meat, fish and seafood" (group 3.1); "In our opinion, the regular consumption of dietary supplements [in vegan diets] would amount to a questionable need, at least in terms of everyday convenience" (group 2.2).
3. *Cultural justifications*: alluding to the cultural context, the prestige of meat and the social difficulties for adopting a vegetarian diet. In 5 reports: 1O, 4 O-. The report of group 4.1 illustrates the role of the cultural and symbolic value of animal-based food in their decision:

   > Most of the gastronomic festivals that we know highlight animal-based products, especially meat and fish. Food is part of our tradition and it is very difficult to decouple it from it, because, in its own way, they are a component of our identity [...] with this kind of diet [omnivorous], we would not renounce to any cultural asset.

These results suggest that the conflicting dimensions of the issue were generally considered by the participants, especially by those electing intermediate options that would involve moderate changes, although not always explicitly. The shortcomings of their elections were generally recognized. The transparency in the acknowledgment of the underlying conflict and in the limitations of one's choices is, we argue, a sign of intellectual honesty and critical thinking. The analysis of the justifications for overcoming the conflict in the participants' arguments reveal the considerable weight of social and cultural influence on their decisions.

12.4.3 Critical Action: Change Towards Sustainability and Identification of Obstacles

Although the task was an argument about dietary choices, which did not demand to explain their personal choice or why would be difficult to carry on with it, the participants framed it as a real-world decision. In some cases, the proposal is considered in the wider social context, rather than just as an academic exercise.

We examine the component of critical action by analysing the existence of spontaneous references to the need for modifying social or individual behaviours, and the identification of obstacles preventing change in the written reports.

Table 12.3 summarizes these results. Seven of the 20 groups spontaneously highlighted the need or convenience for change in their arguments. These references are identified only in groups proposing O-, V or Vg, in other words, any option that involves modifying eating behaviours, appearing in 7 out of the 13. No references to the need for change appear in the 7 groups proposing O, which seems coherent with their conclusion, since these proposals would not entail an implementation of individual or social changes.

The convenience for change appears in different forms. In some cases, as direct appeals to act, involving public persuasion: "We encourage you all to reduce the presence of meat in your diets" (group 2.1). In others, is presented as the result of the reflection on the issue: "We consider that a restructuring should take place for avoiding the overconsumption of meat and in order to not generate as much polluting gas" (group 4.2). Specific changes in particular contexts are also suggested: "We think that a good option would be to include in school canteens, in order to educate kids on food, the Mondays without meat, in which meals, for a day, would not include any kind of meat" (group 4.5).

In contrast, we interpret signs of lack of empowerment in three groups. These include acknowledgments of the weight of social and cultural influence, which would prevent individual agency: "If a majority had a vegetarian diet we would choose it, so we can say that a diet is adopted because of heritage, fashion" (group 1.2); and reveal reluctances to act, which might be interpreted as a refusal to take responsibility on the issue: "There are many other pollution causes without having to directly influence on the diet we should follow" (group 2.3).

Thirteen groups pointed to the presence of obstacles preventing dietary changes. Six of these made no proposals to overcome them (4 O, 2 O-); while five of them did so (2 O-, 2 V, 1 Vg), and two (2 O-) included both obstacles with and without possible solutions for them. The characterization of these obstacles and the inclusion of ways to surmount them appears to be coherent, or at least related, to the options chosen. The lack of proposals was more frequent in groups with options that involved little or no change (4 O), while the inclusion of proposals appeared in all groups aiming at noticeable changes (2 V; 1 Vg). Groups electing O- fall somewhere in the middle, with more dispersion across categories.

Table 12.3 Analysis of dimensions related to agency in participants' discourse

Category	Sub-category	Number of reports (N = 20)
Disposition to action/ change	Highlighting the need or convenience for change	7 (6O-, 1V)
	Expressing lack of empowerment or disposition to act	3 (2O, 1O-)
Identification of obstacles for change	Identifying obstacles that prevent change with no proposals to overcome them	8 (4O, 4O-)
	Identifying obstacles and making proposals to overcome them	7 (4O-, 2V, 1Vg,)

Thematically, the most frequent obstacles pointed out in the reports were related to economic constraints, for instance in group 4.3: "in the case that we all started to follow a vegetarian diet, we would experience an enormous decrease in the economic aspect"; or in group 4.2: "quitting meat would be a great economic loss for us". Pragmatic and cultural difficulties are also highlighted. We identified proposals to overcome both economic: "A great new vegetable industry could be developed to substitute the reduction in cattle production" (group 2.5); and cultural issues, for instance highlighting the role of institutions and families in social change: "This would be an important cultural change, that is why it should be supported by food and nutrition education, in which the school plays an essential role, as well as families and institutions" (group 3.1).

Besides the written arguments, the action component was also identified in the individual portfolio reports, which the participants were asked to submit at the end of the course. Seven reports mentioned spontaneously that the task made them reflect on the need for a change of diet, emphasizing the impact of the task in their assumptions about diets. This issue was not the focus of the portfolio report, which was supposed to address pedagogical reflections about the tasks carried out throughout the course.

12.5 Discussion and Significance

The analysis of the participants' reports allows some insights into their dispositional traits of CT and their thoughts and perceptions about changing ways of life, in particular changing ways of eating. Although the task did not explicitly require them to justify personal choices, the context of the argumentation sequence promoted that choices needed to be justified, not only in terms of evidence or premises supporting them, but also in terms of evidence or premises contradicting them. Environmental responsibility and animal suffering were generally explicitly considered, but had limited impact on the conclusion. However, we argue that the analysis of the way in which participants use and take –or not– ownership of these data provided us with an opportunity to examine their critical character in terms of their consideration or dismissal of inconvenient, lifestyle-threatening evidence; confirmation biases; and intellectual honesty, captured in the accuracy of their recollection and interpretation of the information presented to them by their partners and the task design.

Conflict was present and recognized in most reports, although not always explicitly. This acknowledgment indicates the capacity of understanding the ill-structured nature of the task and its embedded conflict, which lays the ground for engaging in the practice of making an independent decision. Furthermore, these results show that students made their decisions transparent, at least to some degree, avoiding the impulse of disregarding uncomfortable, inconvenient evidence. The awareness of potential dissonances might trigger dissatisfaction with previous ideas and assumptions, which could facilitate the revision of one's worldviews and, thus, informs us of dispositional traits related to critical character.

The participants' reports point to some reasons perceived as difficulties hindering the conclusion that was coherent with their evaluation of evidence. The way in which participants justified their decisions and overcame conflicts in their arguments reveals how they valued and prioritised different dimensions of the issue, which has implications regarding their own perceptions about personal and social responsibility. The analysis of participants' arguments, in terms of the self-reported social influence in their opinions and decisions, points to a substantial weight of socially established ideas, and to the difficulties, experienced or anticipated, for taking non-mainstream options. The general attribution of such influence to society and culture may be interpreted also as a shifting of responsibility about sustainable and healthy diets onto society, although diets might be considered as personal options to much greater extent that, for instance, energy choices, which are more dependable on government regulations. The choice of mainstream diets was also influenced by socioeconomic concerns, and hence it was, at least to some extent, built on the basis of perceived responsibilities towards social well-being and traditional lifestyles of Galician population. However, we must observe that from an ESD point of view, the promotion of *ecocentric* perspectives (Kopnina, 2012) and a greater degree of commitment towards the environment would be more desirable in order to guarantee future sustainability.

To be clear, we do not think that being a critical thinker means opting for non-mainstream options, or for a specific option within a specific dilemma for specific reasons, perhaps those coinciding with the ideas which the teacher or, more generally, people in charge, have in mind. Quite the opposite: it is possible that students arguing for mainstream options show high CT skills and dispositions, particularly if they perceive a consensus in the audience, and, in that particular context, still decide to argue for a different option or perspective. However, we argue that choosing non-conventional options usually represents a greater opportunity (at a higher risk) to practice CT, and involves a higher potential of development of independent positions. This does not mean, of course, that if one chooses a radical option, one must be automatically considered a critical thinker. Choices are not enough: we need to take a look at how they are developed and justified. That's why we analyse the existence of conflict in the arguments and how it is solved. So, as an educational implication, we think that it is important that practitioners do not convey the idea that being an "original" or "different" thinker necessarily reveals criticality.

The step from critical thinking to critical action is not a smooth one, even when the action is hypothetical. The task did not required them to take actual action, but only to decide a hypothetical recommendation about which diet would be healthy and sustainable. Even being aware that their decision (in the report) did not mean any commitment to actually change their diet, they experienced difficulties in drawing conclusions coherent with the evidence considered, obstacles similar to those discussed by Macdiarmid et al. (2016) regarding social dietary changes towards plant-based diets. Agency towards sustainability involves new behavioural patterns and, as such, is dependent on the capacity of developing independent positions and challenge social ideas, to go beyond those established. In this sense, and for promoting sustainable agency, it might be worth exploring the role played by students' notion of culture and traditions, and how understanding them not as static entities, but dynamic, permeable and ever-changing, might contribute to open dispositions for facilitating large-scale changes towards sustainability.

Our study adds to the literature about criticality findings revealing explicit or implicit awareness of the conflict between evidence and final claims. Conflict is, we argue, an essential feature of learning environments designed to promote CT. If there are no possible conflicts, no space for controversies stemming from the interplay of values and points of view, there is no room for CT to be clearly enacted –no space of coherent options to be critical about. Our analysis presents some methodological limitations: it studies CT at the level of groups, and it is, in a way, an indirect approach, focusing on products derived from argumentative practice. Future research may develop more refined analytical approaches comprising detailed discourse analysis of oral interactions, focusing, for instance, on the influence of peer pressures on the capacity to develop independent opinions, challenge established ideas and increase dispositions towards agency.

Acknowledgements Work supported by the Spanish Ministry of Science, Education and Universities, partly funded by the European Regional Development Fund (ERDF). Contract grant PGC2018-096581-B-C22.

References

Abrami, P. C., Bernard, R. M., Borokhovski, E., Wade, A., Surkes, M. A., Tamim, R., & Zhang, D. (2008). Instructional interventions affecting critical thinking skills and dispositions: A stage 1 meta-analysis. *Review of Educational Research, 78*, 1102–1134.

Abrami, P. C., Bernard, R. M., Borokhovski, E., Waddington, D. I., Wade, C. A., & Persson, T. (2015). Strategies for teaching students to think critically: A meta-analysis. *Review of Educational Research, 85*(2), 275–314. https://doi.org/10.3102/0034654314551063

Acar, O., Turkmen, L., & Roychoudhury, A. (2010). Student difficulties in socio-scientific argumentation and decision-making research findings: Crossing the borders of two research lines. *International Journal of Science Education, 32*(9), 1191–1206.

Aikenhead, G. S. (1985). Collective decision making in the social context of science. *Science Education, 69*(4), 453–475.

Breiting, S., & Mogensen, F. (1999). Action competence and environmental education. *Cambridge Journal of Education, 29*(3), 349–353.

Brocos, P., & Jiménez-Aleixandre, M. P. (2020a). What to eat here and now: Contextualization of scientific argumentation from a place-based perspective. In I. Sánchez-Tapia (Ed.), *International perspectives on the contextualization of science education* (pp. 15–46). Springer.

Brocos, P., & Jiménez-Aleixandre, M. P. (2020b). El impacto ambiental de la alimentación: argumentos de alumnado de magisterio y secundaria. *Enseñanza de las Ciencias, 38*(1), 127–145.

Colucci-Gray, L., Camino, E., Barbiero, G., & Gray, D. (2006). From scientific literacy to sustainability literacy: An ecological framework for education. *Science Education, 90*, 227–252.

Creswell, J. W. (2013). *Qualitative inquiry & research design* (3rd ed.). Sage.

Davies, M., & Barnett, R. (2015). Introduction. In M. Davies & R. Barnett (Eds.), *The Palgrave handbook of critical thinking in higher education* (pp. 1–25). Palgrave Macmillan.

de Haan, G. (2010). The development of ESD-related competencies in supportive institutional frameworks. *International Review of Education, 56*(2), 315–328.

Food and Agriculture Organization. (2012). *Sustainable diets and biodiversity: Directions and solutions for policy, research and action*. Proceedings of the International Scientific Symposium, Biodiversity and Sustainable Diets United Against Hunger. FAO, Rome, Italy.

Garrison, J., Östman, L., & Håkansson, M. (2015). The creative use of companion values in environmental education and education for sustainable development: Exploring the educative moment. *Environmental Education Research, 21*(2), 183–204.

Glaser, B. G., & Strauss, A. L. (1967). *The discovery of grounded theory: Strategies for qualitative research*. Aldine.

Heijltjes, A., Van Gog, T., & Paas, F. (2014). Improving students' critical thinking – Empirical support for explicit instructions combined with practice. *Applied Cognitive Psychology, 28*(4), 518–530.

Ideland, M., & Malmberg, C. (2015). Governing 'eco-certified' children through pastoral power: Critical perspectives on education for sustainable development. *Environmental Education Research, 21*(2), 173–182.

Intergovernmental Platform on Biodiversity and Ecosystem Services. (2020). *IPBES workshop on biodiversity and pandemics*. Retrieved from: https://ipbes.net/pandemics

Jiménez-Aleixandre, M. P., & Brocos, P. (2018). Shifts in epistemic statuses in argumentation and in conceptual change. In T. Amin & O. Levrini (Eds.), *Converging perspectives on conceptual change: Mapping an emerging paradigm in the learning sciences* (pp. 171–179). Routledge. ISBN 978-1-138-20540-6.

Kolstø, S. D. (2001). "To trust or not to trust, . . ." – Pupils' ways of judging information encountered in a socio-scientific issue. *International Journal of Science Education, 23*(9), 877–901.

Kolstø, S. D. (2005). *The relevance of values for coping with socioscientific issues in science education*. Paper presented at the European Science Education Research Association (ESERA) conference, Barcelona, Spain.

Kolstø, S. D. (2006). Patterns in students' argumentation confronted with a risk-focused socio-scientific issue. *International Journal of Science Education, 28*(14), 1689–1716.

Kopnina, H. (2012). Education for sustainable development (ESD): The turn away from 'environment' in environmental education? *Environmental Education Research, 18*(5), 699–717. https://doi.org/10.1080/13504622.2012.658028

Laurie, R., Nonoyama-Tarumi, Y., Mckeown, R., & Hopkins, C. (2016). Contributions of education for sustainable development (ESD) to quality education: A synthesis of research. *Journal of Education for Sustainable Development, 10*(2), 226–242.

Lotz-Sisitka, H., Wals, A. E., Kronlid, D., & McGarry, D. (2015). Transformative, transgressive social learning: Rethinking higher education pedagogy in times of systemic global dysfunction. *Current Opinion in Environmental Sustainability, 16*, 73–80.

Macdiarmid, J. I., Douglas, F., & Campbell, J. (2016). Eating like there's no tomorrow: Public awareness of the environmental impact of food and reluctance to eat less meat as part of a sustainable diet. *Appetite, 96*, 487–493.

Merriam, S. B., & Tisdell, E. J. (2016). *Qualitative research: A guide to design and implementation* (4th ed.). Wiley.

Mezirow, J. (2000). *Learning as transformation: Critical perspectives on a theory in progress*. Jossey-Bass.

Morin, O., Simonneaux, L., Simonneaux, J., Tytler, R., & Barraza, L. (2014). Developing and using an S^3R model to analyze reasoning in web-based cross-national exchanges on sustainability. *Science Education, 98*(3), 517–542.

Oliveira, A. W., & Ackerson, V. (2015). *Environmental agency in classroom argumentation: Crossing sociocultural and sociocognitive boundaries*. Paper presented at the ESERA conference, Helsinki, August 31–September 4.

Orr, D. W. (2017). Foreword. In B. Jickling & S. Sterling (Eds.), *Post-sustainability and environmental education: Remaking education for the future* (pp. vii–x). Springer.
Rieckmann, M. (2012). Future-oriented higher education: Which key competencies should be fostered through university teaching and learning? *Futures, 44*(2), 127–135.
Slavich, G. M., & Zimbardo, P. G. (2012). Transformational teaching: Theoretical underpinnings. Basic principles, and core methods. *Educational Psychology Review, 24*(4), 569–608.
Springmann, M., Godfray, H. C. J., Rayner, M., & Scarborough, P. (2016). Analysis and valuation of the health and climate change cobenefits of dietary change. *Proceedings of the National Academy of Sciences, 113*(15), 4146–4151.
Stehfest, E., Bouwman, L., van Vuuren, D. P., den Elchen, M. G. J., Eickhout, B., & Kabat, P. (2009). Climate benefits of changing diet. *Climatic Change, 95*, 83–102.
Stevenson, R., & Stirling, C. (2010). Environmental learning agency in diverse cultural contexts. In R. B. Stevenson & J. Dillon (Eds.), *Engaging environmental education: Learning, culture and agency* (pp. 219–238). Sense Publishers.
Thompson, S., Gower, R., Darmon, N., Vieux, F., Murphy-Bokern, D., & Maillot, M. (2013). *A balance of healthy and sustainable food choices for France, Spain, and Sweden*. World Wildlife Fund.
Tilbury, D. (2011). *Education for sustainable development: An expert review of processes and learning*. UNESCO. Retrieved from: http://unesdoc.unesco.org/images/0019/001914/191442e.pdf
Tilman, D., & Clark, M. (2014). Global diets link environmental sustainability and human health. *Nature, 515*(7528), 518–522.
Tytler, R. (2012). Socio-scientific issues, sustainability and science education. *Research in Science Education, 42*(1), 155–163.
United Nations. (2015). *Transforming our world: The 2030 Agenda for Sustainable Development*. Resolution adopted by the General Assembly on 25 September 2015. Retrieved from: http://www.un.org/ga/search/view_doc.asp?symbol=A/RES/70/1&Lang=E
United Nations Educational, Scientific and Cultural Organization. (2012). *The education for sustainable development sourcebook. Education for Sustainable Development in Action, Learning and Training Tools N° 4*. Paris: UNESCO. Retrieved from: http://unesdoc.unesco.org/images/0021/002163/216383e.pdf
United Nations Educational, & Scientific and Cultural Organization. (2017). *Education for sustainable development goals: Learning objectives*. UNESCO.
Wiek, A., Withycombe, L., & Redman, C. L. (2011). Key competencies in sustainability: A reference framework for academic program development. *Sustainability Science, 6*(2), 203–218.

Chapter 13
Epistemic Beliefs as a Means of Understanding Critical Thinking in a Socioscientific Environmental Debate

Kévin De Checchi, Gabriel Pallarès, Valérie Tartas, and Manuel Bächtold

13.1 Dealing Critically with Unfriendly Epistemic Contexts

Environmental and sustainability issues are of decisive importance for our society. As future citizens, students need to be able to take part in an informed way in debates on environmental socioscientific issues (SSIs) and to think and argue critically. Developing students' critical thinking (CT) about science and its links to societal issues has thus become a major challenge (Hazelkorn et al., 2015). Environmental SSIs are complex (Morin et al., 2017), as students need to combine knowledge from different disciplines with values and other people's opinions, in order to adopt an enlightened position and engage in critical argumentation. Learners also need to deal with knowledge uncertainties (Kampourakis, 2018), as these are a distinctive feature of SSIs. Lastly, students need to be aware of the openness of these issues: there are numerous reasonable answers to an SSI, none of them is self-evident and all must be argued (Oulton et al., 2004).

Students therefore need to be able to problematize, conceptualize, question, analyze, and argue on SSIs. These skills can be developed during the teaching of specific topics and domains, but only if teachers allow sufficient room for argumentation in their teaching (Schwarz & Baker, 2017). Argumentation is a key component of CT (Facione, 2000, 2011), and some authors consider the two to be somewhat similar (Groarke & Tindale, 2013; Kuhn, 2019). Nevertheless, as other authors

K. De Checchi (✉) · G. Pallarès · M. Bächtold
LIRDEF, University of Montpellier and University of Paul-Valéry Montpellier 3, Montpellier, France
e-mail: kevin.de-checchi@umontpellier.fr; gabriel.pallares@univ-reims.fr; manuel.bachtold@montpellier.fr

V. Tartas
CLLE, University of Toulouse, CNRS, Toulouse, France
e-mail: valerie.tartas@univ-tlse2.fr

B. Puig, M. P. Jiménez-Aleixandre (eds.), *Critical Thinking in Biology and Environmental Education*, Contributions from Biology Education Research,
https://doi.org/10.1007/978-3-030-92006-7_13

(Ennis, 2018) have pointed out, CT is a complex construct that involves not only argumentative skills, but also dispositions to use them. In other words, in order to apply these skills, students must not only master them, but also be disposed to use them (Facione, 2000; Kuhn, 2019).

Several studies have found that students' epistemic beliefs (i.e. their representations of knowledge and knowing) influence their argumentation (Iordanou et al., 2019; Kuhn et al., 2000; Mason & Scirica, 2006; Nussbaum et al., 2008). As SSIs involve various kinds of knowledge and opinions, we assumed that if we wanted to examine epistemic beliefs in this context, we would need to take account of students' representations not only of knowledge, but also of opinions. We expect this investigation to shed light on students' CT skills and dispositions to argue in the context of environmental education. Highlighting these links is important, as it can help teachers improve their teaching strategies for developing students' argumentation and their critical dispositions in relation to environmental SSIs. It would also enable us to ask about the nature of the links between CT and epistemic beliefs in the context of socioscientific argumentation, and more specifically in the context of environmental SSIs.

The main purpose of this chapter is to develop a theoretical framework that could connect CT to epistemic beliefs, defined as representations of both knowledge and opinions. We test this theoretical framework with an empirical study, looking at the arguments produced by six students during debates about environmental SSIs, and transcripts of interviews undertaken to elicit their epistemic beliefs. In the following sections, after setting out our theoretical framework, we describe this empirical study and analyze the two sets of data and their possible interconnections. We then discuss how our study opens up new avenues for developing environmental education and fostering students' critical argumentation on environmental SSIs.

13.2 Towards a Theoretical Framework Connecting Critical Thinking and Epistemic Beliefs

13.2.1 Critical Thinking

CT has been conceived by Ennis (2018) as "judging in a reflective way what to do or what to believe" (p. 136). As this is a rather vague description, CT has been given a variety of definitions, ranging from a very broad set of skills and dispositions to a list of specific behaviors (Ennis, 2011). There is nonetheless a consensus among many authors that there is a strong link between CT and argumentation. For example, each of the six *core critical thinking skills* highlighted by Facione and colleagues in their Delphi project (American Philosophical Association APA, 1990), namely *interpretation*, *inference*, *evaluation*, *explanation*, *analysis* and *self-regulation*, are closely linked to argumentation. By the same token, argumentation scholars often define an ideally *good* argumentation as one that contains *critical discussions* (Van Eemeren

& Grootendorst, 2004) or *critical questions* (Walton, 1989, 1996). This link can also be found in instructional contexts. For instance, Kuhn (2019, p. 147) observed that "inquiry and argument [. . .] get us closer to empirically identifiable skills or behaviors than does the term critical thinking, while capturing much of what critical thinking is envisioned to encompass." Following Kuhn's methodological concerns, we chose to consider CT through the lens of argumentation in this study, despite considering that CT cannot be reduced to argumentative skills and behaviors.

CT can be seen both as a set of skills and as a set of dispositions (Ennis, 2011). As Kuhn (2019, p. 148) noted, researchers currently hold that "critical thinking is at least as much a disposition as it is a skill or ability," as students may have the ability to put forward arguments to explain their opinion, but may not necessarily be disposed to use it. CT dispositions can be seen as the difference between *critical thinking* and *critical thinker*: the former is an activity that can be achieved with the use of specific skills, while the latter is the individual who can decide whether or not to use these CT skills. Accordingly, *dispositions* are linked to willingness to engage in CT. The Delphi project (APA, 1990, p. 6) listed some of the many dispositions of an *ideal critical thinker*. We consider this list relevant but to be viewed with caution, as dispositions are broad and can manifest themselves in a variety of ways in the context of argumentation. For example, in a debate, a student who is *open-minded* and *willing to reconsider* may be inclined to acknowledge that his or her opinion is not self-evident and attempt to argue in favor of it, but may also tend to take another person's opinion into account, try to understand it better, and either challenge it or agree with it.

Another issue is whether students develop all these critical dispositions at the same time. Some dispositions may be easier to develop, if they make more sense to students with regard to the current activity or topic. In the context of environmental SSIs, which is a favorable one for CT practice or development (Simonneaux, 2007), it may be useful to identify which kinds of epistemic beliefs can influence both the development and use of certain critical dispositions. So let us turn now to epistemic beliefs and the most accurate way of defining them.

13.2.2 Epistemic Beliefs

Epistemic beliefs can be defined as "beliefs that might be more or less independent, rather than existing in integrated fashion and developing in a coordinated sequence" (Hofer, 2004, p. 45). This definition implies that "there are multiple dimensions to be considered and these dimensions can be thought of independently, as well as together" (Schommer, 1990, p. 301).

In a developmental approach, Kuhn et al. (2000) identified three stages. At the *absolutist* stage, individuals view knowledge as a certain, objective entity supported by external sources. At the *multiplist* stage, they no longer view knowledge as an objective entity that can be acquired, believing instead that having a given item of knowledge is as a matter of choice. In other words, all individuals have the right to their opinions, and all opinions have equal value. Finally, at the *evaluativist* stage, knowledge is assumed to contain elements of uncertainty because it is constructed by individuals, but there are objective criteria for evaluating and comparing it, which allow this uncertainty to be reduced. More broadly, the developmental approach considers that all individuals move through the same increasingly elaborated stages that reflect the development of the criteria and/or strategies expressed by students to deal with their awareness of uncertainty (King & Kitchener, 2002).

In a dimensional approach, Hofer and Pintrich (1997) described the content of epistemic beliefs in terms of four dimensions separated into two components. The first component concerns the *nature of knowledge (i.e.* what an individual believes knowledge is) which includes the dimensions *certainty of knowledge* and *simplicity of knowledge*. The second component concerns the *nature of the process of knowing* (i.e. how an individual comes to know), and contains the dimensions *justification for knowing* and *source of knowledge* (Hofer & Pintrich, 1997). Moreover, Chinn et al. (2011) proposed to study epistemic beliefs more comprehensively and argued "for a fine-grained, context-specific analysis" in terms of five dimensions: epistemic aims and epistemic values; the structure of knowledge and other epistemic achievements; the sources and justification of knowledge, and the related epistemic stances; epistemic virtues and vices; and reliable and unreliable processes for achieving epistemic aims.

Context-dependency has been supported by several empirical studies, which have shown that epistemic beliefs vary notably according to the academic discipline (Kuhn et al., 2000) or SSI (e.g. Khishfe et al., 2017). Zeidler et al. (2009) observed, for example, that students may be at different stages of epistemic beliefs, depending on which SSI they are being asked about. In this regard, we chose to focus our attention on epistemic beliefs in the context of argumentation, and more specifically in the context of environmental SSIs.

13.2.3 Epistemic Beliefs and Socioscientific Argumentation

Many studies investigating the influence of epistemic beliefs on argumentation have shown that the more elaborated these are, the better individuals argue (e.g. Kuhn et al., 2000; Mason & Scirica, 2006). More specifically, elaborated epistemic beliefs have been observed to lead to better reasoning (Zeineddin & Abd-El-Khalick, 2010), or more alternative arguments and better coordination of facts and hypotheses (Kuhn, 1991). Nussbaum and Bendixen (2003) also observed epistemic beliefs influence the way students engage in argumentative activities. More precisely these authors pointed out, for example, that students who consider knowledge to be certain and simple state that "arguments were anxiety-promoting" (p. 3) and tend to avoid dealing with them. A correlation has been found between certain dimensions of epistemic beliefs (according to Hofer & Pintrich, 1997) and argumentative skills: *justification from knowing* develops in the same direction as the argumentative quality of written productions (Ferguson & Bråten, 2013; Mason & Scirica, 2006), while *certainty of knowledge* develops in the opposite direction to willingness to engage in argumentation (Nussbaum & Bendixen, 2003). Moreover, the nature of the topic being discussed influences the way in which students argue: the latter do not argue in the same way if the debate concerns a scientific issue or an SSI (Simonneaux, 2007; Zeidler et al., 2009; Zeineddin & Abd-El-Khalick, 2010). The particular features of the topic under discussion should therefore be taken into account in order to describe epistemic beliefs in a situated way in the context of socioscientific argumentation.

Links between epistemic beliefs and students' argumentation have been brought to light through the use of interviews or questionnaires within the framework of SSIs (Barzilai & Weinstock, 2015; Kuhn, 1991; Mason & Scirica, 2006). In these studies, the authors presented students' epistemic beliefs as concerning only knowledge. However, the interview questions aimed at eliciting epistemic beliefs about knowledge also focused on opinions: "How sure are you of your view, compared to an expert?" (Kuhn, 1991, p. 175), "Can you say that one opinion is better and one is worse?" (King & Kitchener, 2002). The same is true for students' responses, in which knowledge and points of view were interwoven. For example, in reply to the question "Could someone prove that you were wrong?" (Kuhn, 1991, p. 175), students responded "No, they couldn't prove it [...] because it's my opinion [...]" (p. 181), "I was wrong, but I would probably not change my opinion. It's the result of lifelong personal experience and quite frankly, I think it is right. [...]" (p. 182). The first answer refers to an opinion, and the second answer concerns statistical knowledge that can be given to refute the student's proposal, lending more weight to one opinion than to another based on personal experience. It should be noted that these excerpts can contain different terms, such as *view*, *point of view*, *opinion*, *belief* and *position*. However, these terms are not always defined or explicitly considered in the literature on students' epistemic beliefs. In our research, we took them to be synonymous, and chose to use the term *opinion*. It is also apparent from these few examples that students' responses involved knowledge, opinions, and the connections between the two. This suggests that students may endorse different beliefs about knowledge and about opinions. Even though both kinds of beliefs appear to be expressed in students' responses, the latter have been less studied in the literature and defined as part of the epistemic beliefs to be taken into account in the context of argumentation on SSIs. The description of epistemic beliefs in this context remains restricted to beliefs about knowledge.

13.2.4 Research Questions

Our study aims to achieve a better understanding of the factors at play in high-school students' development of CT on environmental SSIs. In line with other authors (Facione, 2000; Groarke & Tindale, 2013; Kuhn, 2019), we take argumentation to be a major component of CT. More specifically, we assume that CT about environmental SSIs relies on both skills and dispositions to argue on them. As CT dispositions influence the use of CT skills (Facione, 2000), we chose to specifically study CT dispositions. As pointed out in other studies (Nussbaum et al., 2008; Zeidler et al., 2009), the quality of students' argumentation on SSIs is related to their epistemic beliefs. This leads us to regard epistemic beliefs as a key to understanding how students develop their CT on environmental SSIs. Moreover, as SSIs involve both knowledge and opinions, we investigate epistemic beliefs as representations of knowledge, but also opinions. Based on the literature, we predict that the more elaborated the students' epistemic beliefs are, the more they have developed CT dispositions to argue. Therefore, our research questions are: How precisely are students' epistemic beliefs related to their CT, especially to their dispositions to argue, in the context of environmental SSIs? Which features of their epistemic beliefs about knowledge and opinions are most important in this respect?

We begin exploring our research questions by separately describing the argumentation of six students during an environmental socioscientific debate and their epistemic beliefs on the same topic. We then cross-analyzed our data to show how some features of epistemic beliefs can help us better understand students' critical dispositions to argue.

13.3 Methodology

13.3.1 Context: The AREN Project and the Participants

This study is part of the *Argumentation et Numérique* (AREN) French project (the French word "*numérique*" means "digital"). The purpose of this project is to design an online debate platform (also called AREN) that promotes the development of students' argumentative skills and CT on SSIs, following a design experiment method (Sandoval, 2013). We developed a teaching sequence consisting of three phases: (1) a preparatory phase where students acquire knowledge on the topic; (2) an online debate on an SSI, mediated by the AREN platform; and (3) a synthesis phase, where students undertake a reflective analysis of the quality of the arguments produced during the debate.

The data were collected in two Grade 10 biology classes (mean age: 16 years) in two high schools located in the center of a city from south of France (around 250,000 inhabitants). The first is attended by students from mixed socio-economic backgrounds, and the second by students from middle and low socio-economic backgrounds. We analyzed the productions of six students, three from each class. The sample was composed of four girls (Silène and Hibiscus from the first high school, and Azalée and Crocus from the second high school) and two boys (Jonquille from the first high school and Muguet from the second high school). All original first names have been changed here. All six students were volunteers and were selected with the help of their teachers to reflect varying levels of involvement in class activities. In each class, the teaching sequence was implemented twice during the school year. We examine the second debate, which focused on an environmental SSI, use of renewable energy and/or fossil fuel.

AREN-mediated argumentation has several specific features. Argumentation on the platform is based on a text, which appears on the left side of the screen. Students can debate by posting comments on the right side of the screen. To do it, a student has to select some words, generally a full sentence, in the text or in a peer's comment. This triggers an argumentation pop-up asking the student to reformulate them, give an opinion on this sentence (color-coded: *Tend to agree* in blue/*Tend to disagree* in red/*Do not understand* in grey), and justify this opinion through argumentation. Students are free to fill the *argumentation* box as they wish: the platform induces arguments with a Toulminian structure (Toulmin, 1958), comprising a thesis (here, the opinion) and grounds (in the argumentation box), but this is the only extent to which students' arguments are structured.

As students can react to any part of the text or their peers' comments, argumentation on AREN is not linear, and can take an arborescent structure. It should be noted that there is no guarantee that all the students will actually read all the arguments of the debate, as they may limit themselves to reading only parts of the arguments that are developed in parallel. The reflective synthesis phase, at the end of the teaching sequence, ensures that students have read all the kinds of arguments produced during the debate.

13.3.2 Data Analysis

For our analysis, we first examine the arguments students produced during a debate on environmental SSIs. Second, we describe their epistemic beliefs about knowledge and opinions, based on thematic analysis. Third, we subject the features of both their epistemic beliefs and their argumentative practices to a cross-analysis.

13.3.2.1 Analysis of Students' Argumentation

Assuming that CT is instantiated in argumentation and that argumentation is methodologically the easiest way to evaluate CT (Kuhn, 2019), we chose here to determine students' CT by considering the arguments they produced during debates. To this end, we used a coding scheme developed in the frame of a previous study of the AREN project and applied to analyse about 2500 arguments (Pallarès, 2020) to evaluate the quality of students' socioscientific argumentation. In order to assess CT dispositions in students' argumentation, we link the dispositions listed in APA (1990) to items in the coding scheme. This scheme was based on the view that argumentation is both a dialogical process, in the context of a debate, and a monological process, in relation to students' reasoning (Jiménez-Aleixandre & Erduran, 2007). It was composed of what we called *argumentative moves*. For each of them, we also assessed whether students tried to justify their affirmation, or thesis, for instance using empirical data, examples or personal values.

Concerning CT dispositions, the frequent use of justifications can be linked to the disposition to inquiry, where data are a core component. These argumentative moves, and the precise ways in which they could be related to CT dispositions, are described in Table 13.1, where the last column show examples (underlined) of the argumentative moves, in their context of enunciation. It should be noted that an argument, treated here as the product of an argumentative process (Jiménez-Aleixandre & Erduran, 2007), could contain more than one argumentative move.

We also analyzed the monological aspects of socioscientific argumentation, namely the content of the arguments. We recorded the occurrence of a domain of validity, awareness of uncertainties relative to knowledge, and the socioscientific domains taken into account in the arguments. Each of these indicators, examples for them, and the precise way in which they could be related to CT dispositions, are described in Table 13.2.

13.3.2.2 Analysis of Epistemic Beliefs

The interviews served to elicit students' epistemic beliefs, that is, beliefs about knowledge, opinions and the link between the two. These interviews were conducted after a preparatory phase and before a debate in class. The preparatory phase allowed the students to study definitions and knowledge related to environment in a biology class. They were therefore prepared in terms of knowledge content and knew that they would be debating in a future session on a theme related to what they had studied in biology lessons.

Before the interviews, the researcher explained to students that the aim was not to judge or evaluate what they said, and there were no right or wrong answers. A statement related to the socioscientific theme seen during the preparatory phase: "Human activities that enable economic and social development should not be changed just because they might cause the disappearance of animal or plant species" was then shown to the students, who were asked to express their agreement or disagreement with it. This statement had been previously tested within the AREN project. Each interview lasted about 15 min, was audio-taped and transcribed. We prepared an interview guide featuring nine questions, developed by the first author and their validity discussed with the third and fourth authors. To ensure that the questions were well formulated and understood, the interviews were tested on eight students. Their responses ensured that the questions were well understood and had the potential to elicit students' beliefs about knowledge (Q6 & Q8), opinions (Q2, Q7 & Q9), and the relationship between the two (Q3, Q4 & Q5) (Appendix).

We ran a multistep thematic analysis of the interview transcripts. The first step consisted in describing for each student her or his beliefs about knowledge, opinions, and the link between the two. In the second step, these analyses were compared so as to identify common areas and specific themes. We chose to conduct a thematic analysis first, based on the students' responses, instead of an analysis based on the dimensions established by Hofer and Pintrich (1997) or Chinn et al. (2011). This choice was justified by the fact that we did not know beforehand whether the

Table 13.1 Description of argumentative moves (dialogical aspects of argumentation)

Argumentative moves	Description	Possible links to APA (1990) CT dispositions	Example (translated from French; the relevant parts of the excerpts are underlined)
Concession	Involves the acceptance of another's justification or thesis.	"Flexibility" and "willingness to reconsider" which may concern the thesis or justification one is ready to accept.	Accidents can happen in nuclear power plants (even if it's rare) Nuclear accidents are rare but may be more frequent in the future because nuclear plants grow old
Refutation of the thesis	Counterargumentative move, focused on another's thesis and intended to undermine it.	"Inquiry process," in rebutting with sound data erroneous hypotheses, use of "reasonable criteria," which may be the kind of processes involved when evaluating another's thesis with the aim of refuting it.	I think solar energy is the best, because it does not pollute and is infinite, do not emit greenhouse gas, however we need solar panels and it's expensive Solar energy is not efficient enough, we can't even power a city without another energy
Refutation of the justification	Counterargumentative move aimed at denying a justification put forward by another student.	"Inquiry process," "prudence in making judgments," use of "reasonable criteria," which may be the kind of processes involved when evaluating another's justification with the aim of refuting it.	People's mind is changing, thanks to recycling people care about the planet There's nothing to do with recycling, in any case recycling doesn't prevent millions of people to litter plastic or metallic trash
Nuance	Partial refutation of another's thesis or justification aimed at by pointing out its limitations.	"Fair-mindedness in evaluation," "prudence in making judgments," which may concern another's thesis or justification.	Accidents can happen in nuclear power plants (even if it's rare) Nuclear accidents are rare but may be more frequent in the future because nuclear plants grow old
Development	Intended to complete or extend another student's thesis or justification, by proving further justification or clarification.	"Trustfulness of reason" which may lead to develop another's reasoning expressed in an argument.	[Nuclear accidents] are rare but may be more frequent in the future because nuclear power plants grow oldSo we have to repair them or build new ones

(continued)

Table 13.1 (continued)

Argumentative moves	Description	Possible links to APA (1990) CT dispositions	Example (translated from French; the relevant parts of the excerpts are underlined)
New idea	Consists in considering an idea or point of view which was not discussed before during the debate.	"Open-mindedness" and "inquisitiveness," which may concern a new idea or point of view concerning the topic being disputed.	[In a discussion about nuclear waste and what to do with it] Nuclear waste are generally buried deeply

Table 13.2 Description of content of arguments (monological aspects of argumentation)

Content of the arguments	Description	Possible links to APA (1990) CT dispositions	Example (translated from French; the relevant parts of the excerpts are underlined)
Awareness of a domain of validity for assertions	Identification of the cases in which the argument/thesis can be applied or clarification of the degree of trust in the conclusion.	"Fair-mindedness in evaluation" and "prudence in making judgments," which may consists in identifying the degree of trust and the domain of validity.	Yes [nuclear energy] is one of the best energies from a climatic point of view but not a good energy for its local consequences which are terrible
Awareness of the uncertainties	Expression of specific reservations about the certainty of knowledge or showing prudence in considering the development of technologies.	"Prudence in making judgments," "reasonableness in the selection of criteria" to evaluate "results which are as precise as the subject and the circumstances of inquiry permit," what might amount to taking into consideration the uncertainties related to the situation.	For now there is no energy which both respects the environment and sustainable Which is why we need to find energies like this and if it doesn't exist we'll have to use other means!
Socioscientific domains	Domain(s) which are involved in an argument. Eight domains have been identified: Scientific, Technical, Economic, Political, Social, Axiological (values), Sanitary and Environmental.	"Orderliness in complex matters," which can consist in tackling the SSI systematically in all its complexity by considering its different domains.	We have to [*Axiological*, moral imperative] find other energies with similar capacities as nuclear [*Technical* features] but without being dangerous! [risks for *Health* and *Environment*]

dimensions identified in the literature to describe beliefs about knowledge would be equally suitable to describe beliefs about opinions and the links between both. The thematic analysis allowed us to identify four themes: *opinion and knowledge*, *certainty of knowledge*, *certainty of opinion*, and *possibility and means of obtaining a better opinion*. For each theme, we categorized the types of responses given by students. Moreover, in each theme, we distinguished between students' epistemic beliefs according to the richness of their elaboration. Based on King and Kitchener (2002), we judged the relative elaboration of epistemic beliefs on two main criteria: richness of the awareness of uncertainty, and complexity of criteria and/or strategies for obtaining the best opinion available.

13.4 Results Concerning Students' Arguments and Epistemic Beliefs

In this section we first examine the arguments students produced during a debate on environmental SSIs. Second, we describe their epistemic beliefs about knowledge and opinions, based on thematic analysis. Third, we subject the features of both their epistemic beliefs and their argumentative practices to a cross-analysis.

13.4.1 Analysis of Students' Argumentation and Epistemic Beliefs

The results of the analysis of students' argumentation are summarized in Table 13.3. One "argument" is defined as corresponding to one posted comment. It may consist of several argumentative moves or no argumentative move at all (e.g. "I completely agree").

The analysis of students' epistemic beliefs allowed identifying four themes: *opinion and knowledge*, *certainty of knowledge*, *certainty of opinion*, and *possibility and means of obtaining a better opinion*. The main results regarding epistemic beliefs are summarized in Table 13.4.

13.4.2 Cross-Analysis of Argumentation and Features of Epistemic Beliefs

The interviews indicated different profiles of epistemic beliefs among students, based on the four themes (*opinion and knowledge*, *certainty of knowledge*, *certainty of opinion*, and *possibility and means of obtaining the best opinion*). By the same token, concerning argumentation in the socioscientific computer-mediated debate,

Table 13.3 Analysis of students' argumentative moves and content of arguments (numbers in parentheses correspond to the fraction of argumentative moves with a justification)

Argumentativemoves	Azalée	Crocus	Hibiscus	Jonquille	Muguet	Silène
Arguments (Total)	5	8	6	9	12	7
Concessions	2 (1)	2 (0)	2 (0)	0 (0)	1 (0)	0 (0)
Nuances	0 (0)	2 (1)	4 (1)	1 (0)	3 (2)	0 (0)
Refut. thesis	1 (0)	1 (1)	0 (0)	0 (0)	3 (3)	0 (0)
Refut. justif.	1 (1)	0 (0)	0 (0)	0 (0)	1 (1)	1 (1)
Developments	1 (0)	5 (3)	1 (1)	6 (3)	5 (4)	3 (1)
New ideas	1 (0)	0 (0)	1 (1)	2 (1)	1 (0)	1 (0)
Domain of validity	2	0	3	1	3	2
Uncertainties	0	0	1	0	0	3
Socioscientific domains	4 Technical 1 Scientific 1 Environ.	5 Scientific 4 Technical 2 Environ. 2 Health 1 Social 1 Without domain	3 Environ. 3 Social 2 Health 2 Technical 1 Scientific	4 Technical 3 Environ. 2 Health 1 Political 1 Axiological 2 Without domain	7 Technical 5 Environ. 3 Social 1 Scientific	5 Technical 3 Axiological 3 Environ. 2 Health 1 Without domain

Table 13.4 Types of student responses by epistemic beliefs theme

Epistemic beliefs theme	Opinion and knowledge	Certainty of knowledge	Certainty of opinion	Possibility and means of obtaining the best opinion
Azalée	Different nature of opinion and knowledge	Source uncertainty	Source uncertainty	Possible if we do some research and ask someone who knows
Crocus	Explicit reference to a link between opinion and knowledge	Source uncertainty	Source uncertainty	Impossible but some opinions are better than others, depending on the arguments
Hibiscus	Explicit reference to a link between opinion and knowledge	Uncertainty related to learning, source uncertainty	Source uncertainty	Impossible but some opinions are better than others, depending on whether we ask a specialist
Jonquille	Explicit reference to a link between opinion and knowledge	Almost certain	Uncertainty owing to the nature of opinion	Impossible right now, but time will tell
Muguet	Different nature of opinion and knowledge	Almost certain	Uncertainty owing to the nature of opinion	Possible if we take the most likely opinions in relation to scientific theory
Silène	Different nature of opinion and knowledge	Almost certain	Source uncertainty	Possible if we ask a specialist

we found various kinds of critical arguments, in terms of argumentative moves. The first theme, *opinion and knowledge*, allowed us to make a clear distinction between two groups of students. This led us to analyze the features of the arguments produced by each group. In order to make sense of the specificities of their respective arguments, we considered the degree of elaboration of students' epistemic beliefs.

The two groups differed on their belief about a link between knowledge and opinions: The students Azalée, Muguet and Silène in Group 1 did not mention any link, whereas the students Crocus, Hibiscus and Jonquille in Group 2 explicitly acknowledged and described a link between knowledge and opinions: Crocus, "your opinion is formed from your knowledge"; Hibiscus, "I will form a opinion based on what I know"; Jonquille, "an opinion is formed from what we have seen, what we have heard" (knowledge being identified here by the student to personal experience). These two groups produced similar numbers of arguments during the debate, 24 arguments for Group 1 and 23 for Group 2. However, they differed on the nature of these arguments. Group 1 produced more diverse argumentative moves, whereas Group 2 seemed to focus mainly on developments and nuances. Moreover, Group 2's arguments featured a better combination of the socioscientific domains than Group 1's.

Looking more closely at the diversity of the argumentative moves produced by Group 1, each of the students in this group produced a refutation of another student's justification, which is a complex critical argumentative move (Sampson & Clark, 2008), especially as these three refutations of justifications were each accompanied by a justification of their own. For example, answering another student's claim "if we stop nuclear power plants before an alternative is found we'll run out of energy" the refutation produced by Azalée was: "there are a lot of different energy sources [refutation of the necessity of finding an alternative], there is not only nuclear, but also solar or photovoltaic energy [justification of this refutation: examples of alternatives which already exist]".

By contrast, only two of the students in Group 2 (Crocus and Hibiscus) produced four or more different kinds of argumentative moves, namely developments, concessions, nuances, and refutation of the thesis (Crocus) and new idea (Hibiscus). Another difference was that the arguments of students in Group 2 were more focused on developments and nuances than on counter-arguments and refutations. For example, Crocus produced an argument combining a concession and a justified nuance (even if it was on erroneous grounds): "Each industry has risks, but nuclear is one of the less lethal energy sources (less than solar or wind)."

Regarding the exploration of SSI complexity, students in Group 2 mostly produced arguments tackling more than one socioscientific dimension, whereas students in Group 1 mostly made arguments tackling only one dimension at a time. The specificity of the domains tackled (e.g., if a student tackled more technical or social matters) did not seem to differ across the two groups of students. For example, Hibiscus produced an argument that simultaneously considered the technical, environmental and social domains: "today there is no BEST energy, namely productive enough [for the needs of our society], cheap, risk-free AND nonpolluting."

For Group 1, we also analyzed the complexity of the criteria and/or strategies expressed by students to obtain the best opinion available and their awareness of uncertainty, in order to determine which epistemic beliefs were more elaborated (King & Kitchener, 2002). Muguet set out not only criteria (e.g., "better from someone who knows") but also a genuine strategy for obtaining the best opinion available: for him, opinions had to be compared with scientific theory to decide which one was the most trustworthy. For Azalée, to obtain the best opinion, it was important to do research and ask a knowledgeable person. We interpret that this process was less elaborated than the one described by Muguet, as it referred more to criteria than to an actual strategy. Silène considered that the best opinion must be one that came from a specialist, and she therefore described the least elaborated way of obtaining the best opinion. Muguet produced more advanced argumentative moves than Azalée and Silène did. Furthermore, Muguet produced more refutations than Azalée, who made more refutations than Silène. Muguet was also the only one of the six students to question another student.

Regarding Group 2, it was awareness of uncertainty that appeared to be decisive in differentiating between the degrees of elaboration of students' epistemic beliefs. Hibiscus seemed to have the most elaborated epistemic beliefs: "there are things that I know, well maybe it is not true [...] with more advanced knowledge [...]. So I think there are some knowledge that are safe, that everybody learns and it's a reality and some others that aren't necessarily. [...] you really need to be specialized in a field to have more advanced knowledge." Crocus seemed to have less elaborated epistemic beliefs than Hibiscus, as she only considered uncertainties about the source of knowledge. Jonquille had even less elaborated epistemic beliefs, as for him, knowledge was always almost certain. Incidentally, Hibiscus's argumentation was more critical than that of Crocus and Jonquille: she produced fewer arguments, but was more focused on nuances (four nuances on six arguments, which is a lot considering that nuancing is a complex critical move), associating them with concessions, and showing considerable awareness of the domain of validity in her arguments (three arguments out of six). Furthermore, Hibiscus tackled more than one socioscientific domain in almost all her arguments (five out of six). Similarly, Crocus's arguments were better than those of Jonquille: her nuances were combined with concessions, she produced a refutation of the thesis, and explored SSI complexity more.

Overall, regarding argumentation and epistemic beliefs, Muguet (Group 1) and Hibiscus (Group 2) were the ones who produced the most critical arguments and who had the most elaborated epistemic beliefs. However, these two students did not argue in the same manner, and their epistemic beliefs differed in one important respect (i.e., expression or not of a link between knowledge and opinions). Muguet was the one who produced the greatest variety of argumentative moves, including justified refutations and questions. We consider that Muguet had the most highly elaborated epistemic beliefs, based on the complexity of the strategy he described for obtaining the best opinion. By contrast, Hibiscus produced a great many nuances and developed arguments related to several domains. We consider that Hibiscus produced the most highly elaborated epistemic beliefs, based on her awareness of uncertainty.

13.5 Discussion and Conclusion

The goal of this study is to highlight links between epistemic beliefs and CT dispositions. As such, our research questions were: How exactly are students' epistemic beliefs related to their CT, and more specifically to their dispositions to argue, in the context of environmental SSIs? Which features of their epistemic beliefs about knowledge and opinions are the most important components in this respect?

In the wake of findings that epistemic beliefs seem to be linked to the way in which students argue, the study provided a new and more fine-grained analysis, offering a mean of defining the relationships between specific features of epistemic beliefs and ways of participating in a computer-mediated discussion on an environmental SSI. Previous studies had found that the more elaborated individuals' epistemic beliefs are, the better they argue (e.g. Kuhn, 1991; Mason & Scirica, 2006). Our study shed further light on this influence. Students may produce arguments focused more on nuances, or focused more on refutations. Furthermore, it seems that when epistemic beliefs are elaborated, with respects to King and Kitchener's (2002) criteria, students' arguments tend to become more critical, with more nuances (than developments) and more refutations.

Our cross-analysis yields two main results. First, students could be categorized according to whether they ignored the link between knowledge and opinions, or whether they acknowledged and explicitly described it. Indeed, students produced different arguments, depending on whether or not they drew this link: students who ignored it made various argumentative moves and were the only ones to refute justifications, while students who explicitly described this link focused on developments and nuances, and produced more complex arguments from a socioscientific perspective. Second, these specific argumentative features seemed to be related to students' awareness of uncertainties of knowledge and/or their strategy for obtaining the best opinion. The more elaborated students' epistemic beliefs regarding these two aspects, the better they argued. Hibiscus and Muguet, the students with the most elaborated epistemic beliefs, exhibited the most critical argumentation, but in different ways, as Hibiscus saw a link between knowledge and opinions, whereas Muguet did not.

Concerning CT, it should be noted that argumentative moves, be they in the form of nuances or refutations of justifications, can be linked to the same evaluative dispositions, namely *reasonableness of the selection of criteria*, *fair-mindedness in evaluation*, and *prudence in making judgments* (APA, 1990). This is in line with the literature, which indicates that evaluation is a crucial component of critical argumentation (Facione, 2000, 2011; Groarke & Tindale, 2013; Van Eemeren & Grootendorst, 2004; Walton, 1996). Moreover, even if Hibiscus and Muguet differed on their epistemic beliefs, they expressed the same CT dispositions and had the most elaborated epistemic beliefs. It therefore seems that the more elaborated their epistemic beliefs, the more students were disposed to CT. However, the nature of students' epistemic beliefs about the link between knowledge and opinions led them to operationalize these same dispositions in different ways. For instance, by contrast with other students who explicitly drew a link between knowledge and opinions, Hibiscus's arguments contained a high proportion of "nuances" moves. Meanwhile, Muguet's arguments contained a greater variety of argumentative moves and more refutations than those of other students who neglected the link between knowledge and opinions. There were other characteristic features of students' arguments, namely the use of justification for students who ignored the link between knowledge and opinions, and socioscientific complexity for students who explicitly drew such a link. Concerning CT, these features referred to different dispositions, namely *focus in inquiry* (i.e., a set of procedures and criteria appropriate for making reasonable judgments), and *orderliness in complex matters* (i.e., dealing with and organizing complexity in specific issues) (APA, 1990). It seems that when students argue, they are more focused either on inquiry or orderliness, depending on whether they neglect or consider respectively the link between knowledge and opinions.

Furthermore, epistemic beliefs appear to be more elaborated when students become aware of the uncertainty of knowledge, viewing it as a flawed product that does not reflect reality as it is, but rather an approximation of it (King & Kitchener, 2002; Kuhn et al., 2000). In this regard, it should be noted that, as in the study by Mason et al., (2011), most of our participants seemed primarily focused on the credibility of the source, rather than uncertainties inherent to knowledge itself. This may be explained by the fact that such uncertainties, to be acknowledged, require a developed epistemological view on the nature of knowledge. Our results argue in favor of considering epistemic beliefs not only about knowledge, but also about opinions, as well as the link between them, in order to highlight links with CT in the context of environmental SSIs. This would provide means to investigate precisely how the link between knowledge and opinions drawn by students influences the way they perceive content relating to different SSI domains. For example, depending on their epistemic beliefs, do students perceive arguments relating to the technical and scientific domain as involving only knowledge, and arguments relating to both the social and moral domains as involving only opinions (Kuhn et al., 2000)? The link between epistemic beliefs and CT seems to be a complex one, and needs to be studied with regard to the features of both knowledge and opinions, in order to highlight their influence on students' argumentation. Our cross-analysis focused on two specific aspects of epistemic beliefs: the link between opinion and knowledge, and the elaboration of epistemic beliefs regarding the *certainty of knowledge* and the *possibility and means of obtaining the best opinion*. However, epistemic beliefs can be described from many other aspects (e.g., dimensions proposed by Chinn et al., 2011). Furthermore, the operationalization of critical dispositions appears to differ across contexts: previous studies showed that students' arguments vary according to the SSI being debated (Pallarès, 2020; Pallarès et al., 2020), as well as students' epistemic beliefs (Zeidler et al., 2009). Implementing teaching sequences in the context of other SSIs or in nonsocioscientific debates could yield more detailed data on the link between epistemic beliefs and CT. Finally, it should be noted that our study focuses on computer-mediated argumentation, which may have induced a very different operationalization of CT dispositions from oral argumentations, and further investigation is also needed in that direction.

Despite these limitations and the need for further research, as far as the implications for teaching are concerned, our study highlighted specific epistemic beliefs that should be fostered in environmental education, in order to improve the relevant CT dispositions and thereby students' socioscientific argumentation, making it more critical. This might provide an answer to the problem of students being uncritical of environmental issues that has been identified in previous research (Barthes & Jeziorski, 2012). First, to improve students' critical argumentation, the first set of CT dispositions that need to be fostered are linked to evaluation: *reasonableness of the selection of criteria*, *fair-mindedness in evaluation*, and *prudence in making judgments*. These three dispositions seem to be linked to the critical argumentative moves we observed, notably in students Hibiscus and Muguet. In this regard, one way of improving CT dispositions may be to develop some aspects of epistemic beliefs related to these critical dispositions. It might be useful for environmental education to foster the sort of epistemic beliefs exhibited by Hibiscus and Muguet, namely the link between knowledge and opinions and either uncertainties (Hibiscus) or the criteria for obtaining the best opinion (Muguet). Before or during a socioscientific debate, teachers could help students ask themselves about the links between knowledge and opinions, the uncertainties of knowledge, and strategies to articulate knowledge and opinions and to deal with these uncertainties. As one of the aims of environmental education is to foster students' CT about SSIs (Morin et al., 2014, 2017; Simonneaux, 2007), this focus on epistemic beliefs would be in line with its objectives.

Second, CT dispositions to *focus in inquiry* and *orderliness in complex matters* also appear important for fostering critical argumentation. Considering that the latter seems to induce a more complex argumentation from a socioscientific perspective, it might be preferable to focus specifically on this in the context of environmental education, which has to deal with complex SSIs. As Leung (2020) pointed out, students who only consider uncertainties about inquiry, and not about the nature of knowledge, may work well when they have to deal with well-established and reliable knowledge. However, this may be more problematic in a context where they have to argue about environmental SSIs, which are complex and uncertain (Morin et al., 2014, 2017).

Overall, this discussion shows that the relationship between students' epistemic beliefs and students' dispositions to argue about environmental SSIs remains a very complex question. However, it also points out that explicitly considering both knowledge and opinions in this respect opens up new avenues that deserve to be explored in future research.

Appendix: Questions of the Interview Guide

- Q1: What do you think about the statement? Do you agree?
- Q2: Would you say your opinion about this subject is certain?
- Q3: Who might have the best opinion on this?
- Q4: How can we obtain the best opinion/least bad opinion?
- Q5: What are the differences and similarities between an opinion and knowledge? What is knowledge?
- Q6: Is knowledge certain or uncertain?
- Q7: In comparison, is an opinion certain or uncertain?
- Q8: Does knowledge change over time?
- Q9: Does an opinion change over time?

References

American Philosophical Association. (1990). *Critical thinking: A statement of expert consensus for purposes of educational assessment and instruction*. ERIC document ED 315–423.

Barthes, A., & Jeziorski, A. (2012). What kind of critical university education for sustainable development? A comparative study of European students and social representations. *Journal of Social Science Education, 11*(4), 62–77.

Barzilai, S., & Weinstock, M. (2015). Measuring epistemic thinking within and across topics: A scenario-based approach. *Contemporary Educational Psychology, 42*, 141–158.

Chinn, C. A., Buckland, L. A., & Samarapungavan, A. L. A. (2011). Expanding the dimensions of epistemic cognition: Arguments from philosophy and psychology. *Educational Psychologist, 46*(3), 141–167.

Ennis, R. H. (2011). Critical thinking: Reflection and perspective, Part I. *Inquiry, 26*(1), 4–18.

Ennis, R. (2018). Critical thinking across the curriculum. *Topoi, 37*(1), 165–184. https://doi.org/s11245-016-9401-4/s11245-016-9401-4

Facione, P. A. (2000). The disposition towards critical thinking: Its character, measurement, and relationship to critical thinking skill. *Informal Logic, 20*(1), 61–84.

Facione, P. A. (2011 [2006]). *Critical thinking: What it is and why it counts*. Measured Reasons and The California Academic Press.

Ferguson, L. E., & Bråten, I. (2013). Student profiles of knowledge and epistemic beliefs: Changes and relations to multiple-text comprehension. *Learning and Instruction, 25*, 49–61.

Groarke, L. A., & Tindale, C. W. (2013). *Good reasoning matters! A constructive approach to critical thinking*. Oxford University Press Canada.

Hazelkorn, E., Ryan, C., Beernaert, Y., Constantinou, C., Deca, L., Grangeat, M., ... Welzel-Breuer, M. (2015). *Science education for responsible citizenship*. Report to the European Commission of the expert group on science education. European Union.

Hofer, B. K. (2004). Epistemological understanding as a metacognitive process: Thinking aloud during online searching. *Educational Psychologist, 39*(1), 43–55.

Hofer, B. K., & Pintrich, P. R. (1997). The development of epistemological theories: Beliefs about knowledge and knowing and their relation to learning. *Review of Educational Research, 67*(1), 88–140.

Iordanou, K., Muis, K. R., & Kendeou, P. (2019). Epistemic perspective and online epistemic processing of evidence: Developmental and domain differences. *The Journal of Experimental Education, 87*(4), 531–551.

Jiménez-Aleixandre, M.-P., & Erduran, S. (2007). Argumentation in science education: An overview. In S. Erduran & M. Jiménez-Aleixandre (Eds.), *Argumentation in science education* (pp. 3–27). Springer Science.

Kampourakis, K. (2018). Science and uncertainty. *Science & Education, 27*(9–10), 829–830.

Khishfe, R., Alshaya, F. S., BouJaoude, S., Mansour, N., & Alrudiyan, K. I. (2017). Students' understandings of nature of science and their arguments in the context of four socio-scientific issues. *International Journal of Science Education, 39*(3), 299–334.

King, P., & Kitchener, K. (2002). The reflective judgement model: Twenty years of research on epistemic cognition. In B. Hofer & P. Pintrich (Eds.), *Personal epistemology: The psychology of beliefs about knowledge and knowing* (pp. 37–61). Lawrence Erlbaum Associates.

Kuhn, D. (1991). *The skills of argument*. Cambridge University Press.

Kuhn, D. (2019). Critical thinking as discourse. *Human Development, 62*, 146–164.

Kuhn, D., Cheney, R., & Weinstock, M. (2000). The development of epistemological understanding. *Cognitive Development, 15*(3), 309–328.

Leung, J. S. C. (2020). Students' adherences to epistemic understanding in evaluating scientific claims. *Science Education, 104*(2), 164–192.

Mason, L., & Scirica, F. (2006). Prediction of students' argumentation skills about controversial topics by epistemological understanding. *Learning and Instruction, 16*(5), 492–509.

Mason, L., Ariasi, N., & Boldrin, A. (2011). Epistemic beliefs in action: Spontaneous reflections about knowledge and knowing during online information searching and their influence on learning. *Learning and Instruction, 21*(1), 137–151.

Morin, O., Simonneaux, L., Simonneaux, J., Tytler, R., & Barraza, L. (2014). Developing and using an S3R model to analyze reasoning in web-based cross-national exchanges on sustainability. *Science Education, 98*(3), 517–542.

Morin, O., Simonneaux, L., & Tytler, R. (2017). Engaging with socially acute questions: Development and validation of an interactional reasoning framework. *Journal of Research in Science Teaching, 54*(7), 825–851.

Nussbaum, E. M., & Bendixen, L. D. (2003). Approaching and avoiding arguments: The role of epistemological beliefs, need for cognition, and extraverted personality traits. *Contemporary Educational Psychology, 28*(4), 573–595.

Oulton, C., Dillon, J., & Grace, M. (2004). Reconceptualizing the teaching of controversial issues. *International Journal of Science Education, 26*(4), 411–423.

Pallarès, G. (2020). *Développer les compétences argumentatives de lycéens par des débats numériques sur des questions socio-scientifiques. Vers une didactique de l'argumentation et de l'esprit critique*. Unpublished doctoral dissertation, University of Montpellier, France.

Pallarès, G., Bächtold, M., & Munier, V. (2020). Des débats numériques pour développer les compétences argumentatives des élèves sur des questions socio-scientifiques? Recherches en Didactique des Sciences et des Technologies.

Sampson, V., & Clark, D. B. (2008). Assessment of the ways students generate arguments in science education: Current perspectives and recommendations for future directions. *Science Education, 92*(3), 447–472.

Sandoval, W. (2013). Educational design research in the 21st century. In R. Luckin, J. Underwood, N. Winters, P. Goodyear, B. Grabowski, & S. Puntambekar (Eds.), *Handbook of design in educational technology* (pp. 388–396). Taylor & Francis.

Schommer, M. (1990). Effects of beliefs about the nature of knowledge on comprehension. *Journal of Educational Psychology, 82*(3), 498.

Schwarz, B., & Baker, M. (2017). *Argumentation, dialogue and education: History, theory and practice*. Cambridge University Press.

Simonneaux, L. (2007). Argumentation in socio-scientific contexts. In S. Erduran & M. Jiménez-Aleixandre (Eds.), *Argumentation in science education* (p. 179–199). Springer.

Toulmin, S. E. (2008 [1958]). *The uses of argument* (18th ed.). Cambridge University Press.

van Eemeren, F. H., & Grootendorst, R. (2004). *A systematic theory of argumentation: The pragma-dialectical approach*. Cambridge University Press.

Walton, D. (1989). Dialogue theory for critical thinking. *Argumentation, 3*, 169–184.

Walton, D. (1996). *Argumentation schemes for presumptive reasoning*. Lawrence Erlbaum Associates.

Nussbaum, E. M., Sinatra, G. M., & Poliquin, A. (2008). Role of epistemic beliefs and scientific argumentation in science learning. *International Journal of Science Education, 30*(15), 1977–1999.

Zeineddin, A., & Abd-El-Khalick, F. (2010). Scientific reasoning and epistemological commitments: Coordination of theory and evidence among college science students. *Journal of Research in Science Teaching, 47*(9), 1064–1093.

Zeidler, D., Sadler, T., Applebaum, S., & Callahan, B. (2009). Advancing reflective judgment through socioscientific issues. *Journal of Research in Science Teaching, 46*(1), 74–109.

Chapter 14
Primary School Teachers' Understanding of Critical Thinking in the Context of Education for Sustainable Development

Eli Munkebye and Niklas Gericke

14.1 Introduction

Critical thinking (CT) is one of the key skills in the twenty-first century for the education profession to develop among young people to facilitate their success as individuals, citizens, and workers (European Commission, 2016). A common argument for this is that everyone must be able to critically relate to their own beliefs and defend them in a logical way. A healthy critical attitude protects an individual from being manipulated, deceived, and exploited (Vieira et al., 2011). Living and participating in a democratic society places many demands on citizens' critical thinking. Hence, CT is of great importance in many areas of an individual's personal life as well as in their role as a participant in society. This chapter will focus on CT within the context of education for sustainable development (ESD), because ESD has grown in importance within curricula in many countries, such as Norway (Ministry of Education, 2017). In contemporary curricula, ESD is described as a broad teaching approach that often replaces environmental education as a way to connect the human world (society and economy) with the world of other species (the environment). Therefore, ESD includes biology and environmental education but also transcends these domains (Gericke & Ottander, 2016).

E. Munkebye (✉)
Department of Teacher Education, Norwegian University of Science and Technology, NTNU, Trondheim, Norway
e-mail: eli.munkebye@ntnu.no

N. Gericke
Department of Teacher Education, Norwegian University of Science and Technology, NTNU, Trondheim, Norway

Department of Environmental and Life Sciences, Karlstad University, Karlstad, Sweden

B. Puig, M. P. Jiménez-Aleixandre (eds.), *Critical Thinking in Biology and Environmental Education*, Contributions from Biology Education Research,
https://doi.org/10.1007/978-3-030-92006-7_14

The underlying assumption of the importance of CT in ESD is the complexity of sustainability issues with no definite answers, and where values as well as knowledge form the base for decision making, leading to CT. The application of scientific advances has led to increasing complexity in today's society, which makes it more difficult to solve the challenges we face in achieving sustainability. To meet and overcome these challenges, political decisions must often be made between different interests relating to development versus sustainability. For example, a choice often needs to be made about whether to preserve a natural habitat or increase economic development based on the exploitation—and inevitable degradation—of that habitat. These issues do not have one correct answer, and there is often no right or wrong position; instead, plausible, different answers can be made depending on what position or argument people personally validate as being the most important. Such complex issues are suggested to be addressed in ESD by using a pluralistic teaching approach that embraces these different positions (Boeve-de Pauw et al., 2015). To address these challenges in a plural and democratic society, high-quality CT skills are crucial when validating the important choices that must be made (Davies & Barnett, 2015). Hence, CT is a key aspect in ESD (Wiek et al., 2011). However, despite this urge to develop CT among the younger generation to help them cope with complex sustainability issues, very little research has been conducted in the field, and for primary education, it is non-existent.

This chapter presents an interview study with Norwegian primary school teachers that aimed to explore the teachers' views and understanding of CT in an ESD context. The research questions were as follows: (1) How do primary school teachers understand CT?; (2) How do primary school teachers understand CT in an ESD context?

14.2 Critical Thinking—A Complex Idea

CT is often described as dispositions within an individual or as skills, such as judging and evaluating, and it is described as normative because specific criteria can be used in the assessment of its qualities (Bailin et al., 1999). Moreover, the relevance of criteria often depends on the domain or subject of interest in which CT are evaluated (Lai, 2011). One definition of CT that is often used is that of Ennis (1985, p. 45): "Critical thinking is reasonable reflective thinking focused on deciding what to believe or do." This section discusses what characterizes a critical thinker and the extent to which CT can be described as a general or a subject-specific skill.

A critical thinker is often described as someone who possesses the specific cognitive skills and sub-skills necessary for CT. Facione (1990) categorizes these skills into six core categories (*interpretation, analysis, evaluation, inference, explanation,* and *self-regulation*), which this study uses as analytical tools to discern teachers' understanding of CT. In contrast to skills that are manifested in actions, dispositions can be described as the inner motivations that stimulate a critical thinker to apply their critical skills (Facione, 1990). Facione (1990) categorized dispositions into two approaches: *to life and living in general* (e.g., inquisitiveness) and *to specific issues, questions, or problems* (e.g., clarity in stating the question). Herein, Facione's framework will guide our analysis of this aspect of CT. This study is interested in investigating which skills and dispositions teachers relate to when they reflect on CT as well as investigating what criteria teachers use when evaluating claims and arguments in their own thinking and teaching.

If a person merely possesses the basic skills and dispositions that define CT, it does not necessarily mean that this person acts like a critical thinker. Therefore, Davies and Barnett (2015) extended CT through the concept of criticality to include thinking, being, and acting. Through this concept of criticality, CT is defined as a trait (i.e., how a person behaves in the world). A person with this trait is critically oriented toward the outside world and acts in relation to it. In an ESD context, the ability to act is central to achieving a more sustainable world (Mogensen & Schnack, 2010). It is not enough to know what to do; one must also act sustainably based on a critical approach to environmental challenges. Therefore, this study aimed to see if this aspect of CT also appeared in teachers' understanding of CT in the conducted interviews.

Much of the research on teachers' understanding of CT has been at the level of higher education (e.g., Choy & Cheah, 2009), and no studies were found relating to primary schools. However, Alazzi (2008) examined Jordanian secondary and high school teachers' understanding of CT and found that teachers often exemplify CT as a disposition or attitude used in a specific situation, or they describe someone with CT as being inherently sceptical by asking questions. Howe (2004) compared Canadian and Japanese secondary and high school teachers' understanding of CT. The teachers ranked the top 10 of 50 keywords describing common conceptions of CT and answered an open-ended questionnaire. When looking at all the teachers' responses, *analysis* was the most preferred keyword, followed by *reasoning*, *drawing inferences*, *problem solving*, and *analytical skills*. The next five definers were *inductive reasoning*, *creative thinking*, *clarifying ideas*, *logical*, and *thoughtful judgements*. Hence, these teachers rated the cognitive and logical aspects of CT skills as important. Howe (2004) also found significant differences between Japanese and Canadian teachers. Canadian teachers understood CT as a cognitive domain, as shown above, while Japanese teachers understood CT as an affective domain and ranked conscientious judgements (e.g., *objectivity*) the highest, followed by intellectual engagement (e.g., *active participation*). Hence, CT might be viewed differently in various cultural contexts. We must be careful when interpreting and comparing results from different countries, as in this case where the data is drawn from Norway alone.

Although none of the reviewed studies included primary school teachers, similar patterns of varying ideas about CT among these teachers might be expected in this study. However, as mentioned earlier, most of the research on CT stems from higher education, where teachers are closely linked to their respective disciplines, a fact that might influence the results for their understanding of CT. In contrast, primary school teachers teach several school subjects that have less connection to academic disciplines. This might lead to more possible multidisciplinary and multidimensional perspectives focusing more on the learner rather than content, which, as argued below, is important in the context of ESD.

14.3 Critical Thinking in the Context of Education for Sustainable Development

ESD is a policy concept established in international steering documents originally from the UN and UNESCO. Within ESD, CT has become a key competence according to these documents, where it is summarized as "the ability to question norms, practices and opinions; reflect on own one's values, perceptions and actions; and take a position in the sustainability discourse" (UNESCO, 2018, p. 44). Over the years, ESD and CT have trickled down into many national curricula, such as the Norwegian (Ministry of Education, 2017). As policy concepts, both ESD and CT have been given vague definitions and often multiple meanings. Not only are they closely intertwined at the policy level, as shown in the quote above, but CT is described as a foundation of sustainability education according to a research review by Wiek et al. (2011). The following will discuss ESD and how it might relate to CT.

ESD is polysemous and has encompassed different meanings in the literature (Vare & Scott, 2007). In a research review of the ESD field, Eilam and Trop (2010) outlined a framework of four essential components of ESD pedagogy based on accumulated theory and experience in the field. The framework aspires to encompass most prevailing teaching approaches within a set of four basic principles. The first principle, the *traditional academic style of teaching and learning*, takes place within traditional school subjects and is decontextualized from the learning objective. This principle supports the development of analytical-rational modes of intelligence. The second principle is *multidisciplinary learning* (inter and/or multidisciplinary). This approach combines knowledge from a variety of disciplines. It is considered capable of supporting the acquisition of systemic thinking and the formation of linkages between cause and effect within systems. The third principle, *multidimensional learning*, can be achieved when the academic learning of principle one and the multidisciplinary system of principle two are added with time and space dimensions. It facilitates the development of contextual ways of thinking (Hopkins & McKeown, 2002) and the acquisition of abilities to think "out of the box" and investigate systems in their relations to other systems, spaces, and times. The fourth principle, *emotional learning*, involves the emotions in a learning activity that simultaneously activate processes of value and ethics clarification. In this way of learning, students are motivated to do activities that make them feel any type of emotion ranging from enjoyment to distress. In other words, they are led on a path of emotional learning—to care, according to Eilam and Trop (2010).

Further, Eilam and Trop (2010) argue that it is necessary to include all these principles for learning opportunities in ESD teaching to achieve action competence in line with sustainability (Mogensen & Schnack, 2010). However, the principles also represent a hierarchy where each step precedes the other in the development of true ESD. A question emerges: What does CT become in these learning principles? We would assume that CT is enacted differently in these different steps of ESD. This study will relate primary school teachers' understanding and practices of CT to these different learning principles of ESD. First, we will theoretically discern what CT might look like in these steps.

The ideas of CT as generic or subject-specific have been discussed in the field of CT research since the 1980s (e.g., Facione, 1990). McPeck (1981) and Ennis (1989) claimed that CT is expressed differently according to the discipline within which it is practised, and criteria for CT that apply in one discipline do not necessarily apply in another discipline. There is a broad consensus that content knowledge plays an important role when thinking critically, because—naturally—we need something to think critically about (Willingham, 2008). Content knowledge is necessary, as CT takes place within a context characterized by concepts, values, and beliefs. Hence, if teachers recognize subject-specific aspects of CT, we can see that they achieve the first step of ESD, according to Eilam and Trop (2010).

Most of the literature recommends that ESD should be taught in an interdisciplinary and/or multidisciplinary way that promotes holistic and pluralistic ideas for students to address complex problems (e.g., Sund & Gericke, 2020; Munkebye et al., 2020). ESD is commonly defined by environmental, economic, and social dimensions (Gough & Scott, 2003), i.e., referring to many school subjects of various domains. In this context, the role of the subject is not entirely clear because of the interdisciplinary and multidisciplinary nature of ESD. If the teachers recognized that CT should be addressed in many settings (i.e., from an inter- or multidisciplinary perspective), we would see traces of step two in Eilam and Trop's (2010) framework. As a study by Beilin and Bender (2010) suggests, the disciplines can be approached by teachers using a question-based teaching methodology possessing a willingness to transcend the cultural boundaries of the disciplines, as this opens new perspectives on CT. Apart from that study, so far, educational research has not addressed what CT becomes in interdisciplinary teaching settings. Does CT transform into something different in an ESD interdisciplinary setting? This study will address this question by asking the interviewed teachers about their view of CT both from a general perspective and from an ESD perspective and comparing the differences.

In ESD research, holism is referred to as a crucial subconstruct of ESD (Gericke et al., 2019). Holism can be defined as a holistic approach since it includes multiple perspectives on content that emphasise the necessity to include all three dimensions of sustainability (environmental, social, and economic) and focus on their interrelationship, as well as interactions over time and space (Gough, 2002). Holism agrees with step three of Eilam and Trop's framework (2010). Through education for CT, it would be possible to prepare students to deal with these complex problems of holism (Vieira et al., 2011). There are rarely simple answers to sustainability issues like climate change and the fair and equal distribution of natural resources. Instead, these issues often demand that the learner draw from their knowledge, ethics, and values to make decisions that must address goal conflicts, for example, the conflicting goals of sustaining the environment and developing the economy. It follows that CT is a crucial competence for students to discern and address these kinds of problems within ESD according to the third step of Eilam and Trop's (2010) framework.

In a study by Ampuero et al. (2015), it is suggested that empathy is an important addition to cognitive and logic skills and dispositions, as argued in most CT literature. They found a significant advantage in using empathy strategies in teaching to engage students in societal processes to solve environmental problems. This intriguing perspective is in accordance with the suggestions of other scholars that CT is situated and context-dependent (Davies & Barnett, 2015). Moreover, it closely relates to the fourth step or principle of ESD as defined by Eilam and Trop (2010). Hence, if emotions are included in teaching in relation to CT, the last step in ESD will be reached.

Eilam and Trop (2010) argue that ESD teaching often evolves from being characterized by the terms unidisciplinary, unidimensional, and rational to the terms multidisciplinary, multidimensional, and emotional when encompassing all four steps. The latter way of teaching could be denoted as pluralistic teaching. According to Boeve-de Pauw et al. (2015), pluralism is one of the main characteristics of ESD. In education, pluralism emphasises a democratic approach by striving to promote different perspectives, views, values, and emotions when dealing with authentic questions and problems concerning the future of our world (Lijmbach et al., 2002). Moreover, pluralistic teaching has been found to affect students' self-reported pro-sustainability actions (Boeve-de Pauw et al., 2015). We would argue that this teaching approach meets the demand of incorporating all four steps that are asked for in the framework of Eilam and Trop (2010).

Pluralistic teaching involves critically evaluating different perspectives, asking questions, considering other people's arguments, formulating valid arguments, and having an open and fair mindset; these are skills and dispositions that also characterize CT. It is suggested that equipping students with CT enables them to participate in a sustainable society when leaving school (Willingham, 2008). In this study, primary school teachers were asked about how they understand CT and how it relates to ESD. This chapter will discuss in what ways the teachers' understanding of CT reflects the different steps of ESD according to Eilam and Trop (2010).

14.4 Methods, Context, and Participants

A semi-structured interview protocol (Kvale & Brinkmann, 2009) was developed, inspired by previous studies that examined teachers' understanding of CT (Alazzi, 2008; Choy & Cheah, 2009). The first questions inquired about teachers' competences, as well as which subjects and grades they taught. The teachers were then asked to express their understanding of CT. They could speak freely, and afterward, they were asked follow-up questions about skills, dispositions, and what criteria they used when assessing CT. Next, they were asked how they taught about CT and asked to provide examples from their own teaching. Finally, they were asked about their understanding of CT in the context of ESD. This sequence of questions was meant to give teachers an opportunity to obtain a broader picture of CT before linking CT to ESD. For the interview guide, see the Appendix.

The interviews were transcribed and then studied using thematic analysis (Braun & Clarke, 2006). The dataset was encoded with an initial open coding where different understandings of CT were highlighted. The initial codes were then grouped, and new categories were developed. Through repeated, systematic reviews of the dataset considering the categories and codes, themes were gradually developed that represented the content of the dataset (e.g., source criticism, argumentation, and need for knowledge). Then, the themes were categorized as skills or dispositions and organized, where possible, according to Facione's (1990) overview of CT skills and dispositions.

As an analytical framework for CT skills, we used Facione's (1990) categories of skills into six core dimensions. First, *interpretation*: skills such as categorization, decoding significance, and obtaining clarification. Second, *analysis*: skills that can be used to explore ideas and uncover and analyse arguments. Third, *evaluation*: skills to evaluate assertions and arguments by determining relevant criteria for assessment and assessing their relevance when using them. Fourth, *inference*: skills for judging whether one has sufficient information, skills to obtain necessary information, and skills to conjecture alternatives and draw conclusions. Fifth, *explanation*: the ability to state results, justify procedures, and present arguments. Sixth, *self-regulation*: skills to self-monitor cognitive activities and to correct errors and weaknesses that are revealed.

Facione's framework was also used for the analysis of CT dispositions (Facione, 1990); see Table 14.1.

This study focuses on Norwegian primary school teachers' understanding of CT. Norwegian compulsory education is divided into two main levels: primary school (grades 1–7) and lower secondary school (grades 8–10). The compulsory education system is based on the principle of equitable education for all and is primarily financed by the municipality. All schools follow the same national curricula (Ministry of Education, 2020).

Recently, a new Norwegian core curriculum has been developed that includes sustainable development as one of three interdisciplinary themes (Ministry of Education, 2017). CT has also been highlighted; according to the reform's overarching goals, students should "become curious and ask questions, develop scientific and CT, and deal with ethical awareness" (Ministry of Education, 2017, p. 7). CT is interpreted and operationalized in different subject-specific ways. This means that CT is formulated both as a general ability and in more specific ways depending on what subject is to be studied.

Ten primary school teachers teaching grades 5–7 were interviewed about their understanding and practice of CT; see Table 14.2 for an overview of the participants.

Nine of the 10 teachers were selected from a professional development program named Sustainable Backpack programme (further described in Munkebye et al., 2020), because we wanted sustainable development to be a familiar concept to them. This programme aims to increase awareness for sustainable development, understanding, and competencies among teachers and students of primary and secondary schools, although there is little explicit attention given to CT.

Table 14.1 Analysis framework

Disposition approaches to life and living in general	
Subcategories	1. Concerns of becoming and remaining generally well-informed 2. Inquisitiveness regarding a wide range of issues 3. Alertness to opportunities to use CT 4. Trust in the processes of reasoned inquiries 5. Self-confidence in one's own ability to reason 6. Open-mindedness regarding divergent world views 7. Flexibility in considering alternatives and opinions 8. Understanding the opinions of other people 9. Fair-mindedness in appraising reasoning 10. Honesty in facing one's own biases, prejudices, stereotypes, and egocentric or sociocentric tendencies 11. Prudence in suspending, making, or altering judgments 12. Willingness to reconsider and revise views where honest reflection suggests that change is warranted
Disposition approaches to specific issues, questions, or problems	
Subcategories	13. Clarity in stating the question or concern 14. Orderliness in working with complexity 15. Diligence in seeking relevant information 16. Reasonableness in selecting and applying criteria 17. Care in focusing attention on the concern at hand 18. Persistence when difficulties are encountered 19. Precision to the degree permitted by subject and circumstances

Dispositions According to Facione (1990)

Table 14.2 Information of participants

Distribution	Participants teach at different schools in five municipalities, except two who work at the same school.
Geographical location	Central Norway Six of them are in urban locations.
Teaching grades	All teachers work with students aged from 9 to 11 years.
Teaching experience	All teachers have between four and 30 years of teaching experience.
Teaching subjects	One subject: T7[a]; two subjects: T1, T5; three subjects: T8; four subjects: T9; more than four subjects: T2, T3, T4, T6, T10 Natural science: T1–10; Social studies: T4, T6, T10; Language: T2, T3, T6, T10

[a]The letter and numbers in brackets are the identifiers of the teachers

14.5 Results

First, teachers' understanding of CT and the criteria they use for critical assessment is presented. Then the focus is shifted to what teachers do in the classroom to promote student CT. Finally, how teachers understand CT in the specific context of ESD is presented.

14.5.1 Teachers' Understanding of CT

CT as Scepticism Eight of the 10 interviewees understood CT as an attitude people possess where they do not accept everything they are told. A couple excerpts from interviews highlight this, where some teachers said this attitude means to "face things with a certain, it may be wrong to use the word scepticism, but with a little attention to the things you read and hear (1)" and "to meet information with a healthy scepticism (10)". Two teachers pointed out that students must understand the difference between CT and negativity, and one of them said: "They [the students] often associate criticism with something negative, and it does not have to be so. It takes time before they can see that to be a critical thinker is not just to be negative about something". Hence, most of the teachers first and foremost viewed CT as scepticism.

CT Skills and Dispositions When teachers described their understanding of CT, most of them referred to what they considered to be CT skills (Table 14.3).

When it came to dispositions, the teachers described five types that coincided with some categories of Facione's (1990) framework (Table 14.4).

None of the teachers addressed whether CT could be understood diversely when viewed from the perspective of different disciplines or school subjects. The teachers seemed to mostly talk about CT as a generic ability. However, two teachers mentioned CT in specific relation to mathematics. They linked problem solving in mathematics with CT, as students are encouraged to be critical of their own solutions in mathematics and physics and to justify their solution strategies.

When the teachers were asked about what criteria they used when assessing something critical, few answers emerged. Seven teachers expressed the view that this was a difficult question that they had not really thought about. Only three of them answered. Teacher 5 argued that "whatever sounds logical" was a criterion, while teachers 7 and 8 stated that disciplinary aspects of content and methods within the natural sciences formed their standards against which they judged CT. Teacher 7 pointed out that the methodological aspects of science studies can vary and that, for example, an important quality criterion is to evaluate against the sample size of specific studies, where a large sample means better quality.

Six of the teachers stated they often promoted CT in their teaching, while only one said it rarely happens. The teachers mostly referred to activities in which they believe the students get to practise CT. In Table 14.5, the categories found in the analysis are outlined. As can be seen, they are largely in accordance with the understanding of CT (see Table 14.3) and support the notion that teachers teach in accordance with their perception of CT.

Table 14.3 Skills teachers associated with CT

Skills according to Facione (1990)	CT skills mentioned by the teachers[a]
Interpreting	
Clarifying meaning	Asking clarifying questions (1)
Analysing	
Examining ideas	Distinguishing between established knowledge and opinions (1) Seeing a case from multiple perspectives (4)
Detecting and analysing arguments	Looking for more sources that point in the same direction (1) Reasoning in relation to statistical representations (2)
Evaluating	
Assessing claims and arguments	Considering the credibility of the sources (9) Evaluating arguments against what is logical (1)
Inferring	
Drawing conclusions	Making choices based on their own assessments (1)
Explaining	
Justifying procedures	Justifying solutions in maths (1)
Stating results of one's reasoning	Making arguments for their own views (8) Making arguments for and against a topic (1)
Self-regulating	
Self-examining	Being critical of their own answers in maths and physics (1)

[a]The number of teachers providing these responses is presented in brackets

Table 14.4 Dispositions mentioned by teachers

Dispositions according to Facione (1990)[a,b]	Dispositions exemplified by teachers' quotes[c]
1. Inquisitiveness regarding a wide range of issues (6)	A kind of 'driving force' to find an answer. . . . enthusiasm or interest, I do not know what is best. Engagement maybe. (T10) Ask questions about what you hear. (T4)
2. Concerns about becoming and remaining generally well-informed (4)	To be able to judge whether something is true, knowledge is important. (T1)
6. Open-mindedness regarding divergent world views (8)	Being open to change their point of view. (T7) Acknowledge that they have different opinions. (T1) To listen. (T1) Taking someone else's perspective. (T4)
9. Fair-mindedness in appraising reasoning (1)	A sense of justice. (T10)
16. Reasonableness in selecting and applying criteria (2)	Having trust in science as criteria to evaluate against. (T7)

[a]The disposition number refers to the category number in Table 7.1
[b]he number of teachers providing these responses is presented in brackets
[c]The letter and numbers in brackets are the identifiers of the teachers

Table 14.5 Teachers' responses about their practices of CT

Teachers' teaching practices[a]
Approaching different texts and figures critically (10)
Exploring different perspectives through dialogue and/or texts (9)
Practicing CT through dialogue (5)
Problem solving (1)

[a]The number of teachers providing these responses is presented in brackets

Table 14.6 Teachers' different approaches to CT in ESD

Different approaches[a]	Examples of the approaches
CT as an overarching approach to ESD (3)	CT as a pillar of ESD Evaluating today's contribution to a better future
CT skills and teaching approaches included in ESD (5)	Argumentation (5), dialogue (2), discussing complex issues (2), and an exploratory approach (1)
Others (3)	Authenticity (1), environmentally friendly choices (1), and CT across disciplines and subjects (1)

[a]The number of teachers providing these responses is presented in brackets

14.5.2 Teachers' Understanding of CT in the Context of ESD

We now turn to the second research question about teachers' understanding of CT in the context of ESD. The interviewees found it difficult to express their understanding because they had not thought much about the relationship between ESD and CT. Table 14.6 provides an overview of the teachers' different approaches to CT in ESD.

Three of the teachers considered CT to be an overarching approach to ESD, as shown in this excerpt:

> [CT is] the essence of it [ESD]. If you have to think about ESD, it's about the complex issues and problems that do not have an answer. If you want to move forward in any way, you have to think critically and put arguments against each other. So, I think it is very difficult to carry out ESD without CT as a pillar. (T8)

Five of the teachers talked about teaching approaches and CT skills as argumentation. One teacher (T7) said that by approaching different environmental issues, such as the climate crisis and wolf/sheep problem, the students found arguments for and against them and identified and defended different points of view. Another teacher (T5) stated that students worked on approximating contradictory arguments, such as reducing the use of plastic to reduce microplastics in the ocean and packing food in plastic to increase durability and reduce food waste. The skills highlighted by the teachers were similar to those highlighted earlier in the interview.

Three of the teachers referred to teaching methods that form part of the Sustainable Backpack programme, such as presenting students with complex problems related to the local environment and making environmentally friendly choices, like choosing between farmed or wild salmon. According to the teachers, these approaches should help students develop skills related to argumentation. Based on these statements, we can see that primary school teachers teach CT from a broad and general perspective relating to complex problems. This is epitomised by a quote from Teacher 7:

> That you have the ability to think critically across disciplines, that you manage to think critically in a bigger picture. And because sustainability has three dimensions, there are many, three dimensions, and in each dimension, there are incredible amounts of information that you have to decide on. It says something about how difficult it can be to think critically about sustainability then. (T7)

14.6 Discussion

It is important to get insight into how teachers subjectively understand CT, as this is the starting point for equipping teachers to undertake the efforts required to promote students' CT. We will start by discussing teachers' understanding of CT. In our study, both skills and dispositions were mentioned by the teachers, although different aspects dominated. Moreover, none of the teachers included empathy, emotion, intuition, or imagination as aspects of CT. Hence, the more affective, cultural, and context-dependent aspects of CT—as suggested by Ampuero et al. (2015)–were not addressed by the teachers. When looking into what the teachers said about specific skills and dispositions in more detail, many of their comments can be seen to relate to an understanding of CT essentially as scepticism, possessing an inquiring attitude and applying source criticism.

Secondly, in this section, we interpret these results in the context of ESD. Interestingly, teachers thought of CT as the core of ESD, though they gave it more or less the same meaning as they did outside the context of ESD.

14.6.1 Primary School Teachers' Understanding of CT

Six of the teachers understood CT as a questioning and inquiring attitude; one teacher described it as a "healthy scepticism". Hence, it seems that the teachers' notion of CT is close to the concept of criticality (Davies & Barnett, 2015), i.e., that a student should be critically oriented toward the outside world and act accordingly. The teachers also emphasised that the students had to understand the difference between negativity and scepticism, as they believed many students understood being critical as something negative. Moore (2013) pointed out that scepticism can be understood as a propensity to judge in a negative way. This way of applying the term scepticism does not coincide with the teachers' use of the term. The teachers who used the term explicitly pointed out that it must not be understood in this way.

Motivation to find answers to complex questions was identified by some teachers as crucial to becoming a critical thinker at the primary school level. We consider this an important point, because possessing the skill of asking questions does not necessarily mean that one does ask questions. Mastering CT skills does not necessarily produce a critical thinker, since one must practise to be so (Bailin et al., 1999). Therefore, at the primary school level, it is not only about skills and dispositions but also about creating learning environments that enable students to use and further develop these skills and dispositions. Here, we would argue that ESD is an excellent context for cultivating CT skills, as will be discussed in the following section.

Looking at the skills teachers defined as CT skills, they largely coincide with Howe's (2004) findings related to the Canadian teachers at the secondary level, in which cognitive strategies to achieve CT were a dominant notion among teachers.

Source criticism was a CT skill referred to by four of the participants. These teachers reported that they frequently taught source criticism. In the Norwegian curriculum, source criticism is found in several subjects as part of digital skills, which is one of the basic skills that applies to all subjects. However, previous research shows that students often disregard source information and only focus on the content; this applies to all levels of education (Britt & Aglinskas, 2002). Furthermore, Norwegian secondary school students do not master source criticism as well as they should (Bråten et al., 2019). This is concerning, as source criticism is important in the twenty-first century because of the exponential growth and accessibility of information on the internet. It is also a key competence in ESD.

To conclude, it seems as though the teachers had a coherent idea of CT that was also manifested in their teaching practices. This is in line with Alazzi's (2008) study of Jordanian secondary school teachers. However, although the primary school teachers had a coherent idea, this idea represented a quite limited understanding and teaching of CT focusing on attitudinal aspects, such as scepticism, source criticism, and argumentation, as generic skills. Mostly cognitive strategies were used to achieve CT, but no assessment criteria were mentioned, and the influence of the subject was hardly addressed. In the next section, we will discuss how this way of perceiving CT might affect CT as an aspect of ESD.

14.6.2 Primary Teachers' Ability to Enact CT in ESD

In this section, the framework of Eilam and Trop (2010) will be used to analyse how the primary school teachers of this study could enact CT within ESD, what barriers there might be, and what aspects need to be addressed in future professional development and research. In this endeavour, we will explain the results generated from the interviews regarding teachers' general understanding and their practice of CT, as well as how they explicitly understood CT in relationship to ESD. As can be seen in the results section, three of the teachers recognized CT as an overarching feature of ESD, and five of them recognized it as a teaching approach exposing different arguments and perspectives on complex sustainability issues (i.e., similar to a pluralistic view on teaching ESD) (Boeve-de Pauw et al., 2015). Still, the teachers had difficulties outlining this more specifically. From these results, we can conclude that the teachers recognized the centrality of CT in ESD and possessed the willingness to teach accordingly, but they may lack the qualification to enact CT in ESD in line with that belief. That conclusion will be explored by analysing our results using the four-step framework of Eilam and Trop (2010).

The first aspect of ESD, according to Eilam and Trop (2010), is the traditional *academic style of teaching and learning* taking place within traditional school subjects. From our results, we can see that this rarely happens at the primary school level. Only two teachers referred to aspects of math and science where disciplinary analytical-rational modes were mentioned, such as problem-solving strategies. No referents were made to other disciplinary skills or rationales connected to other specific subjects, although source criticism could be argued to be more practiced in some school subjects like civics. Further, the lack of criteria for evaluating CT in the teachers' practices might be related to the fact that the primary school teachers do not connect CT to subject-specific traditions where they are familiar with the teaching traditions, including assessment practices.

The second aspect of Eilam and Trop's framework is *multidisciplinary learning*. This teaching approach combines knowledge from a variety of school subjects supporting the acquisition of systemic thinking. From our results, we can see that only one of the teachers referred to this principle, interdisciplinary, in their teaching practice of CT in ESD. The teachers pointed to aspects that look multidisciplinary and interdisciplinary at a superficial level, and some of them referred to the different dimensions of ESD (environment, society, and economy). However, less is explicated about what this would look like in practice. Instead, references were made to argumentative practices around specific topics, such as climate change, but not how these topics might be informed from different school subjects or disciplinary aspects. It seems that the teachers also ignore this principle to a large degree.

The third principal or step of ESD, according to Eilam and Trop (2010), is *multidimensional learning*, referring to holism, which includes multiple perspectives on content over time and space (Boeve-de Pauw et al., 2015). Several of the teachers mentioned that they worked with complex issues when teaching CT, and they highlighted skills and dispositions related to issues with multiple points of view (i.e., a holistic approach). The teachers also understood CT in ESD as taking and maintaining a stand and making arguments leading to choices, which are actions that can be linked to dealing with complex issues. To conclude, it seems that the primary school teachers participating in this study understood the discourse on CT as a multidimensional learning aspect of ESD.

The fourth principle of ESD relates to *emotional learning*, which is meant to activate the processes of clarifying values and ethics for the learner. In the results of this study, one teacher referred to "fair-mindedness" and one teacher to "respect to others". These responses may be regarded as belonging to this fourth teaching principle; otherwise, there were no references. Hence, the teachers almost leave out emotions entirely as important aspects for CT while emotions are claimed as central by Ampuero et al. (2015) and within sustainability education (Ojala, 2013).

Based on our analysis, we can conclude that these primary school teachers did not seem to follow the progression of CT in ESD building on the framework of Eilam and Trop (2010), as outlined in the background section of this paper. The teachers did not seem to teach CT, neither in their ordinary teaching nor in connection to

ESD, according to the first two steps (i.e., unidisciplinary, unidimensional, or multidisciplinary). Likewise, the emotional aspect of step four was left out. The results indicate that the primary school teachers viewed and enacted CT as a cognitive strategy demanding an inquiring attitude with the aim of outlining the multidimensional third aspect of ESD in a way that addresses different viewpoints of complex sustainability issues. An interesting question then emerges: What consequences might such a teaching approach have for the students?

The overall aim of ESD is to establish an action competence for sustainability in the learner (Mogensen & Schnack, 2010; Olsson et al., 2020; Sass et al., 2020). Therefore, the teaching approach cannot only address cognitive decontextualized learning, as in traditional teaching, but also contextualised, multidisciplinary, multidimensional, and emotional learning (Eilam & Trop, 2010). The teachers of this study only focused on the multidimensional aspects, and this does not necessarily lead to action competence. The primary school teachers did not seem to have the knowledge about or tools for what CT might be in a disciplinary or school subject context, apart from some statements relating to mathematics and science. Without the unidisciplinary perspective, the teachers could not articulate multidisciplinary aspects either. Thus, the teachers could not teach the analytical-rational modes of CT in ESD, as linked to the first two principles of Eilam and Trop's framework. This is also indicated by the teachers' inability to point out what criteria they used in assessing CT. The emotional principle, including value-laden aspects, is possibly left out by the teachers because this principle is outside the teachers' usual teaching traditions. This is potentially problematic because these aspects of teaching may be seen to comply with an indoctrination of the learner (see discussion with Jickling and Wals (2012) on the matter).

Therefore, the overall conclusion is that these teachers could have difficulties in teaching CT in the context of ESD in a way that promotes action competence due to their limited coverance of the teaching approaches needed. This was not expected considering several of the teachers understood CT as the essence of ESD, in line with research (Wiek et al., 2011) and policy (UNESCO, 2018). Our interpretation is that teachers have been following the international and national policy discourse around ESD, as, for example, outlined by UNESCO, which mainly addresses the multidimensional aspect of CT. Note that all but one of the teachers in this study participated in the Sustainable Backpack programme; therefore, they could have encountered perspectives and teaching methods related to ESD that influenced their understanding, although CT was not particularly expressed in that programme.

If the teachers of this study are representative of primary school teachers at large, we can conclude that there is a great need for professional development relating to CT in the context of ESD. Such professional development needs to first address what CT might be in different school subjects, such as biology, geography, mother tongue, and civics. The next step should address how multidisciplinary aspects of CT can be taught using tools and aspects of the different subjects. In this endeavour, CT as a rational practice based on scientific knowledge and argumentative practices of the disciplines should be developed into teaching approaches and pertinent evaluation criteria. Only thereafter would we recommend connecting these teaching practices to the multidimensional and holistic aspects of CT in ESD that today seem to dominate primary school teaching. This way, the learning of CT can be built upon deep insights from subject knowledge rather than societal discourse alone. As discussed by Gericke et al. (2020) and Sund and Gericke (2020), multidisciplinary teaching that builds on societal discourse and everyday knowledge would be pointless because all the subjects would provide the same perspectives. Finally, the emotional dimension, including values and ethics at both the individual and societal levels, needs to be addressed. Here, it would be fruitful to follow the advice of Davies and Barnett (2015) to shift from CT's individual focus (for example, a person's ability to relate to a situation critically to avoid being misled) toward a socio-cultural dimension. In an ESD context, CT's socio-cultural dimension seems particularly relevant, as it is about preserving the common good and includes a social focus and critical virtue in the form of ethics.

Acknowledgements This study was conducted within the project CriThiSE (https://www.ntnu.no/ilu/crithise), which is supported by The Research Council of Norway, project number 302774.

Appendix

Interview guide, inspired by Alazzi (2008) and Choy and Cheah (2009)
How many years have you worked as a teacher? What age levels do you teach for? What subjects do you teach?
What does CT mean to you? *Follow-up question:* What are the skills related to CT? *Follow-up question:* What are the dispositions related to CT? What standards/criteria do you use when evaluating the thinking of others?
Is there anything you do in the classroom that you believe fosters CT? How often do you involve students in CT? Do you think CT teaching is important?
What is, for you, critical thinking in an education for sustainable development context?

References

Alazzi, K. F. (2008). Teachers' perceptions of critical thinking: A study of Jordanian secondary school social studies teachers. *The Social Studies, 99*(6), 243–248.

Ampuero, D., Miranda, C. E., Delgado, L. E., Goyen, S., & Weaver, S. (2015). Empathy and critical thinking: Primary students solving local environmental problems through outdoor learning. *Journal of Adventure Education & Outdoor Learning, 15*(1), 64–78.

Bailin, S., Case, R., Coombs, J. R., & Daniels, L. B. (1999). Conceptualizing critical thinking. *Journal of Curriculum Studies, 31*(3), 285–302.

Beilin, R., & Bender, H. (2010). The getting of interdisciplinarity: The everyday practice of environmental curriculum design. In *Interdisciplinary higher education: Perspectives and practicalities* (pp. 181–193). Emerald Group Publishing Limited.

Boeve-de Pauw, J. B. D., Gericke, N., Olsson, D., & Berglund, T. (2015). The effectiveness of education for sustainable development. *Sustainability, 7*(11), 15693–15717.

Bråten, I., Brante, E. W., & Strømsø, H. I. (2019). Teaching sourcing in upper secondary school: A comprehensive sourcing intervention with follow-up data. *Reading Research Quarterly, 54*(4), 481–505.

Braun, V., & Clarke, V. (2006). Using thematic analysis in psychology. *Qualitative Research in Psychology, 3*(2), 77–101.

Britt, M. A., & Aglinskas, C. (2002). Improving students' ability to identify and use source information. *Cognition and Instruction, 20*(4), 485–522.

Choy, S. C., & Cheah, P. K. (2009). Teacher perceptions of critical thinking among students and its influence on higher education. *International Journal of teaching and learning in Higher Education, 20*(2), 198–206.

Davies, M., & Barnett, R. (2015). Introduction. In M. Davies & R. Barnett (Eds.), *The Palgrave handbook of critical thinking in higher education* (pp. 1–25). Springer.

Eilam, E., & Trop, T. (2010). ESD pedagogy: A guide for the perplexed. *The Journal of Environmental Education, 42*(1), 43–64.

Ennis, R. H. (1985). A logical basis for measuring critical thinking skills. *Educational Leadership, 43*(2), 44–48.

Ennis, R. H. (1989). Critical thinking and subject specificity: Clarification and needed research. *Educational Researcher, 18*(3), 4–10.

European Commission. (2016). *Communication from the Commission to the European Parliament, the Council, the European Economic and Social Committee and the Committee of the Regions: EU Action plan against wildlife trafficking*. European Commission.

Facione, P. (1990). *Critical thinking: A statement of expert consensus for purposes of educational assessment and instruction. The Delphi report*. The California Academic Press.

Gericke, N., & Ottander, C. (2016). On the issue of 'research in didactics of biology': Definitions and demarcations. In T. Tal & A. Yarden (Eds.), *The future of biology education research; A selection of papers presented at the 10th conference of European Researchers in Didactics of Biology* (pp. 155–162). The Technion Israel Institute of Technology and The Weitzmann Institute of Science.

Gericke, N., Boeve-de Pauw, J., Berglund, T., & Olsson, D. (2019). The Sustainability Consciousness Questionnaire: The theoretical development and empirical validation of an evaluation instrument for stakeholders working with sustainable development. *Sustainable Development, 27*(1), 35–49.

Gericke, N., Huang, L., Knippels, M. C., Christodoulou, A., Van Dam, F., & Gasparovic, S. (2020). Environmental citizenship in secondary formal education: The importance of curriculum and subject teachers. In A. C. Hadjichambis, P. Reis, D. Paraskeva-Hadjichambi, J. Činčera, J. Boeve-de Pauw, N. Gericke, & M. C. Knippels (Eds.), *Conceptualizing environmental citizenship for 21st century education* (pp. 193–212). SpringerOpen.

Gough, S. (2002). Right answers or wrong problems? Towards a theory of change for environmental learning. *The Trumpeter, 18*(1), 1.

Gough, S., & Scott, W. (2003). *Sustainable development and learning: Framing the issues*. Routledge.

Hopkins, C., & McKeown, R. (2002). Education for sustainable development: An international perspective. In *Education and sustainability: Responding to the global challenge* (p. 13). IUCN.

Howe, E. R. (2004). Canadian and Japanese teachers' conceptions of critical thinking: A comparative study. *Teachers and Teaching, 10*(5), 505–525.

Jickling, B., & Wals, A. E. (2012). Debating education for sustainable development 20 years after Rio: A conversation between Bob Jickling and Arjen Wals. *Journal of Education for Sustainable Development, 6*(1), 49–57.

Kvale, S., & Brinkmann, S. (2009). *Interviews: Learning the craft of qualitative research interviewing*. Sage.

Lai, E. R. (2011). Critical thinking: A literature review. *Pearson's Research Reports, 6*, 40–41.

Lijmbach, S., Van Arcken, M. M., Van Koppen, C. A., & Wals, A. E. (2002). 'Your view of nature is not mine!': Learning about pluralism in the classroom. *Environmental Education Research, 8*(2), 121–135.

McPeck, J. E. (1981). *Critical thinking and education-issues and ideas in education*. Martin Robertson Co.

Minestry of Education. (2017). *Overordnet del – verdier og prinsipper for grunnopplæringen* [Overarching part – Values and principles of primary and secondary education]. https://www.regjeringen.no/contentassets/53d21ea2bc3a4202b86b83cfe82da93e/overordnet-del%2D%2D-verdier-og-prinsipper-for-grunnopplaringen.pdf

Minestry of Education. (2020). *Facts and analysis of kindergarten, primary and secondary education in Norway*. http://utdanningsspeilet.udir.no/2016/en/

Mogensen, F., & Schnack, K. (2010). The action competence approach and the 'new' discourses of education for sustainable development, competence and quality criteria. *Environmental Education Research, 16*(1), 59–74.

Moore, T. (2013). Critical thinking: Seven definitions in search of a concept. *Studies in Higher Education, 38*(4), 506–522.

Munkebye, E., Scheie, E., Gabrielsen, A., Jordet, A., Misund, S., Nergaard, T., & Øyehaug, A. B. (2020). Interdisciplinary primary school curriculum units for sustainable development. *Environmental Education Research, 26*(6), 795–811.

Ojala, M. (2013). Emotional awareness: On the importance of including emotional aspects in education for sustainable development (ESD). *Journal of Education for Sustainable Development, 7*(2), 167–182.

Olsson, D., Gericke, N., Sass, W., & Boeve-de Pauw, J. (2020). Self-perceived action competence for sustainability: The theoretical grounding and empirical validation of a novel research instrument. *Environmental Education Research, 26*(5), 742–760.

Sass, W., Boeve-de Pauw, J., Olsson, D., Gericke, N., De Maeyer, S., & Van Petegem, P. (2020). Redefining action competence: The case of sustainable development. *The Journal of Environmental Education*, 1–14.

Sund, P., & Gericke, N. (2020). Teaching contributions from secondary school subject areas to education for sustainable development—A comparative study of science, social science and language teachers. *Environmental Education Research, 26*(6), 772–794.

UNESCO. (2018). Issues and trends in education for sustainable development. https://unesdoc.unesco.org/ark:/48223/pf0000261445

Vare, P., & Scott, W. (2007). Learning for a change: Exploring the relationship between education and sustainable development. *Journal of Education for Sustainable Development, 1*(2), 191–198.

Vieira, R. M., Tenreiro-Vieira, C., & Martins, I. P. (2011). Critical thinking: Conceptual clarification and its importance in science education. *Science Education International, 22*(1), 43–54.

Wiek, A., Withycombe, L., & Redman, C. L. (2011). Key competencies in sustainability: A reference framework for academic program development. *Sustainability Science, 6*(2), 203–218.

Willingham, D. T. (2008). Critical thinking: Why is it so hard to teach? *Arts Education Policy Review, 109*(4), 21–32.

Part IV
Concluding Remarks

Chapter 15
The Integration of Critical Thinking in Biology and Environmental Education. Contributions and Further Directions

Blanca Puig and María Pilar Jiménez-Aleixandre

15.1 Introduction

The purpose of this volume is to bring together research on critical thinking (CT) in biology and environmental education from international scholars working in this area of research. The volume seeks to broaden current ideas about the role of CT in biology and environmental education taking into account educational challenges in the post-truth era. To this end, the chapters are distributed in three sections, perspectives of a theoretical character (part I), empirical research about CT in the context of biology and health education (part II), and empirical research on CT in the context of environmental and sustainability education (part III). The chapters focus on studies reporting students' engagement in the practice of CT, and display how CT can be integrated in biology and environmental education and why biology and environmental issues are privileged contexts for the development of CT. Biology and environmental education are overlapping fields: some biology topics such as ecology have been for many years suffused with environmental education, as the chapters by Hufnagel, González-Weil and colleagues, or Sezen-Barrie and colleagues evidence. Recently other topics as nutrition have shown potential for been approached through a sustainability education perspective, as the chapter by Brocos and Jiménez-Aleixandre shows. We argue that biology and environmental education provide complex and multifaceted problems in which the practice of CT would be relevant in order to take responsible actions. Furthermore, emotions and identities emerge in the practice of CT when dealing with biology and environmental topics (Hufnagel, Chap. 3, this book). The examination of evidence is a substantial part of CT, but understanding

B. Puig (✉) · M. P. Jiménez-Aleixandre
Facultade Ciencias da Educación, Universidade de Santiago de Compostela, Santiago de Compostela, Spain
e-mail: blanca.puig@usc.gal; marilarj.aleixandre@usc.gal

B. Puig, M. P. Jiménez-Aleixandre (eds.), *Critical Thinking in Biology and Environmental Education*, Contributions from Biology Education Research,
https://doi.org/10.1007/978-3-030-92006-7_15

CT as only this examination is limited, as argued in the chapters by Jiménez-Aleixandre and Puig and by Duncan and colleagues. The volume is an effort to put together theoretically and empirically founded perspectives on how to promote CT in biology and environmental education contexts, approaches that, we believe, can be transferred to other science education contexts. Furthermore, we aim to examine how the components of CT presented in our characterization (Jiménez-Aleixandre and Puig, Chap. 1, this book) are articulated in diverse contexts.

This chapter aims to connect the main ideas and challenges that emerge across the volume and to provide some reflections about future directions for the integration of CT in biology and environmental education based on the studies reported in the book.

15.2 Integration of Critical Thinking in Biology and Environmental Education: Main Ideas and Contributions

Meeting the global challenges of the twenty-first century will require citizens who possess the capacity to apply CT to find solutions to problems that have not been yet anticipated and to navigate through uncertainty, as the pandemic of COVID-19 is showing. In the approach suggested in this book, CT is oriented to action, as evidenced for instance in the chapter by González-Weil and colleagues. CT for action in biology and environmental education implies, among other issues, to engage students in making reasonable decisions and developing appropriate behaviors regarding controversial issues. Decision making and taking action are challenging in the post-truth era, in the contexts of the rise of science denial on issues of urgency, such as climate change or anti-vaccines movements, which advocate for alternative health therapies (e.g., Dillon & Avraamidou, 2020; Puig et al., 2021).

Although the body of research in the domain of CT in biology and environmental education has been growing, and the literature points to the crucial role of CT in order to face the post-truth era challenges (Willingham, 2008), little is known on students' engagement in CT and on how the development of CT interacts with knowledge building dynamics. Consequently, the book seeks to address the questions of what is the role of CT in biology and environmental education and how can we engage students in order to develop CT, to make appropriate use of science in social contexts, and to face post-truth and misinformation.

The chapters report a range of studies that investigate the practice of CT by students' learning biology and environmental issues, and how CT interacts with their knowledge, epistemics beliefs and other factors, as identities and emotions. Four main ideas connect the chapters and are contributions of the book:

(a) *CT oriented towards responsible actions and behaviors is crucial in the post-truth era and biology and environmental education are privileged environments to develop it.*

We are experiencing a rise of post-truth and consequently the harmful effects on the public of misinformation and disinformation, such as people's loss of confidence in fact-based science (Saribas & Çetinkaya, 2021). Climate change, vaccination, gene therapy, sustainable diets, are among the environmental and health topics affected by science denial movements addressed in the chapters. The studies report on the role of CT as a tool to face some of the challenges posed by post-truth in biology and environmental education, and on ways to engage students in CT through the practice of argumentation. Biology and environmental topics are considered privileged contexts to engage students in CT for action since they offer opportunities to: (1) explicitly reflect on the ways media can (intentionally or unintentionally) expose people to inaccurate scientific information; (2) engage them in argumentation interactions, as in chapters by Puig and Ageitos, and Uskola. As discussed by Munkenbye and Gericke there is a range of factors that may shape what people perceive as true. Thus, for instance, the socio-cultural contexts, people's personal experiences and their own judgements that can be biased as Colucci-Gray and Gray, and García Franco and colleagues report; (3) make responsible decisions based on evidence, as reported by Duncan and colleagues, and acting upon conflicting evidence, as discussed by Brocos and Jiménez-Aleixandre; (4) develop appropriate behaviors and actions to protect the environment, as reported by González-Weil and colleagues, and teach climate change, acknowledging the tensions between CT and emotions related to students' identities, as discussed by Hufnagel; and (5) think about their own CT and epistemic beliefs when facing socio-scientific biology and environmental related issues, as discussed by De Checchi and colleagues.

(b) *CT is a dynamic activity that can be developed through engagement in the practice.*

This perspective, drawing from Kuhn (2019), proposed in the book as a framing of CT, is illustrated in the chapters and supports the view that CT is not a static attribute, but rather that it can be developed through engagement in the practice. Its exercise might vary in complexity depending on the topic and the context. For instance, assessing the arguments produced by classmates in peer negotiation, as Brocos and Jiménez-Aleixandre report, may pose challenges different from evaluating information provided by media, as in the chapters by Puig and Ageitos, and Uskola. On the other hand, applying CT to assess evidence as reported in the chapter by Duncan and colleagues, might prove more difficult than using CT to provide arguments based on evidence, as in the chapter by Ergazaki. The view of CT as an activity that is developed with engagement in the practice plays a central role in Kuhn's (2019) dialogic view, and informs several chapters, as Iordanou's.

(c) *CT development in biology and environmental education might be affected by a range of factors, besides scientific knowledge.*

SSIs in biology and environmental education deal with complex issues, demanding the consideration of different dimensions and multiple perspectives (Jiménez-Aleixandre & Brocos, 2021). These topics may be emotionally laden, as health issues and environmental problems addressed in the book, in which emotions, identities, values and ethical concerns are mobilized. Several chapters point to the importance of these dimensions when designing activities for CT development: thus Elizabeth Hufnagel highlights the role of identities to CT, as reference points for emotions; Brocos and Jiménez Aleixandre discuss the obstacles that affect the adoption of decisions challenging culturally established ideas; De Checchi and colleagues report about the influence of epistemic beliefs; and the role of traditional knowledge in designing science education and critical thinking that are culturally relevant is addressed by García Franco and colleagues.

(d) *CT includes a set of components related to purposeful judgement and another set related to civic participation and social justice, related to the criticality approach.*

Engagement in CT is not easy to assess, thus some chapters aim to analyze diverse components of CT, from the characterization proposed by Jiménez-Aleixandre and Puig, as indicative of engagement in CT for action. We argue that CT has two main roles in biology and environmental instruction: one linked to the promotion of rational arguments, cognitive skills and critical dispositions; and other related to the idea of critical action and civic participation, that involves commitment to independent thinking (Jiménez-Aleixandre & Puig, Chap. 1, this book). In the complex biology and environmental problems these components may be articulated. Climate change, genetically modify food, vaccination, homeopathy, food options, are multifaceted problems that involve informal reasoning and elements of critique. Decisions about them entail direct consequences for the well-being of human society and of the environment. As Aikenhead (1985) pointed out people would need to balance subject matter knowledge, personal values and societal norms when making decisions on socio-scientific issues. However, Bencze et al. (2020) argue that they also need to be critical of the discourses that shape their own beliefs and practices in order to act responsibly. Considering these two sets of components involves a shift from a view of CT as a set of skills to a broader notion of CT that expands its meaning to critical action. This volume provides some examples about what CT for "action" means in the practice, as the chapters by Corina González-Weil and colleagues, in particular, and by Brocos & Jiménez-Aleixandre illustrate. It also provides examples about how teachers understand CT and support its practice in the context of biology and education for sustainable development, as Iordanou and Mukenbeye and Gericke report.

The integration of CT in biology and environmental education can be achieved in diverse ways, however all chapters concur on the importance of "teaching for" and "teaching about" CT in order to face post-truth challenges. There is also agreement on that CT can be embedded in biology and environmental instruction as a dialogic practice in a way that help students to take critical actions, as discussed by García Franco and colleagues, and Jiménez-Aleixandre and Brocos. However, teachers might face some difficulties in the integration of CT in their instruction. Some obstacles are related with post-truth, and others with additional pressures from their educational institutions as the curriculum, the number of teaching hours, the number of students, and their own training and knowledge about CT instruction, among others. The consideration of CT as a practice that requires time to be developed, and the acknowledgement that CT can be taught, might stimulate educational institutions and teachers to dedicate more time for their students to cultivate CT through the practice (Vincent-Lacrin et al., 2019). Teachers may face difficulties to put CT into practice, since there is not a clear understanding about what it means (e.g., Abrami et al., 2008; Thomas & Lok, 2015), and how it can be articulated with specific subject domains in a way that make students aware of their own competencies and progression in CT development (Puig et al., 2021).

This volume seeks to provide examples to teachers on how to create the conditions that encourage the exercise of CT. Thus, for instance, engaging students in its practice supports making visible two aspects: a) the components involved in CT practice; b) and elements that affect it being successful. For instance, Iordanou's chapter proposes to engage primary pre-service teachers in dialogic activities on the topic of Genetically Modified Food with the goal of making teachers aware of a fundamental component of CT: individual's ability to take in consideration multiple factors involved in an issue. Puig and Ageitos engage primary pre-service teachers in the assessment of claims from the anti-vaccination movement, with the objective of making them aware of their own knowledge and bias when assessing diverse premises supporting anti-vaccination. Hufnagel's chapter points to the intersections between emotions and CT in a course developed with pre-service teachers about climate change. On the other hand, there are chapters that may be useful in helping teachers to recognize CT in students' arguments and actions, as Brocos and Jiménez-Aleixandre illustrate in the analysis of the dimensions related to taking actions towards sustainability in the participants' discourse.

Drawing from the research presented in the volume we propose three main aspects that can support teachers in engaging students in CT development.

1. *Biology and environmental education instruction centred on SSIs and their social impacts can facilitate students' engagement in CT.* It can also help students to cultivate CT consciousness, as Colucci-Gray and Gray show. To do so, teachers need to provide opportunities to critically reflect on on our own place in the world, particularly on how our actions can affect the environment (negatively or positively). Using empathy strategies to engage students in socioenvironmental processes, as Mukenbye and Gericke suggest, might be beneficial for this purpose.
2. *The creation of learning communities and collaborative work in the classroom, in which CT is considered a situated practice, is of crucial significance in the promotion of CT.* Understanding CT as a situated practice implies identifying its exercise not only as a way of thinking, but also as a way of acting which could be related with community commitments (García Franco and colleagues, Chap. 4, this book). To do so, biology and environmental teaching need to consider students' experiences and to make use of their local contexts, with the goal to promote social and environmental awareness. For instance, regarding the adequate use of natural resources and health supplies and taking steps towards prevention.

3. *Task design that embeds conflict creates fruitful learning opportunities to engage students in the practice of CT*, as Brocos and Jiménez-Aleixandre discuss. To do so, biology and environmental education instruction needs to engage students in dialogic processes, but also in critical actions. For instance, taking actions in order to address environmental problems as food selection, climate change; or taking responsible decisions regarding health problems as vaccination, involves conflicts, both social and personal, as they suggest.

15.3 Future Directions for the Integration of Critical Thinking in Biology and Environmental Education

Taking into consideration the studies presented in the volume and the educational challenges of post-truth, some key ideas regarding future directions for the successful integration of CT are:

Key Idea 1. *CT entails awareness of one's own thinking and knowledge as well as reflection on the thinking on the self and others*. Therefore, biology and environmental instruction should provide learning opportunities in which CT is integrated as a social practice that is shaped by the discourse and dependent on the knowledge domain.

Key Idea 2. *Evidence on how students develop CT might be necessary to make them aware of their own competencies in it*. This suggests that CT could be explicitly mentioned in the learning goals as well as during the development of CT activities. When engaging in CT, students could monitor their thinking (Puig & Ageitos, Chap. 7, this book), but also the skills performed during the activity. Therefore, future directions for the integration of CT could take this aspect into consideration and design activities for this purpose.

Key Idea 3. *Designing learning environments in which students have to deal with uncertainties provide opportunities for CT development*, as Hufnagel and De Checchi et al. point out. Environmental and health issues related with post-truth era challenges, as climate change and health issues, do not present clear cut solutions. They involve some degree of uncertainty, so CT requires not only understanding of the scientific notions behind these topics, but also of the processes of knowledge construction in science. Consequently, future directions in the integration of CT should make an effort in explaining the nature of uncertainty in environmental and biology topics (Kampourakis, 2018; Kampourakis & McCain, 2019).

Key Idea 4. *Epistemic beliefs may help students to plan and evaluate their own and other people's actions* (De Checchi & colleagues; Iordanou, Chap. 13, this book). In the context of environmental issues, it can be helpful to identify which kinds of epistemic beliefs can influence both the development and use of certain critical dispositions, as De Checchi and colleagues suggest. When epistemic beliefs are elaborated, students' arguments become more critical. Hence, future directions for the integration of CT in biology and environmental instruction should attend to these connections in order to improve students' critical argumentation and dispositions to become critical thinkers.

Key Idea 5. *Attention to "Critical consciousness" in biology and environmental education is necessary in order to make students aware of the impacts of our actions in the world* (Colucci-Gray & Gray, Chap. 2, this book). Future directions could integrate "CT awareness" of "critical consciousness" in teaching designs as a way to empower students to take critical actions and to make proper decisions in order to protect the environment and to take action against social inequalities as these authors suggest and the current pandemic and post-truth era demands.

This volume has been completed during the COVID-19 pandemic, a period of time that has promoted collaboration among researchers, teachers and teachers' educators to find new ways to make education relevant for today's circumstances. We hope that the studies in it will encourage teachers and teachers' educators to introduce CT in biology and environmental instruction and to make educational institutions reconsider their priorities in terms of CT for a better society and a healthy environment.

Acknowledgements Work supported by the Spanish Ministry of Science, Education and Universities, partly funded by the European Regional Development Fund (ERDF). Contract grant PGC2018-096581-B-C22. http://www.rodausc.eu

References

Abrami, P. C., Bernard, R. M., Borokhovski, E., Wade, A., Surkes, M. A., Ramim, R., & Zhang, D. (2008). Instructional interventions affecting critical thinking skills and dispositions: A stage 1 meta-analysis. *Review of Educational Research, 78*(4), 1102–1134.

Aikenhead, G. S. (1985). Collective decision making in the social context of science. *Science Education, 69*(4), 453–475.

Bencze, L., Halwany, S., & Zouda, M. (2020). Critical and active public. Engagement in addressing socioscientific problems through science teacher education. In M. Evagorou, J. A. Nielsen, & J. Dillon (Eds.), *Science teacher education for responsible citizenship* (pp. 63–83). Springer.

Dillon, J., & Avraamidou, L. (2020). Towards a viable response to COVID-19 from the science education community. *Journal for Activist Science and Technology Education, 11*(2), 1–6. https://doi.org/10.33137/jaste.v11i2.34531

Jiménez-Aleixandre, M. P., & Brocos, P. (2021). Emotional tension as a frame for argumentation and decision-making: Vegetarian vs. omnivorous diets. *Frontiers in Psychology*. https://doi.org/10.3389/fpsyg.2021.662141

Kampourakis, K. (2018). Science and uncertainty. *Science & Education, 27*(9), 829–830. https://doi.org/10.1007/s11191-018-0019-3

Kampourakis, K., & McCain, K. (2019). *Uncertainty: How it makes science advance*. Oxford University Press.

Kuhn, D. (2019). Critical thinking as discourse. *Human Development, 62*(3), 146–164. https://doi.org/10.1159/000500171

Puig, B., Blanco-Anaya, P., & Pérez-Maceira, J. J. (2021). "Fake news" or Real Science? Critical Thinking to Assess Information on COVID-19. 2021. *Front. Educ*. 6:646909. https://doi.org/10.3389/feduc.2021.646909

Saribas, D., & Çetinkaya, E. (2021). Pre-service teachers' analysis of claims about COVID-19 in an online course. *Science & Education, 30*, 235–266.

Thomas, K., & Lok, B. (2015). Teaching critical thinking: An operational framework. In M. Davies, R. Barnett, et al. (Eds.), *The Palgrave handbook of critical thinking in higher education* (pp. 93–105). Palgrave Macmillan.

Vincent-Lacrin, S., González-Sancho, C., Bouckaert, M., de Luca, F., Fernández-Barrera, M., Jacotin, G., Urgel, J., & Vidal, Q. (2019). *Fostering students' creativity and critical thinking: What it means in school, educational research and innovation*. OED Publishing. https://doi.org/10.1787/62212c37-en

Willingham, D. T. (2008). Critical thinking: Why it is so hard to teach? *Arts Education Policy Review, 109*(4), 21–32. https://doi.org/10.3200/AEPR.109.4.21-32

The manufacturer's authorised representative in the EU is Springer Nature Customer Service Centre GmbH, Europaplatz 3, 69115 Heidelberg, Germany. If you have any concerns regarding our products, please contact ProductSafety@springernature.com

Printed and bound by CPI Group (UK) Ltd, Croydon, CR0 4YY
22/09/2026
02220982-0011